73rd American Helicopter Society International Annual Forum Special Sessions 2017

Fort Worth, Texas, USA
8 - 11 May 2017

ISBN: 978-1-5108-4407-0

Printed from e-media with permission by:

Curran Associates, Inc.
57 Morehouse Lane
Red Hook, NY 12571

Some format issues inherent in the e-media version may also appear in this print version.

Printed by Curran Associates, Inc. (2017)

American Helicopter Society International
2701 Prosperity Ave, Suite 210
Fairfax, VA 22031

Phone (703) 684-6777
Fax: (703) 739-9279

staff@vtol.org

Additional copies of this publication are available from:

Curran Associates, Inc.
57 Morehouse Lane
Red Hook, NY 12571 USA
Phone: 845-758-0400
Fax: 845-758-2633
Email: curran@proceedings.com
Web: www.proceedings.com

TABLE OF CONTENTS

ADVANCED AIR VEHICLE DESIGN

TRANSFORMATIVE URBAN MOBILITY

LATE TECHNICAL PAPERS

Program Executive Office Air ASW, Assault & Special Mission Programs

8 May 2017

Presented to:
AHS International

Presented by:
RDML Dean Peters
PEO(A)

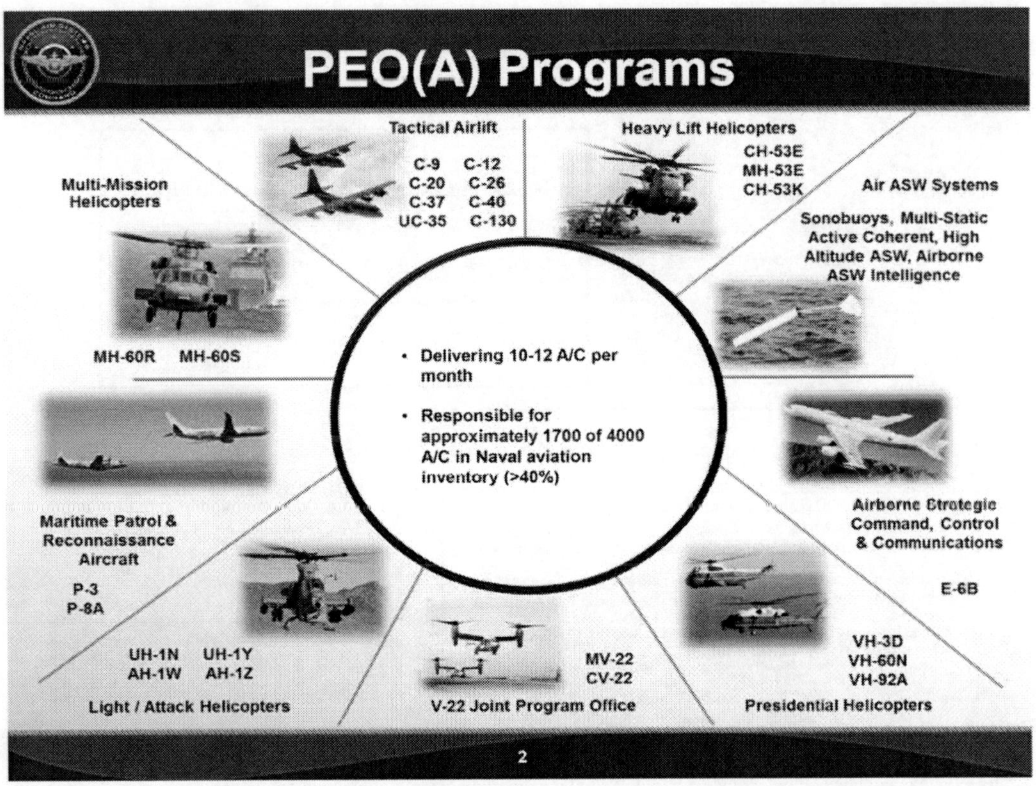

Suppliers, Contractors and Field Activities

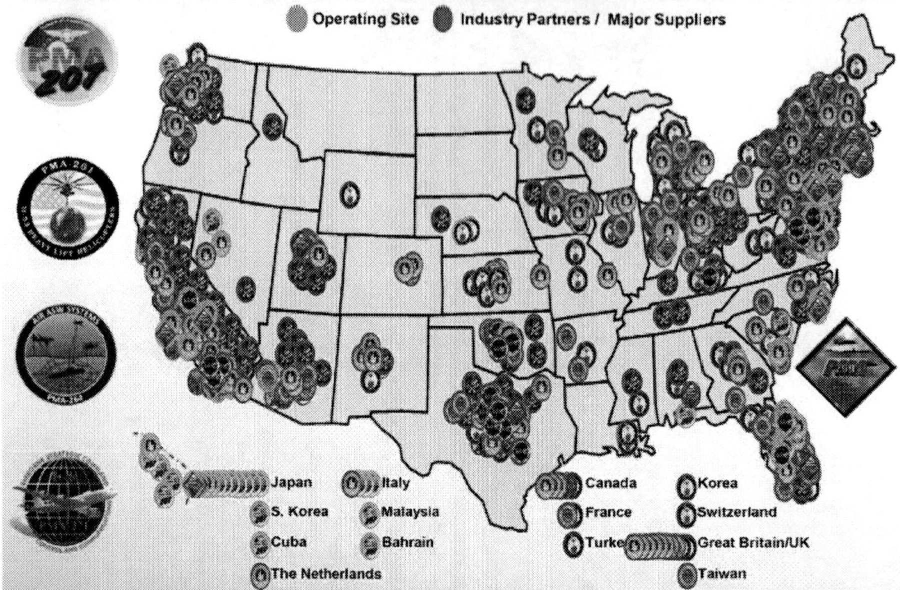

3

Presidential Helicopter VH-92A (PMA-274)

UNITED STATES GOVERNMENT ACCOUNTABILITY OFFICE
Report to Congressional Committees, March 2016
DEFENSE ACQUISITIONS: Assessments of Selected Weapon Programs

The most recent GAO Report (Q2 FY16) recognized the positive progress of the VH-92A program, its consistency with acquisition best practices and the program's drive to affordability.

Reference: GAO-16-329SP

MISSION & REQUIREMENTS

- **VH-92A (Qty. 23):** 21 operational aircraft & 2 test aircraft.
- **Mission:** Presidential Lift Replacement. One type-model series aircraft will replace both the VH-3D and the VH-60N.
- **Improvements:** Operational performance and communications capabilities, while offering increased passenger payload.
- **Acquisition Strategy:** Integrate government-defined mission systems into an existing air vehicle. Maintain the existing FAA airworthiness certification through-out lifecycle. No critical technology elements.
- **Prime Contractor:** Sikorsky Aircraft Corporation (Lockheed Martin) is responsible for air vehicle delivery and mission systems integration.

4

Presidential In-Service Aircraft VH-3D/VH-60N

MISSION & REQUIREMENTS

Marine Helicopter Squadron One (HMX-1) Mission: "Provide safe and timely helicopter transportation for the President of the United States, Vice President , members of the President' Cabinet and foreign Heads of State as directed by the White House Military Office."

"Provide helicopter contingency support for the President of the United States, Vice President of the United States, and other personnel/agencies, as directed by the Director WHMO."

VH-3D

- Introduced in 1974/75
- Quantity – 11
- Mission – Presidential & VIP transport both within the continental US and overseas (Primary DC aircraft)
- SLEP required to carry VH-3D through VH-92A transition.

VH-60N

- Introduced in 1988: 28 years old
- Quantity – 8 (1 A/C attrite in 1993)
- Mission – Presidential & VIP transport both within the continental US and overseas (Primary travel aircraft)
- SLEP required to carry VH-60N through VH-92A transition.

5

PMA-276

Mission Sets

- Common: Strike Coordination and Reconnaissance, Armed Reconnaissance, Close Air Support, Areal Reconnaissance
- AH-1: Offensive Anti-Air Warfare, Air Interdiction
- UH-1: Combat Assault Support, Air Delivery, Airborne Command and Control, Helicopter Rapid Insert and Extract, Tactical Air Control (Airborne), Electronic Warfare

Operational Status

- UH-1Y IOC: Aug 08, FOC: Mar 16
- AH-1Z IOC: Feb 11; FOC: FY20
- 3 UH-1Ys/4 AH-1Zs each assigned to West Coast MEUs
- 3 UH-1Ys/4 AH-1Ws each assigned to East Coast MEU
- Transition ongoing

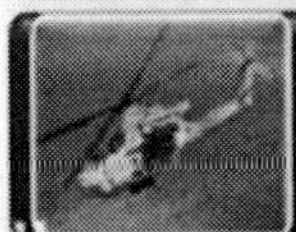

Recent Events

- 8 AH-1Zs permanently deployed to PACOM AOR
- AH-1Z: Program of record: 189; Contracted: 108; Delivered: 63
- UH-1Y: Program of record: 160; Contracted: 160; Delivered: 146

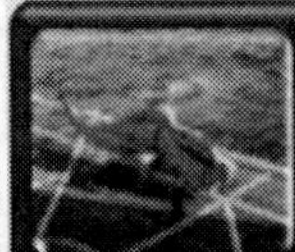

International Sales

- Cases: AH-1W Taiwan (61 a/c and 2 simulators) and support; AH-1W Turkey (11); AH-1Z Pakistan (12)

6

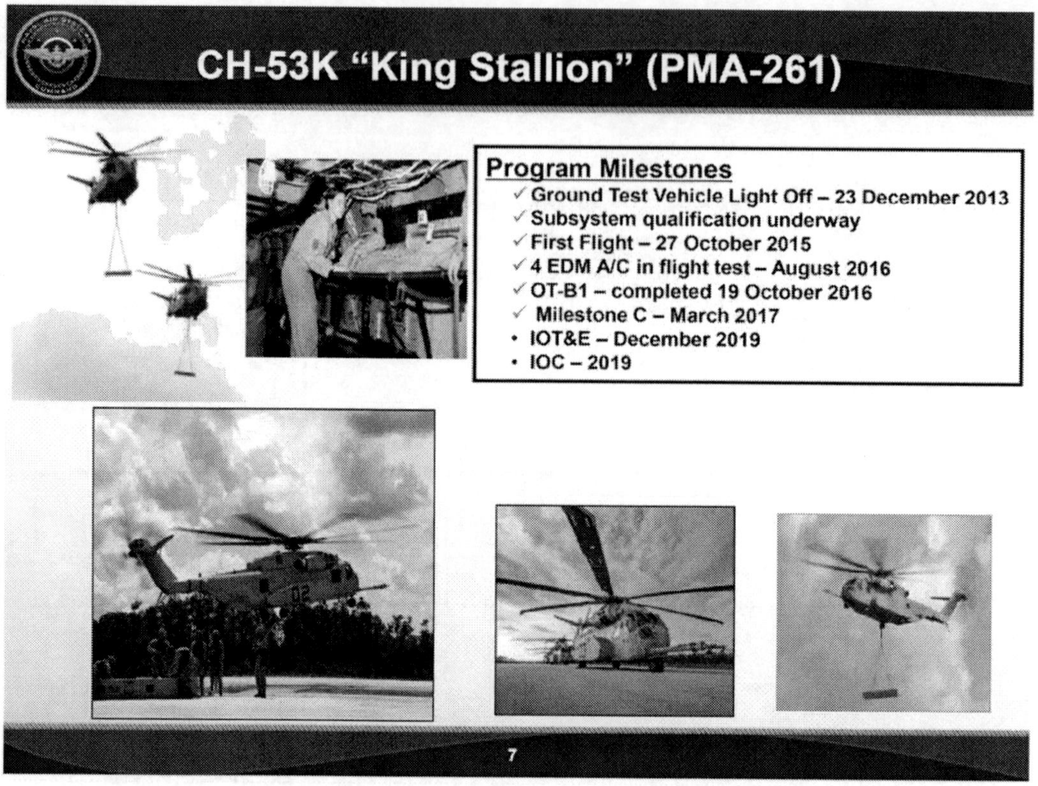
CH-53K "King Stallion" (PMA-261)
Program Milestones
✓ Ground Test Vehicle Light Off – 23 December 2013
✓ Subsystem qualification underway
✓ First Flight – 27 October 2015
✓ 4 EDM A/C in flight test – August 2016
✓ OT-B1 – completed 19 October 2016
✓ Milestone C – March 2017
• IOT&E – December 2019
• IOC – 2019
7

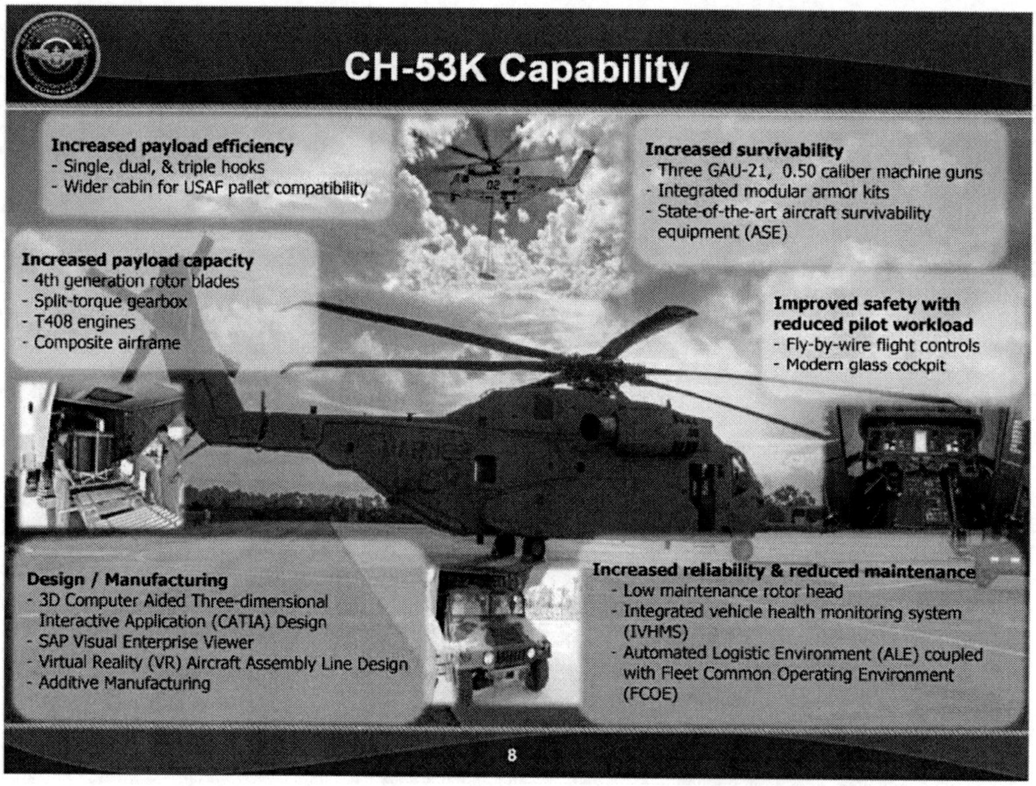
CH-53K Capability
Increased payload efficiency
- Single, dual, & triple hooks
- Wider cabin for USAF pallet compatibility
Increased survivability
- Three GAU-21, 0.50 caliber machine guns
- Integrated modular armor kits
- State-of-the-art aircraft survivability equipment (ASE)
Increased payload capacity
- 4th generation rotor blades
- Split-torque gearbox
- T408 engines
- Composite airframe
Improved safety with reduced pilot workload
- Fly-by-wire flight controls
- Modern glass cockpit
Design / Manufacturing
- 3D Computer Aided Three-dimensional Interactive Application (CATIA) Design
- SAP Visual Enterprise Viewer
- Virtual Reality (VR) Aircraft Assembly Line Design
- Additive Manufacturing
Increased reliability & reduced maintenance
- Low maintenance rotor head
- Integrated vehicle health monitoring system (IVHMS)
- Automated Logistic Environment (ALE) coupled with Fleet Common Operating Environment (FCOE)
8

V-22 Joint Osprey Program (PMA-275)

Program Status

- Currently 340 V-22s delivered (288 MV / 52 CV)
- Program of Record: Marines (360) ◆ Air Force (53) ◆ Navy (48) ◆ Japan (17)

MV-22

- 15 Active Squadrons and HMX-1 FOC Camp Pendleton and Hawaii transition in progress
- Reserves 1 of 2 FOC, planned complete in 2018

CV-22

- 52 aircraft complete: Hurlburt Field, FL, Cannon AFB, NM and RAF Mildenhall, UK
- PACOM transition planned 2021

CMV-22

- First Delivery in 2020; IOC 2021
- First Standup in North Island, VRC-30

IOC

- MV – 2007 ◆ CV – 2009 ◆ Navy – 2021

Operational Significance

- Six units currently deployed (MEUs, SPMAGTFs, National Mission Tasking)

Latest Acquisition Events

- Program meeting production cost, schedule, performance under second multi-year procurement contract
- Currently pursuing a third multiyear procurement to include Navy, Marine and FMS aircraft with pre-negotiated provisions to add additional aircraft

9

Program of Record

USMC	USSOCOM	USN	JAPAN
360 V-22	53 V-22	48 V-22	17 V-22
• Combat Assault & Amphibious Assault • Special Operations Capable	• Long Range Special Operations (Infil / Exfil / Resupply) • Contingency Operations	• Sea Based Logistics • Personnel Recovery • Special Warfare	• Combat Assault & Amphibious Assault • Disaster Relief • Humanitarian Relief

10

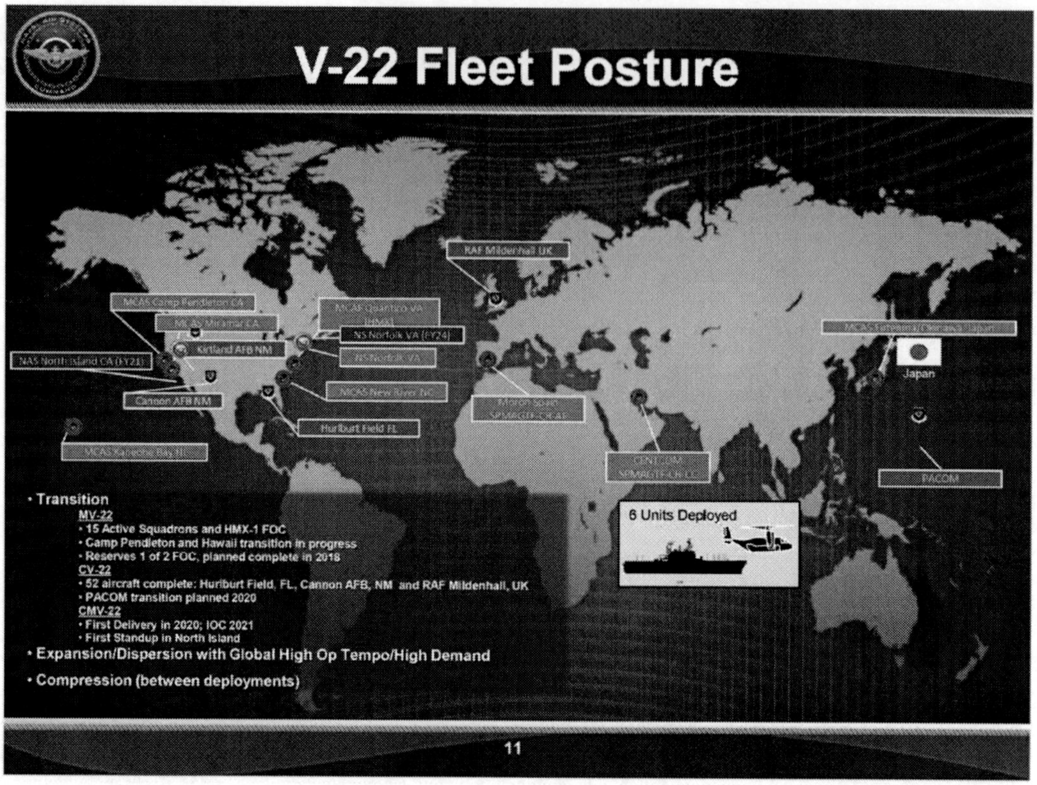

V-22 Significant Events

- Currently 341 V22s delivered (289 MV / 52 CV)
- Surpassed 350,000 Flight Hours
- Continued focus on expanding business base
  - Foreign Military Sales: 2 Japan Cases for 9 a/c
  - Domestic: Working toward MYP III

12

CMV-22 USN Variant: Fleet Experiment
Aug 2016: MV-22 assigned to Marine Operational Test and Evaluation Squadron One (VMX-1) lands aboard aircraft carrier USS *Carl Vinson* (CVN-70), to explore, analyze, and document its operating characteristics within a carrier strike group.
13

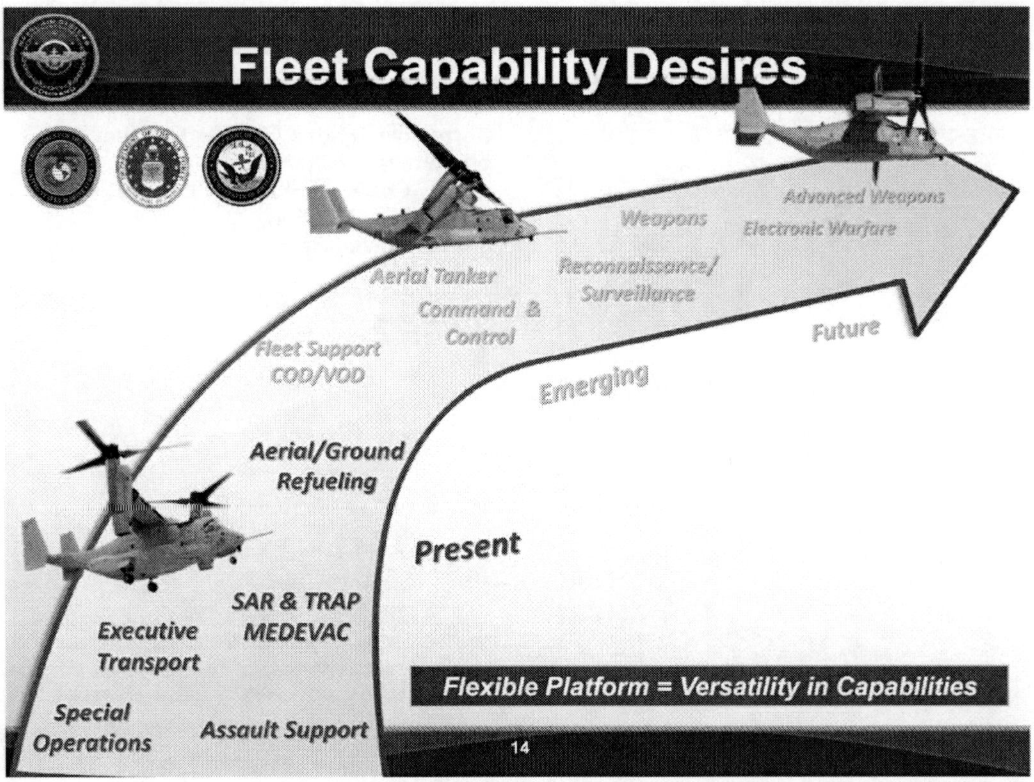
Fleet Capability Desires
Advanced Weapons
Weapons
Electronic Warfare
Aerial Tanker
Reconnaissance/
Surveillance
Command &
Control
Future
Fleet Support
COD/VOD
Emerging
Aerial/Ground
Refueling
Present
SAR & TRAP
MEDEVAC
Executive
Transport
Flexible Platform = Versatility in Capabilities
Special
Operations
Assault Support
14

V-22 Aerial Refueling System

PMA-275 Reliability Improvement Process

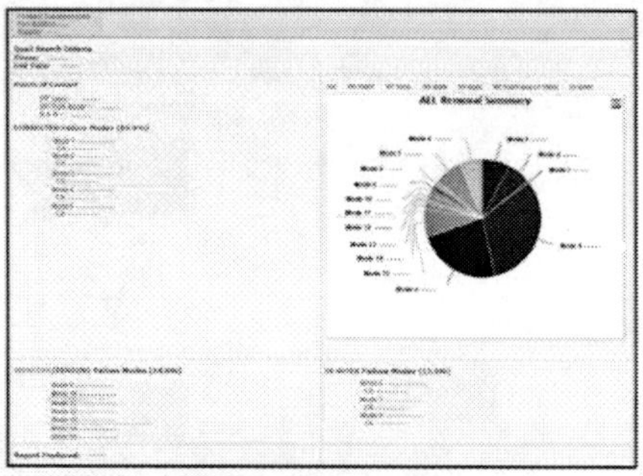

Corrective Actions identified for the predominate failure modes

- Crack
- Delamination
- Crazed/Haze

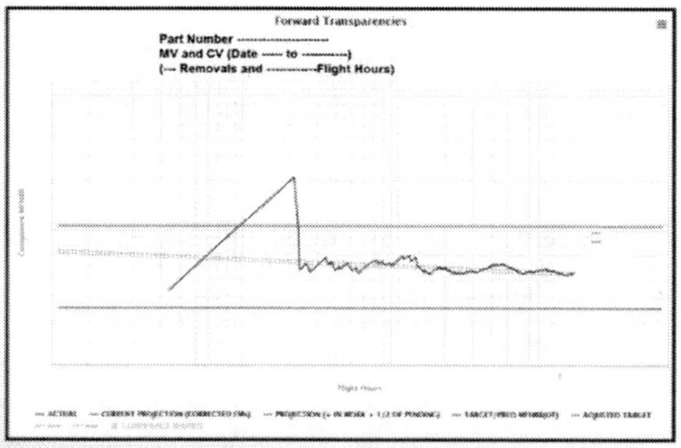

Corrective Action Impact

- Corrective Action: Hard Coating
  - Projected Reliability Improvement of 93%
- Continuing to work remaining faults

OEM repair data vital to effective Corrective Action and Component Improvement

Reliability Trending by V-22 Lot

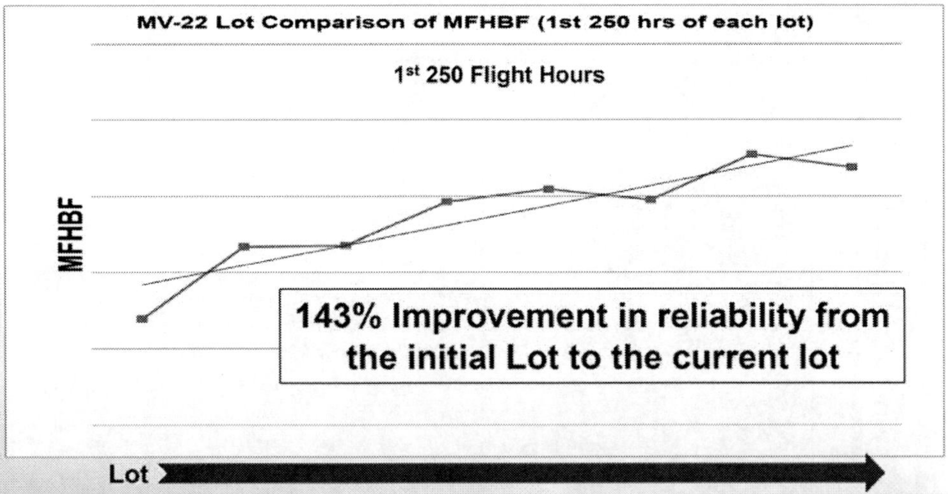

Successive Lots demonstrating improved R&M.

Additive Manufactured Nacelle

- NAVAIR Transformative Goal: rapidly leverage additive manufacturing for delivery of warfighting capability
- Outboard Nacelle Link and Fitting at STA 400 selected for evaluation
- Nacelle link assembly manufactured and instrumented; installed and demonstrated in flight on HX-21 Osprey 70 on 29 July 2016

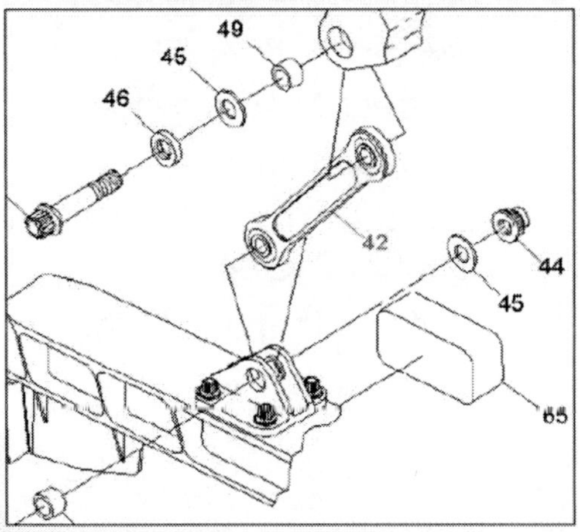

V-22: Additive Manufacturing Test

19

V-22 Planned Readiness Efforts

- Improved Inlet Solution
- Redesigned Variable Frequency Generators
- IR Suppressor
- Nacelle Automatic Vibration Control System
- Additive Manufacturing

20

V-22 Fleet Common Configuration

Single-Configuration Fleet with Reliability and Capability Improvements, Open Architecture, and Technology Foundation for rapid/incremental upgrades

21

H-60 Multi-Mission Helicopter (PMA-299)

22

H-60 Multi-Mission Helicopter (PMA-299)

Program Status

- MH-60R in last lot of domestic production
  - Final fleet delivery summer 2018

Recently Fielded Capabilities

- M197 20mm Gun EOC 2014 (MH-60S)
- DRL/APKWS EOC 2014/2015 (MH-60S/MH-60R)
- Mine Countermeasures Systems IOC November 2016 (MH-60S)

Current Efforts

- Moving map integration
- Helmet Display Tracking System
- Full Motion Video datalink

MH-60S DRL and M197

MH-60R APKWS

MH-60S LMDS

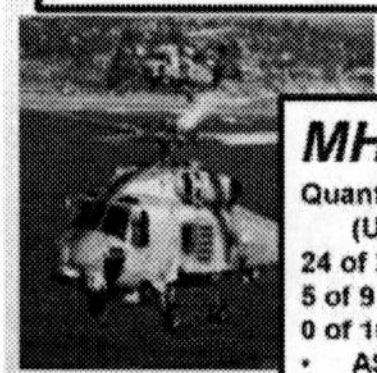

MH-60R

Quantity: 243 of 280 (US)
24 of 24 Australia
5 of 9 Denmark
0 of 10 Saudi Arabia

- ASW, ASUW
- CVN, CRUDES, LCS
- IOC Dec 2005

MH-60S

Quantity: 275 (US)

- VERTREP, VOD, SAR, CSAR, SUW, OAMCM
- CVN, L Class, LCS, Supply Ships
- IOC Aug 2002

HH-60H

Quantity: 15 (US)

- Combat SAR, Spec Warfare Support, ASUW, SAR, MEDEVAC, Log Support
- CVN
- Operational since 1989

MH-60 Availability

- Service Life Assessment Program (SLAP)
  - Examines structural fatigue, dynamic components life, and subsystems aging
  - Defines aircraft service life
- Service Life Extension Program (SLEP)
  - Scope defined by SLAP
  - Ensures MH-60 R/S remain safe, ready, and operationally relevant to meet CNO inventory requirements through FY2035

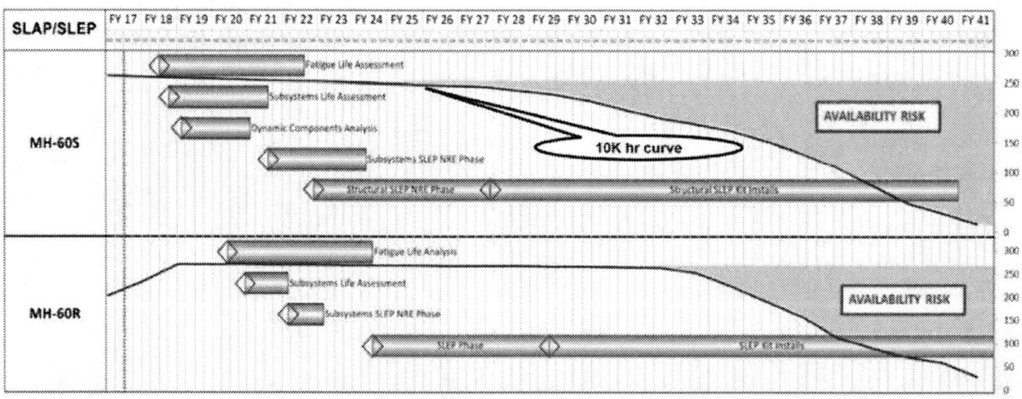

MH-60 Flight Plan

< 5 Years	5-10 Years	10+ years
• Communication • Navigation • Open Architecture – Mission Computer – Displays • Surface Warfare – Fire-and-Forget Weapon – Rapid Targeting • Surveillance – Improved EO/IR Sensors – Digital ESM • Aircraft Survivability Equipment – DAIRCM – Digital ESM • SLAP (Service Life Assessment Program)	• SLEP (Service Life Extension Program) • MLU (Mid-Life Upgrade): Non-Recurring Engineering • Power & Lift Improvements • Avionics and Mission Systems Open Architecture Backbone • Mission Capability Improvements – Non-acoustic ASW sensors – Sensor data fusion improvements – Passive targeting	• SLEP and MLU Production • New Air Vehicle development – FVL Capability Set 2 – Major H-60 redesign • Migrate capabilities without redevelopment

Air vehicle agnostic development of capabilities is key to affordable recapitalization

25

MH-60 Mid-Life Update

- Objectives:
  - Ensure MH-60 remain safe and operationally relevant through 2030s
  - Implement modifications concurrent with planned Service Life Extension Program
  - Establish architectural foundation to migrate capabilities rapidly and affordably into Future Vertical Lift (FVL) replacement for MH-60
- Methodology:
  - Analyze and prioritize current and forecasted capability gaps, material solutions, and constraints (e.g. size, weight, power, cost)
  - Implement broad FACE and HOST based Open Architecture framework for MH-60 avionics and mission systems
  - Prioritize capability suppliers that provide material solutions leveraging FACE integration principles

26

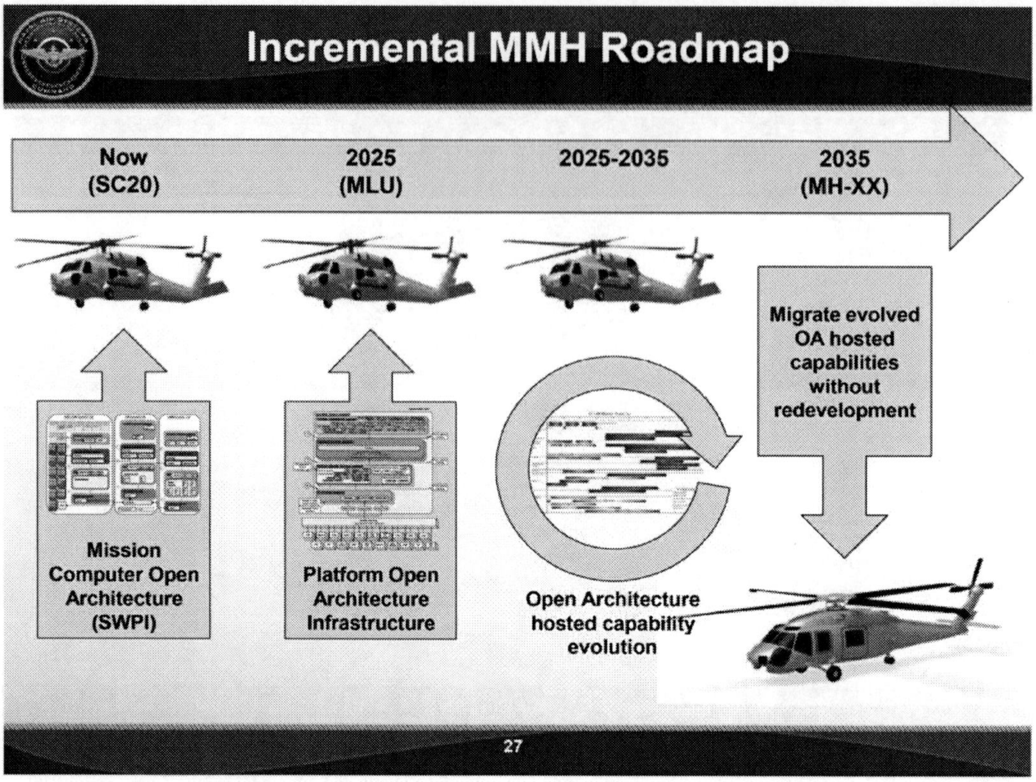
Incremental MMH Roadmap
Now (SC20)
2025 (MLU)
2025-2035
2035 (MH-XX)
Mission Computer Open Architecture (SWPI)
Platform Open Architecture Infrastructure
Open Architecture hosted capability evolution
Migrate evolved OA hosted capabilities without redevelopment
27

Fleet Readiness Initiatives
28

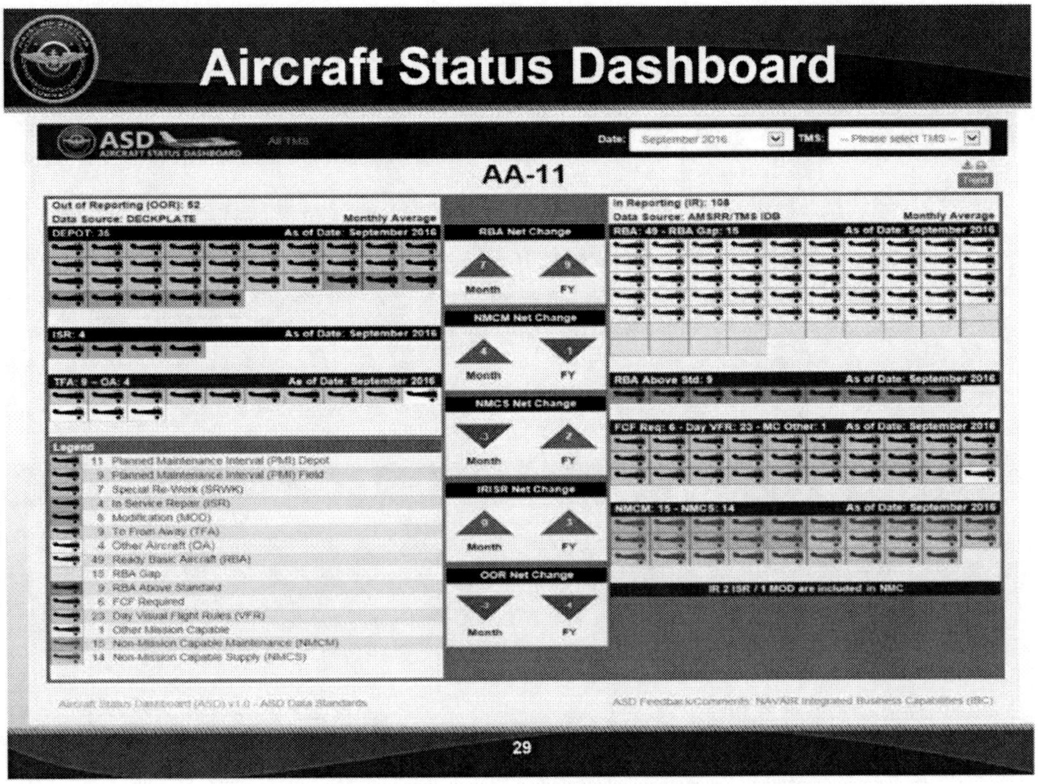
Aircraft Status Dashboard
ASD
AIRCRAFT STATUS DASHBOARD
Date: September 2016
TMS: -- Please select TMS --
AA-11
Out of Reporting (OOR): 52
Data Source: DECKPLATE
Monthly Average
DEPOT: 35
As of Date: September 2016
ISR: 4
As of Date: September 2016
TFA: 9 – OA: 4
As of Date: September 2016
Legend
11 Planned Maintenance Interval (PMI) Depot
9 Planned Maintenance Interval (PMI) Field
7 Special Re-Work (SRWK)
4 In Service Repair (ISR)
8 Modification (MOD)
9 To From Away (TFA)
4 Other Aircraft (OA)
49 Ready Basic Aircraft (RBA)
15 RBA Gap
9 RBA Above Standard
6 FCF Required
23 Day Visual Flight Rules (VFR)
1 Other Mission Capable
15 Non-Mission Capable Maintenance (NMCM)
14 Non-Mission Capable Supply (NMCS)
RBA Net Change
Month
FY
NMCM Net Change
Month
FY
NMCS Net Change
Month
FY
IRISR Net Change
Month
FY
OOR Net Change
Month
FY
In Reporting (IR): 108
Data Source: AMSRR/TMS IDB
Monthly Average
RBA: 49 - RBA Gap: 15
As of Date: September 2016
RBA Above Std: 9
As of Date: September 2016
FCF Req: 6 - Day VFR: 23 - MC Other: 1
As of Date: September 2016
NMCM: 15 - NMCS: 14
As of Date: September 2016
IR 2 ISR / 1 MOD are included in NMC
Aircraft Status Dashboard (ASD) v1.0 - ASD Data Standards
ASD Feedback/Comments: NAVAIR Integrated Business Capabilities (IBC)
29

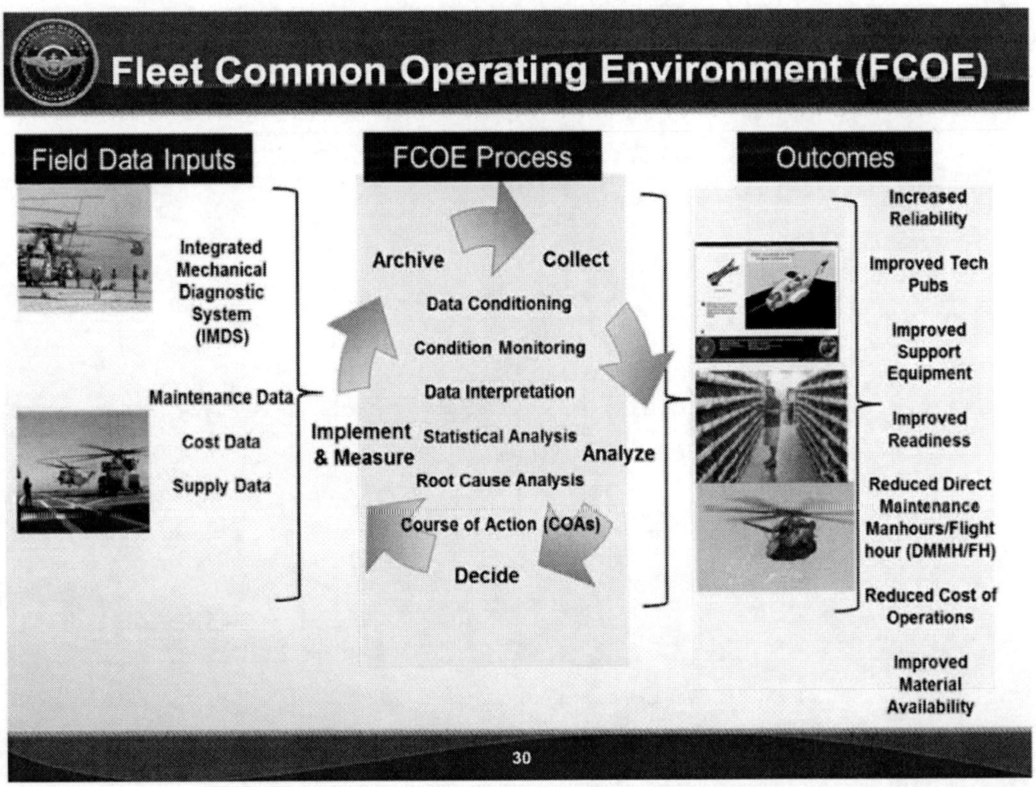
Fleet Common Operating Environment (FCOE)
Field Data Inputs
Integrated Mechanical Diagnostic System (IMDS)
Maintenance Data
Cost Data
Supply Data
FCOE Process
Archive
Collect
Data Conditioning
Condition Monitoring
Data Interpretation
Statistical Analysis
Root Cause Analysis
Course of Action (COAs)
Implement & Measure
Analyze
Decide
Outcomes
Increased Reliability
Improved Tech Pubs
Improved Support Equipment
Improved Readiness
Reduced Direct Maintenance Manhours/Flight hour (DMMH/FH)
Reduced Cost of Operations
Improved Material Availability
30

Logistics Cell (LOGCELL)

Order Fulfillment

	Previous	Current	
Wholesale Fill Rate	81.8%	76.1%	R
UCOs	703	638	Y
Backorders	446	389	G
CWT	18.62	18.67	R
First Pass (DTOs)	94.3%	94.9%	G

Integrated Line of Balance

	Previous	Current	
IPG-1 GWD > 6 Mo.	33	29	G
IPG-1 GWD < 6 Mo.	54	42	R
IPG-2/3 GWD > 6 Mo.	33	26	G
IPG-2/3 GWD < 6 Mo.	88	113	R
Materials with Unfirmed PRs	1,712	1,341	R

Forecast Accuracy

	Previous	Current	
Number of Materials	5,329	5,411	
Materials with Demand	748	956	
Demand Forecast (Understated)	68	0	R
Demand Forecast (On-Target)	751	956	R
Demand Forecast (Overstated)	7	0	G

Retail Allowancing

	Previous	Current	
Net Effectiveness	83.7%	75.6%	R
Gross Effectiveness	70.8%	65.7%	G
Range	68.5%	79.8%	Y
Depth	64.6%	75.9%	Y
Open Dues	786	663	G

Readiness

	Previous	Current
Aircraft IR	42	49
Aircraft OOR	0	1
Aircraft RBA	31	28
Aircraft NMC/PMC	36	21
NMCS/PMCS Requisitions	167	127

Inventory Control

	Previous	Current
Physical Inventory Value		
SIT ≤ 90 Days		
SIT > 90 Days		
Carcass >180 Days		
IR/GR mismatches		

Spares Procurement

	Previous	Current	
ALT Performance	67.9%	62.6%	R
PLT Performance	69.6%	56.6%	R
PR's (Open) Late	92	75	Y
PO's (Open) Late	226	212	R
Unfirmed PR's	141	101	G

Repair Performance

	Previous	Current	
Commercial Repair ALT Performance	96.8%	75.6%	R
RTAT Performance	93.3%	76.3%	R
PR's (Open) Late	76	41	R
PO's (Open) Late	228	207	R
Unfirmed PR's	85	62	G

Outfitting Performance

	Previous	Current	
NA/NM Requisitions (FRWQ)	257	249	R
Projected Fills Current FY (Count)	233	236	
Projected Fills Current FY (Dollars)	$31.2M	$32.7M	
Sales Quotes in FYDP (Count)	3,735	3,696	
Sales Quotes in FYDP (Dollars)	$311.2M	$311.2M	

LOGCELL Status

04/14/2017 - Last update of Material, Planning, PO, PR, Unfirmed PR, STO, Sales Quotes, BPR's and Inventory Data

NOTE : LOGCELL is in User Acceptance Testing, functionality and data are being verified

Apr 17, 2017

31

VECTOR Logistics Analysis Tool

Platform XYZ Component Heat Map

Heat Map

Detailed analysis of removal rates, failure rates, Beyond Capability of Maintenance (BCM), projected demands, as well as, Aviation Depot Level Repairable (AVDLR) and Aviation Fleet Maintenance (AFM) cost drivers.

Sort by: Top Readiness - NMCS

1 2 3 4 5 6 7 8 9 10 ... Page size: 10 — 381 items in 39 pages

HOFNIIN	NOMENCLATURE	COG	TOP NMCS RANK	TRENDING NMCS RANK	TOTAL COST	END ITEM COST	AFM COST	TOTAL SCIR HOURS	NMC HOURS	NMCM HO...
LLZ98W331	HIGH STAGE BLEED AIR AUGMENTOR PNEUMATIC CONTROLLE	0R	1	1	$7.25	$0.00	$7.25	32961	32961.0	
014707884	FLOW CONTROL VALVE	7R	2	2	[illegible]	$8,026.00	$4,000.27	30644.4	29120.0	
014721429	BLEED AIR DUCT HOT AIR SHUT-OFF VALVE	7R	3	3	$989.22	$0.00	$989.22	14230.7	15261.0	
LLZ98X366	ICS CENTRAL SWITCHING UNIT	0R	4	4	$0.00	$0.00	$0.00	18256.3	18256.7	
014756203	HIGH STAGE BLEED AIR AUGMENTOR PNEUMATIC CONTROLLE	7R	5	5	$48,500.09	$43,488.00	$12.09	23817.2	23811.5	
014440419	FLOW CONTROL VALVE	0R	6	6	$1,563.98	$0.00	$1,563.98	16120	18067	
014860676	DUAL INDICATOR AIRSPEED ALTIMETER	7R	7	7	$23,535.39	$23,491.00	$44.39	19851.1	19851.5	
007961672	DIVIDER VALVE	7R	8	14	$91,005.55	$87,939.00	$3,066.55	6218.2	5214.2	
014833183	STICK PUSHER ACTUATOR	7R	9	9	[illegible]	$17,342.00	$68.37	3115	3115.0	
013795875	REFRIGERATION UNIT COOLING TURBINE	7R	10	6	$21,691.26	$19,575.00	$2,116.26	12983.3	2816.8	

1 2 3 4 5 6 7 8 9 10 ... Page size: 10 — 381 items in 39 pages

32

VECTOR Logistics Analysis Tool

Platform XYZ Top 20 Non-Mission Capable Readiness Degraders

HOFNIIN	WUC	NOMENCLATURE	COG	TOP NMCS RANK	TRENDING NMCS RANK
LLZ98W331	293BA	HIGH STAGE BLEED AIR AUGMENTOR PNEUMATIC CONTROLLE	0R	1	1
014707884	41315, 41343	FLOW CONTROL VALVE	7R	2	2
014721429	41351, 41354, 41790, 48450	BLEED AIR DUCT HOT AIR SHUT-OFF VALVE	7R	3	3
LLZ98X364	64RE1	ICS CENTRAL SWITCHING UNIT	0R	4	4
014756203	293BA	HIGH STAGE BLEED AIR AUGMENTOR PNEUMATIC CONTROLLE	7R	5	5
014440419	41315, 41343	FLOW CONTROL VALVE	0R	6	6
014840676	56A55	DUAL INDICATOR AIRSPEED ALTIMETER	7R	7	7
007961672	41510	DIVIDER VALVE	7R	8	14
014833183	14Q30	STICK PUSHER ACTUATOR	7R	9	9
013795875	41322	REFRIGERATION UNIT COOLING TURBINE	7R	10	8
015762522	67J11	MULTIPURPOSE COLOR DISPLAY UNIT	7R	11	10
014521355	42W20	ELECTRONIC CIRCUIT BREAKER UNIT (ECBU) NO.1-14	7R	12	11
014977142	293B880	OIL COOLER FLAP ACTUATOR	7R	13	12
014726850	11820	VERTICAL STABILIZER LEADING EDGE ASSY	7R	14	54
LLZ98X003	293BD	HIGH STAGE BLEED AIR AUGMENTOR VALVE	0R	15	13
014725564	14Q10	ELEVATOR CONTROL BOOSTER MECHANISM ASSY	7R	16	18
014078856	47C10, 47C13	LOW PRESSURE REGULATORS	7R	17	50
016043773	42MC0	FUEL MANAGEMENT CONTROL PANEL ASSY	7R	18	15
014723492	293BR00	THROTTLE QUADRANT ASSY	7R	19	19
014978303	3271200	PROPELLER ASSEMBLY	0R	20	48

33

Questions?

34

HELICOPTERS

May 9, 2017
Fort Worth, Texas, USA

AIRBUS

Aviation Safety: Airbus Helicopters' chief priority

Supporting flight safety for the thousands of women and men around the world who are flying in our aircraft every day.

- ❑ **Safety by aviation SMS**
- ❑ **Safety by training & support to flight operations**
- ❑ **Safety by maintenance**
- ❑ **Safety by design**
- ❑ **Safety by cooperation**

AIRBUS

HELICOPTERS

Breakthrough Design…for Game-Changing Value

[Airbus Helicopters Internal]

Blue Edge ® blades

- Neighborly friendly
- Enhanced passenger comfort
- Superior performance

New generation turboshaft engine

- Step change in performance
- Unprecedented fuel efficiency
- Reduced emissions

New canted Fenestron ®

- Neighborly friendly
- Enhanced passenger comfort
- Superior performance

Full composite airframe

- Maximized occupant safety
- Reduced maintenance

Biplane stabilizer ™

- Exceptional handling
- Superior performance

Electrical landing gear ™

- Reduced maintenance
- Increased payload

68 patents

9 May, 2017 CEO Panel @ AHS Forum 3

AIRBUS

HELICOPTERS

[Airbus Helicopters Internal]

Innovations for the future

Speed with CleanSky 2

Urban Mobility with CityAirbus

Protection and autonomy with VSR700

Urban logistics with Skyways

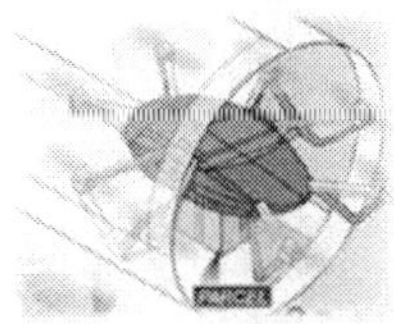

4 9 May, 2017 CEO Panel @ AHS Forum

AIRBUS

HELICOPTERS
[Airbus Helicopters Internal]
Thank you
5
9 May, 2017
CEO Panel @ AHS Forum
AIRBUS

Bell
Helicopter
A Textron Company
Michael Thacker
Executive Vice President,
Technology and Innovation

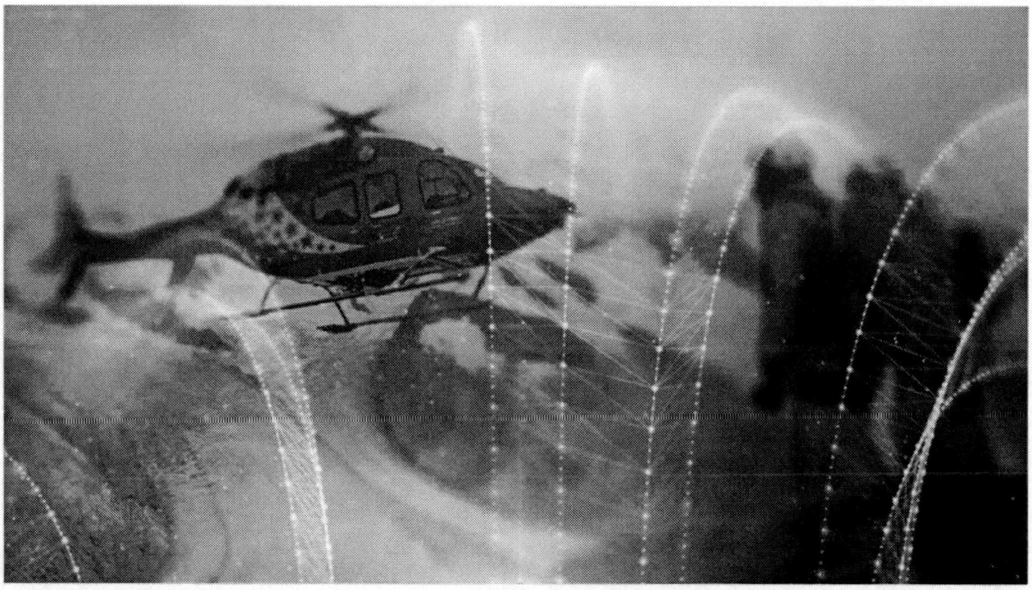

Bell
Helicopter
A Textron Company

Vertical Lift **Products**

V-22 Osprey

AH-64 Apache

H-47 Chinook

Advanced Technologies

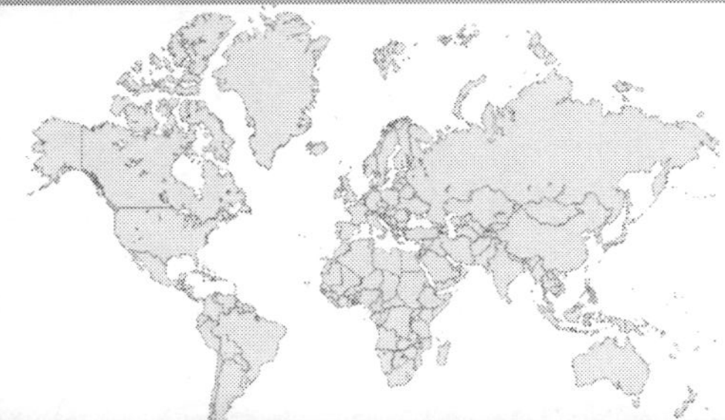

AH-6 Little Bird

Future Vertical Lift

MH-139

The right capability, at the right time, and at the right cost for customers worldwide

AHS Straight Talk from the Top: CEO Panel

Daniele Romiti
Managing Director – Leonardo Helicopters

delivered on his behalf by
Roberto Garavaglia
SVP Competitive Analysis & Strategy

Fort Worth, 9th May 2017

LEONARDO HELICOPTERS

AgustaWestland Products

LEONARDO HELICOPTERS — INGENUITY AT YOUR SERVICE

Leonardo Helicopters 2016 Key Financial Results

LEONARDO HELICOPTERS

USD[1] Bn

- **4.05 Sales**
- **4.05 Order intake**
- **11.6 Order Book**
- **0.4 EBIT [10.8% ROS]**
- **0.3 Net Profit**

Yesterday Finmeccanica, Today Leonardo

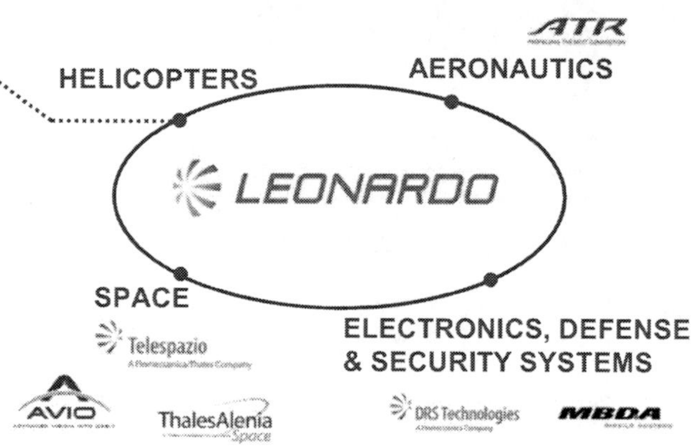

[1] Officially released results in Euros

AgustaWestland Products

LEONARDO HELICOPTERS

INGENUITY AT YOUR SERVICE

Portfolio Pillars: Helicopters, TiltRotor, RUAV

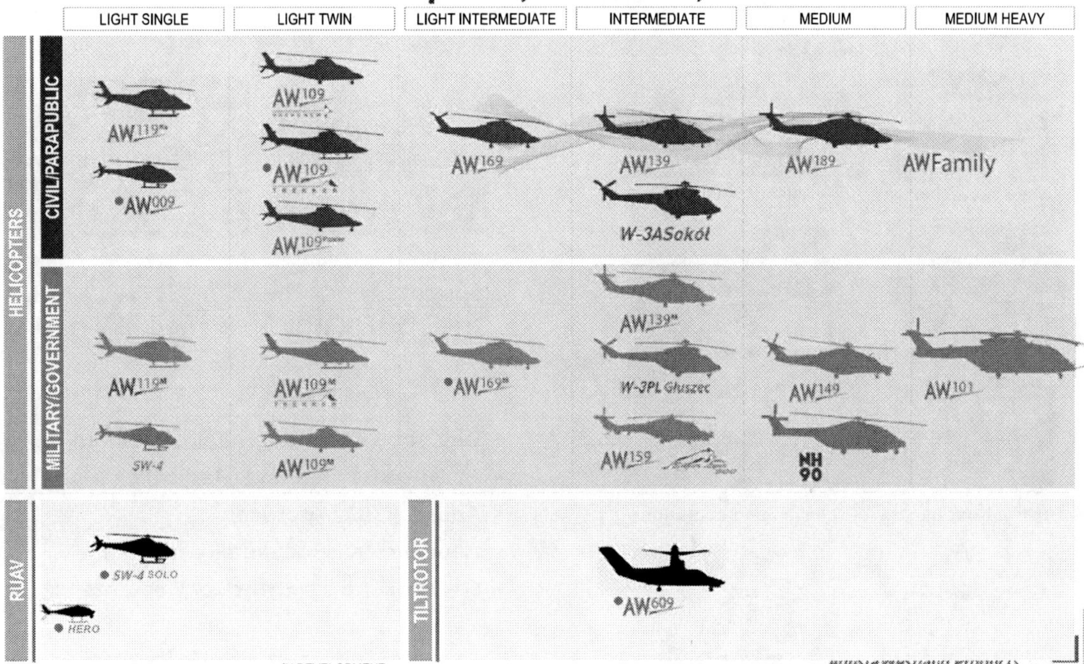

● IN DEVELOPMENT

AgustaWestland Products

LEONARDO HELICOPTERS

INGENUITY AT YOUR SERVICE

Market: Last-Next 10 Years

Total Market – split by segment:

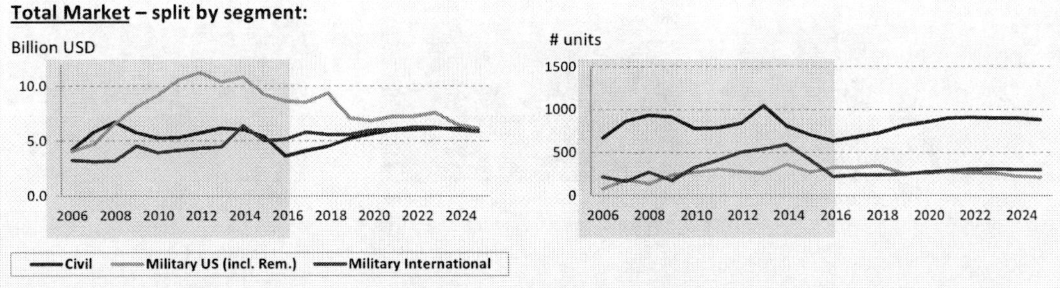

Average 10Y Total Market – split by segment:

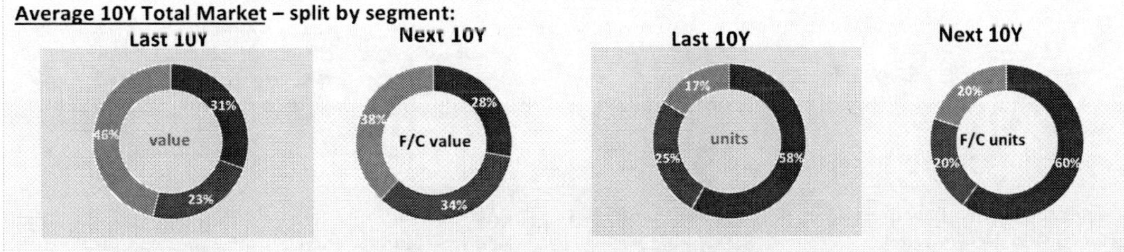

Sources: LHD elaboration of Teal and Ascend database

AgustaWestland Products 4

LEONARDO HELICOPTERS — INGENUITY AT YOUR SERVICE

DRIVERS – Technology & Innovation

Use of **Rotorcraft for Passenger Transport** is still limited compared to Fixed-wing…

public acceptance

resource issues

environmental issues

complexity of operations

infrastructures access

challenging operations

challenging economics

safety concerns

TECHNOLOGY & INNOVATION can change this

- **Develop / Integrate Enabling Technologies**
- **Pursue Incremental & Radical change**
- **Adopt a 360° Approach**
- **Allocate R&T Focused Resources**
- **Leverage on a Network of Competences**

AgustaWestland Products 5

LEONARDO HELICOPTERS — INGENUITY AT YOUR SERVICE

Key Technology Enablers

Multiplatforms NAV/COM/AFCS Integration

All WX, Hostile Environment Capabilities through Advanced System Integration & Integrity

Advanced Technologies, Health/Usage Monitoring, Flexible Maintenance Programs for Cost Reduction

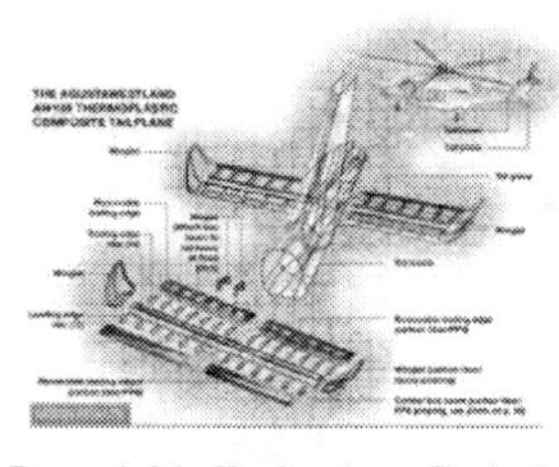
Recyclable Techs, Low Emissions (Noise, CO2, NOx, Waste)

DRY RUN

Intrinsic/Embedded Safety Features

Clean, Silent, Reduced Maintenance Power Plants

Innovative Advanced Configurations

AgustaWestland Products 6

Safe Design – Safe Operations

Embedded Safety

Situational Awareness

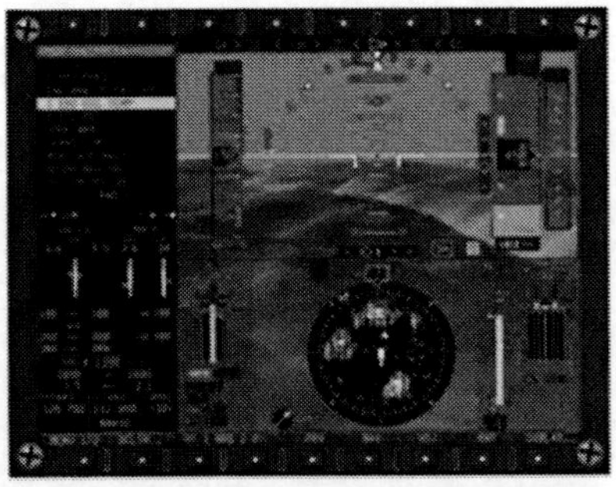

FIPS Ice Protection

PBN Compliance

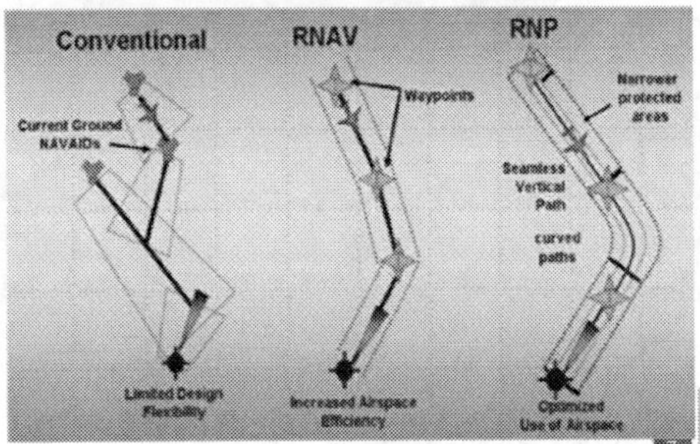

The Future is Here: Fast Rotorcraft & RUAV

AW609 TILTROTOR

SD-150 "HERO"

Next Generation Civil Tiltrotor

SW-4 "SOLO"

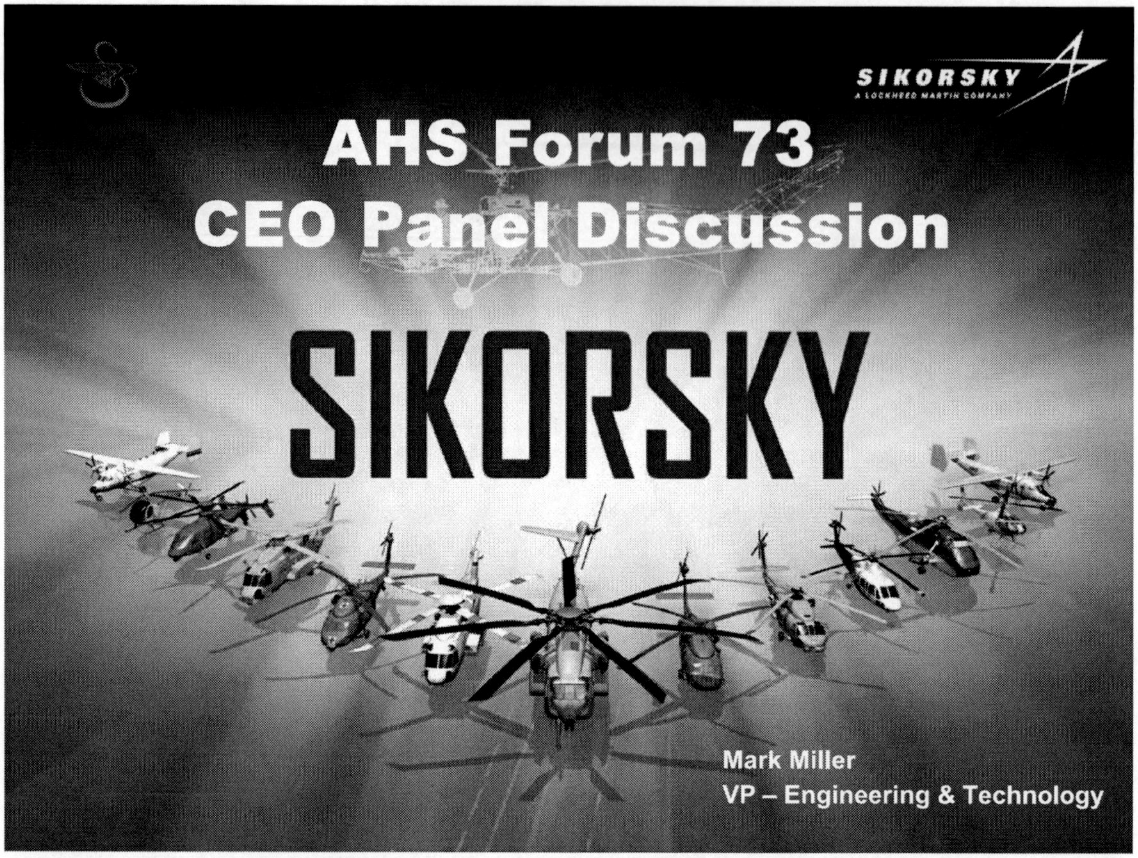
SIKORSKY
A LOCKHEED MARTIN COMPANY
AHS Forum 73
CEO Panel Discussion
SIKORSKY
Mark Miller
VP – Engineering & Technology

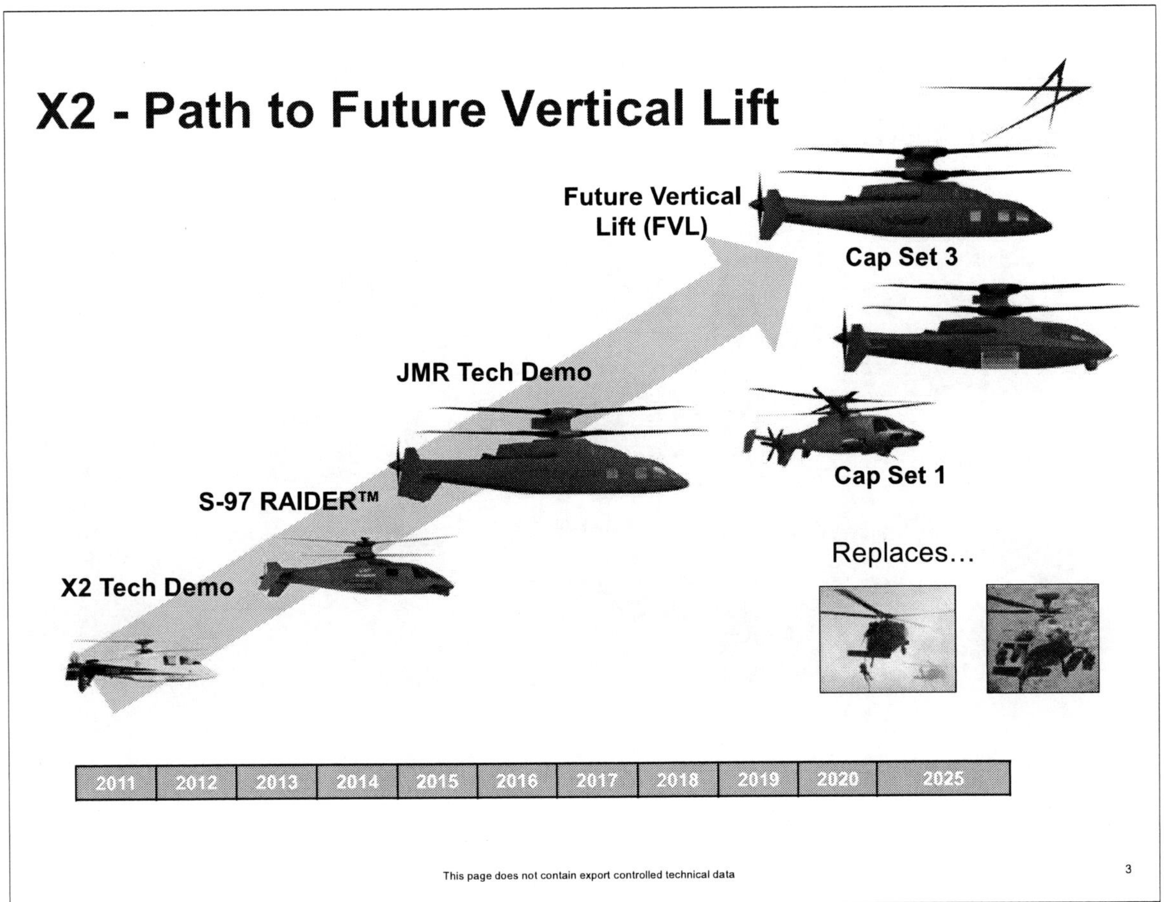

X2 - Path to Future Vertical Lift
Future Vertical Lift (FVL)
Cap Set 3
JMR Tech Demo
Cap Set 1
S-97 RAIDER™
Replaces...
X2 Tech Demo
2011
2012
2013
2014
2015
2016
2017
2018
2019
2020
2025
This page does not contain export controlled technical data
3

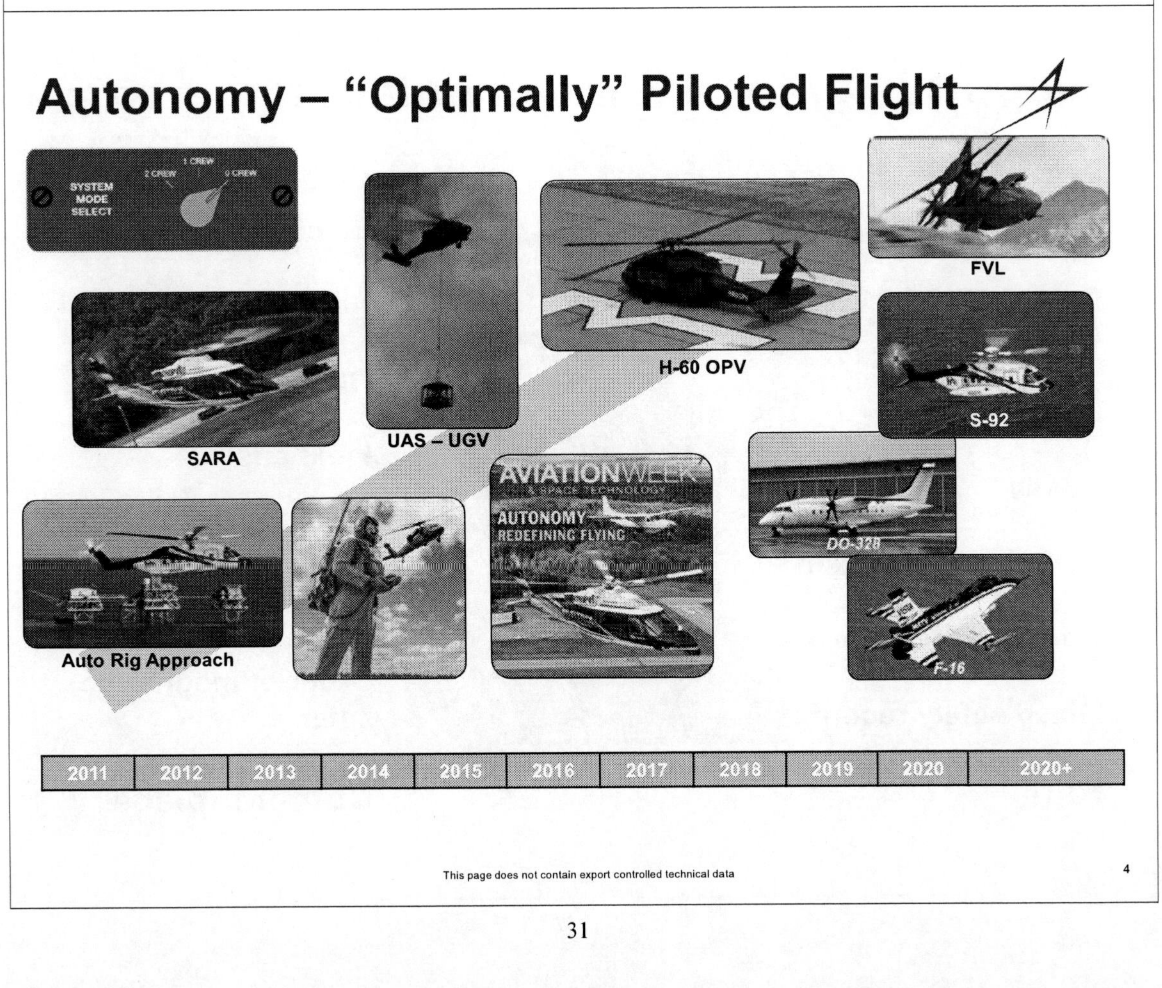

Autonomy – "Optimally" Piloted Flight
SYSTEM MODE SELECT
2 CREW
1 CREW
0 CREW
FVL
H-60 OPV
S-92
SARA
UAS – UGV
AVIATIONWEEK
& SPACE TECHNOLOGY
AUTONOMY
REDEFINING FLYING
DO-328
Auto Rig Approach
F-16
2011
2012
2013
2014
2015
2016
2017
2018
2019
2020
2020+
This page does not contain export controlled technical data
4

Intelligence – Available, Adaptable, & Affordable

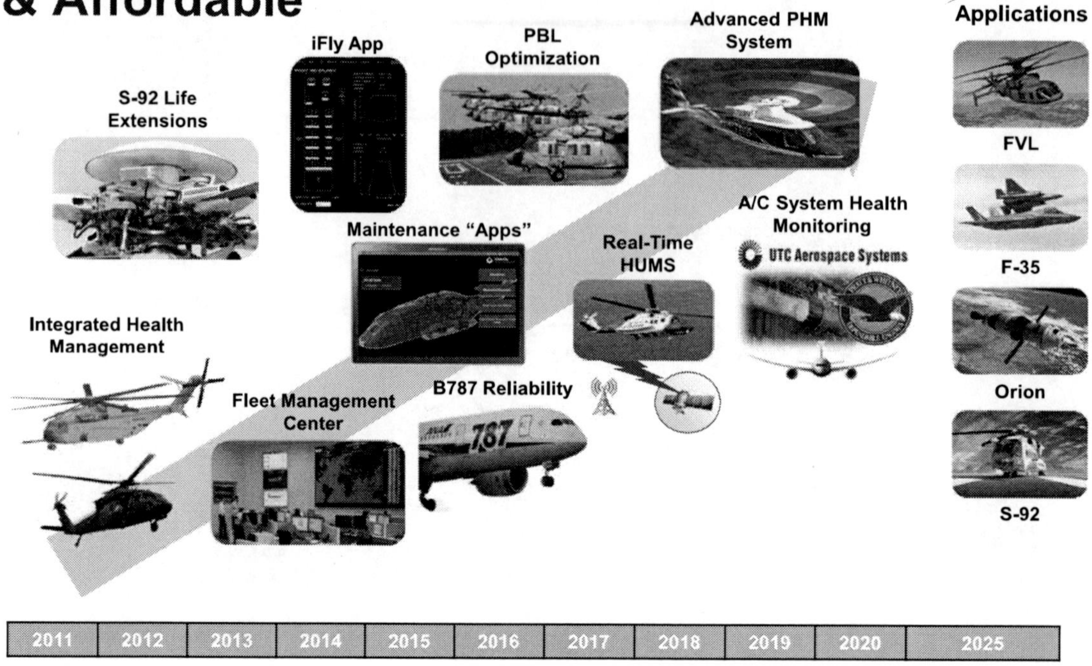

This page does not contain export controlled technical data

5

Opportunity for a Fundamental Shift

Helicopter 1.0

Turbine + Fuel

Conventional Controls

Loud

Costly

Complex to maintain

Operated by experts

Mixed safety record

Rare

Helicopter 2.0 (VERT™)

Electric

Autonomous controls

Quiet

Inexpensive

Reduced complexity

As easy as an elevator

Order of magnitude safer

Commonplace

This page does not contain export controlled technical data

6

NEXTGEN …

Future Vertical Lift

Search and Rescue

Personal Transportation

Commercial

This page does not contain export controlled technical data

Shaping the Future of Vertical Flight…

This page does not contain export controlled technical data

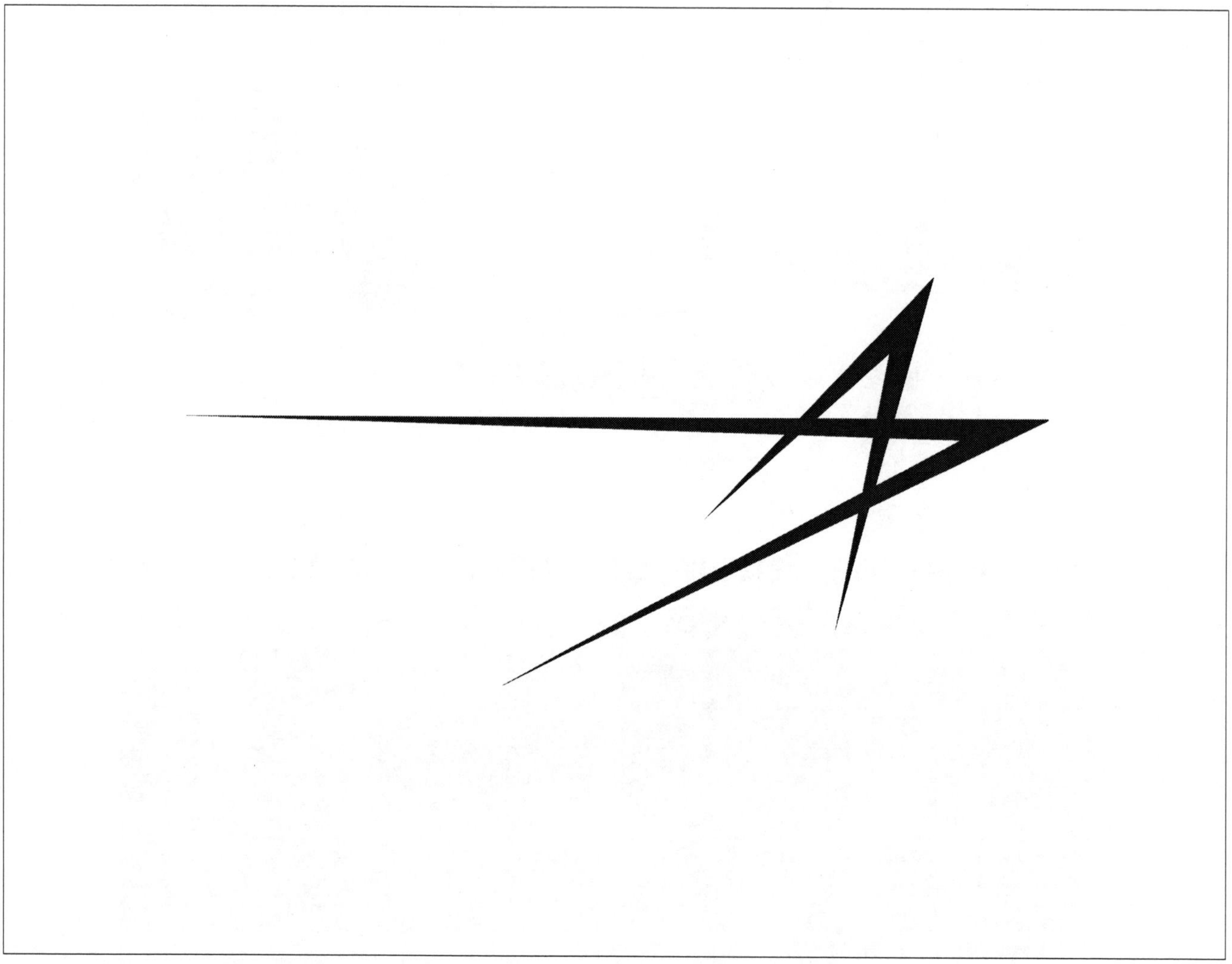

MD HELICOPTERS
MADE IN AMERICA

Disclosures

Confidentiality and Disclosure

This material is also Limited Rights Data or Restricted Rights Data in accordance with FAR subpart 27.4 or Limited Rights Technical Data in accordance with DOD FAR Supplement 227.7102-1(B)(2). Recipient by accepting this document agrees that neither this document nor any part thereof shall be reproduced or transferred to other documents or used by recipient, or disclosed to others, for manufacturing (such as PMA Authority) or any other purpose except as specifically authorized in writing by MD Helicopters, Inc.

Proprietary

This presentation includes information that is the property of MD Helicopters, Inc. (MDHI). This proprietary information, in its entirety, shall not be duplicated, used, or disclosed - directly or indirectly, in whole or in part – for any purpose other than to evaluate this presentation.

2

3

MD HELICOPTERS
MADE IN AMERICA

New Technology Projects

- Open Architecture Glass Cockpits
- MD 6XX Single-Engine Multi-Mission Helicopter
- Safety Enhancing Radar Cocoon

4

MD HELICOPTERS
MADE IN AMERICA

"This year is defined by glass, glass, glass and glass."
- Lynn Tilton, CEO, at HAI

Open Architecture Glass Cockpits

MD 530 F/G & 6XX

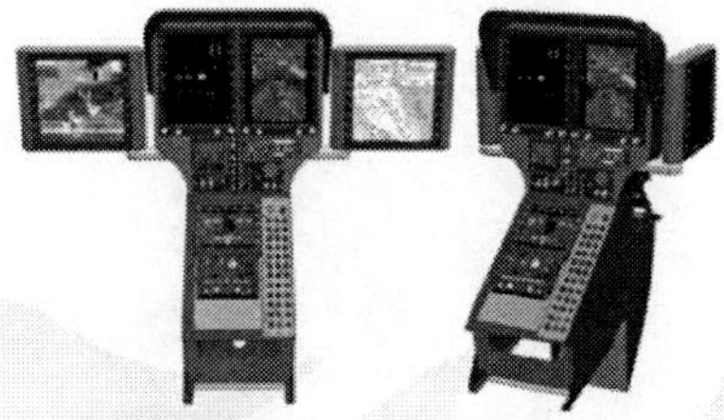

Commercial

Single Engine SPIFR

Fully Digital
100% digital, open architecture enables flexible, aircraft-independent installation

Future Proof
Field-upgradeable via software updates – add new functionality and grow capabilities without replacing hardware components

5

MD HELICOPTERS
MADE IN AMERICA

MD 6XX Single Engine Multi-Mission Helicopter

Projected Performance

MGTOW	5,500 lbs
Range	500 nm
Speed	160 kts
Endurance	4.5 hrs
Passengers	6 or 8 on Planks

6

MD HELICOPTERS
MADE IN AMERICA

Key Enhancements

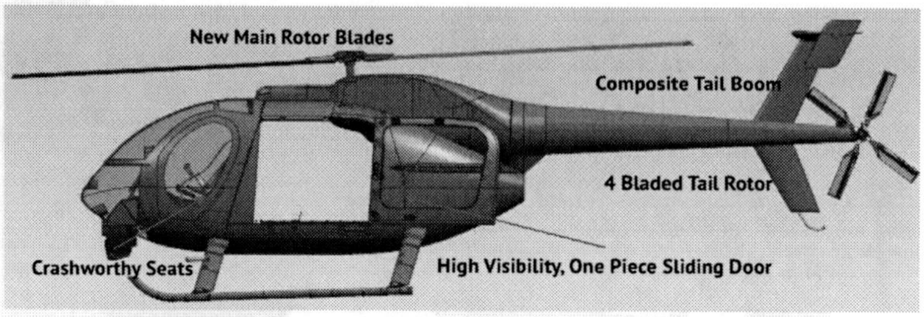

All-Glass Cockpit with PFD/MFD and Dual Tactical Displays

7

MD HELICOPTERS
MADE IN AMERICA

MD 6XX Configuration Variables & Design Considerations

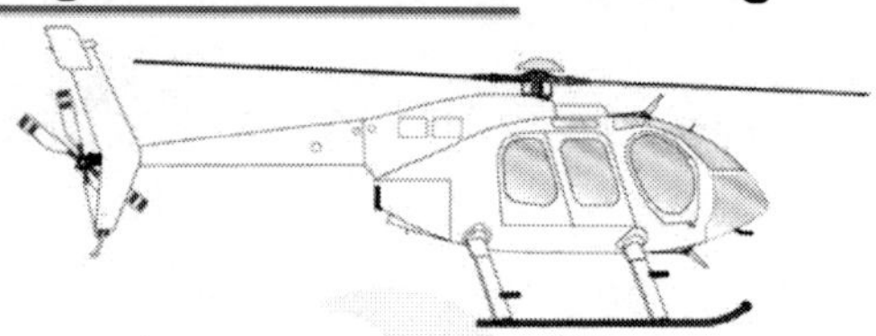

Key Features	Component Design Considerations
Main Rotor Blades	Airfoil Design, Chord, Rotor Solidity
Fuel Capacity	Quantity, Sealing, Location
Tail Rotor	Airfoil Design, Number of Blades
Rotor Head	Life, Number of Blades, Sizing/Solidity
Engine and Shaft Power	SHP, Ratings,
Transmission Rating	SHP, Ratings
Power assist flight controls	Ease of Control, Usage
Tail Boom, Vertical, Structure	Sized for 600 Fuselage
Landing Gear	Increased Capacity

8

MD HELICOPTERS
MADE IN AMERICA

Advanced Technologies: Safety Enhancement & Mission Flexibility

Protective "Radar Cocoon"
Leveraging Low Cost Automotive / Semiconductor Technology

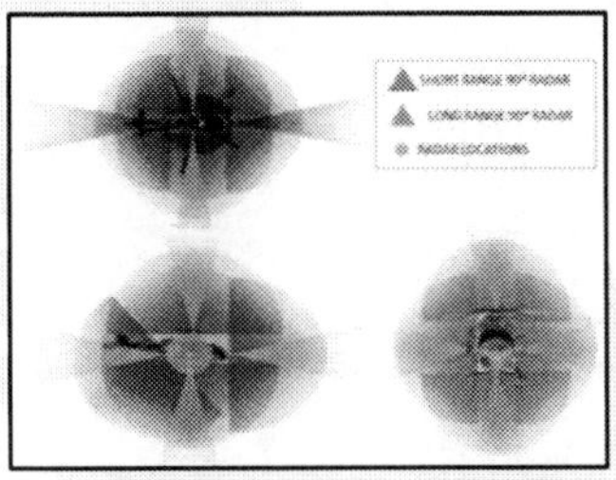

- Prototype Demonstration 1-2Q17
- Production Flight Testing 3Q-4Q17
- **Streamlined FAA Certification 4Q17**
  - Non-Required Safety Enhancing Equipment (NORSEE)

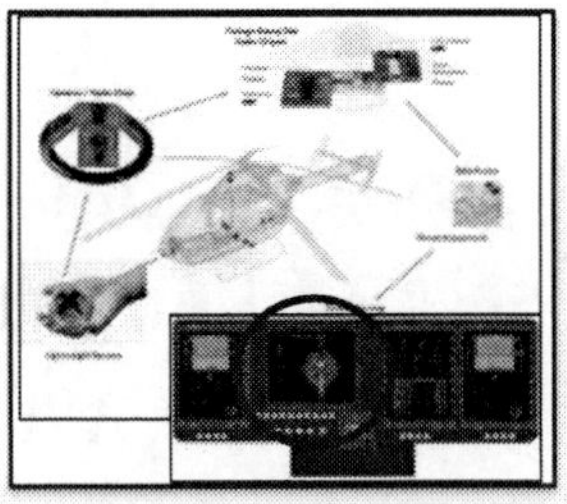

Recoverable Drones

9

*** Formatting Issue - Best Available Paper/Graphic ***

ADVANCES IN ROTORCRAFT CRASHWORTHINESS – TRENDS LEADING TO IMPROVED SURVIVABILITY

Karen E. Jackson, Ph.D.
NASA Langley Research Center
Hampton, VA

Alexander A. Nikolsky Lectureship
73rd American Helicopter Society International Forum
Tuesday, May 9, 2017

OUTLINE

- Introduction and Objectives
- Facilities and Equipment for Conducting Full-Scale Crash Testing
- Updated Crash Certification Requirements
- Crash Modeling and Simulation
- Rotorcraft Structural Design for Crashworthiness
- Recommendations for the Future
- Concluding Remarks

*** Formatting Issue - Best Available Paper/Graphic ***

INTRODUCTION

What is crashworthiness?

- A major goal of crashworthy design is to limit the dynamic loads transmitted to the occupants to survivable levels by incorporating crushable structure, for example
  - crumple zones in cars
  - energy absorbing (EA) landing gear, subfloors and seats in aircraft/rotorcraft

- This structure is designed to fail at a certain load, and then crush in a stable manner through progressive failure, thus removing kinetic energy from the system

- Under compressive loading, the structure is expected to fail locally and crush in a stable manner without global buckling

INTRODUCTION

Examples of Automotive Crumple Zones

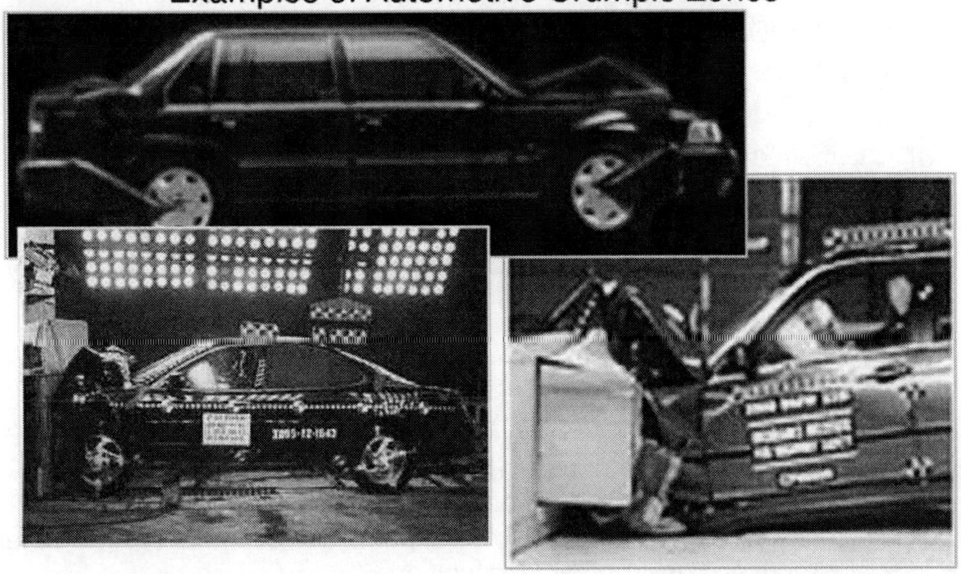

*** Formatting Issue - Best Available Paper/Graphic ***

INTRODUCTION

What factors are involved in rotorcraft crash safety?

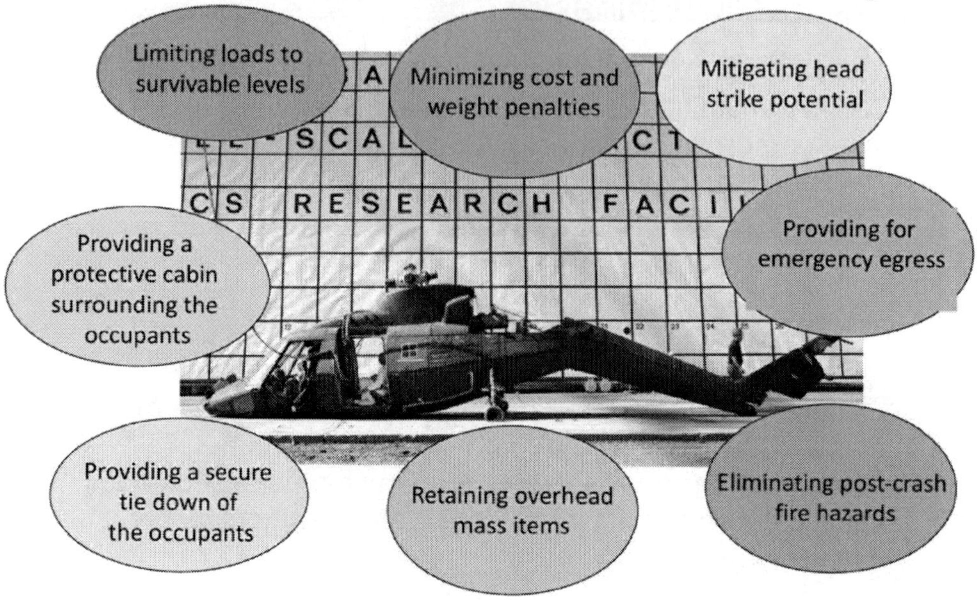

INTRODUCTION

What is the current level of crash survivability?

Study 1: DOD-sponsored Study on Rotorcraft Survivability, sent to the US Congress in October 2009. Reference: Couch M. and Lindell D., "Study on Rotorcraft Safety and Survivability," AHS Forum 66, Phoenix, AZ, May 10-13, 2010.

Conclusions:

- For cargo and utility helicopters, combat hostile action loss rate is nearly 6 times lower, and fatality rate 4 times lower for Operation Enduring Freedom/Operation Iraqi Freedom versus the Vietnam War
- A high percentage of helicopter losses are survivable
- "...to reduce personnel injuries and fatalities for combat threat losses and mishaps, improve aircraft crashworthiness and crash protection for passengers. DOD crashworthiness standards have not been updated since the 1970's and they need to be expanded in scope to cover a wider set of aircraft and environmental conditions."

*** Formatting Issue - Best Available Paper/Graphic ***

INTRODUCTION

What is the current level of crash survivability?

Study 2: US Army AATD sponsored an accident study under the Full Spectrum Crashworthiness Program. Reference: Labun L., "A Study of Rotary-Wing Crashes to Support New Crashworthiness Criteria" AHS Forum 66, Phoenix, AZ, May 10-13, 2010.

Conclusions:

- Crashworthy fuel systems have virtually eliminated fatalities and severe injuries due to post-crash fire
- Onboard systems such as lap belts, shoulder restraints, inertia reels, and load limiting seats are effective in improving crash performance
- Of the crashes studies, fewer that 20% occurred onto prepared surfaces
- The 90th-percentile survivable forward and vertical ground speeds at impact for aircraft designed to crashworthy criteria were generally higher than those for the previous generation of aircraft.

INTRODUCTION

What is the current level of crash survivability?

Study 3: US Navy sponsored a study, Reference: Kent R., "Injury and Fatality Patterns in US Navy Rotary Wing Mishaps: A Descriptive Review of Class A and B Mishaps from 1985 to 2005," AHS Forum 65, Grapevine, TX, May 27-29, 2009.

Conclusions:

- 85-94 vs. 95-05: DON fatality rate (per 100,000 flight hours) reduced from 5.8 to 3.15 and the injury rate (per 100,000 flight hours) reduced from 3.92 to 2.14
- Decadal differences attributed to improved policies, guidelines, training procedures and equipment
- Head injuries are an important cause of morbidity in helicopter mishaps
- Non-pilot personnel appear to be at greater risk for injury

*** Formatting Issue - Best Available Paper/Graphic ***

INTRODUCTION

What is the current level of crash survivability?

Civilian Helicopter Accident Statistics per 100,000 Flight Hours*

Year	Total Accidents	Total Accident Rate	Fatal Accidents	Fatal Accident Rate
2016	106	3.19	17	0.51
2015	121	3.67	17	0.52
2014	138	4.26	21	0.65
2013	146	4.95	30	1.02

* https://ohsonline.com/articles/2017/02/14/faa-helicopter-accident-rates-decreasing.aspx, "FAA: Helicopter Accident Rates Dropping," Feb 14, 2017.

Systems Approach to Crash Safety NASA Schematic 2017

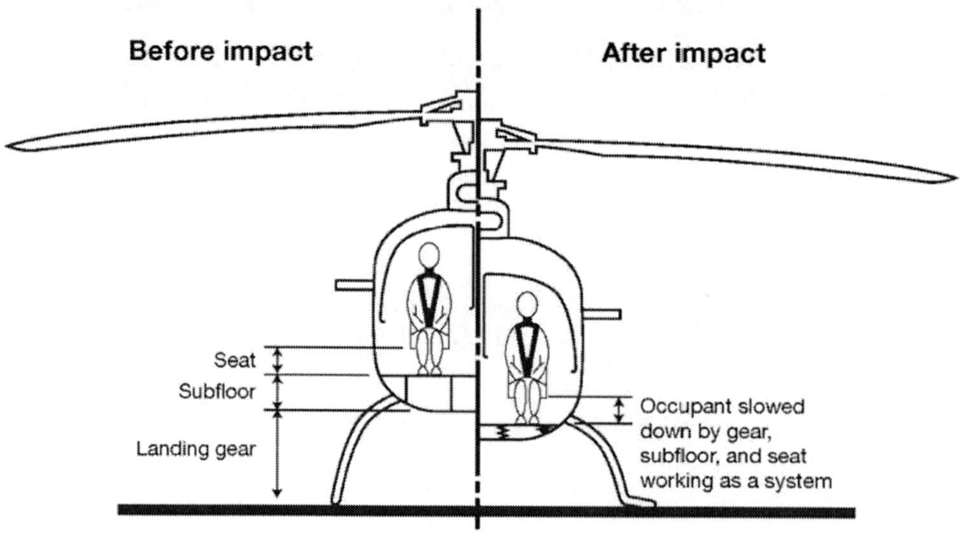

*** Formatting Issue - Best Available Paper/Graphic ***

Schematic Version 1: Aircraft Crash Survival Design Guide

Attributed to Jim Cronkhite, Bell Helicopter 1986

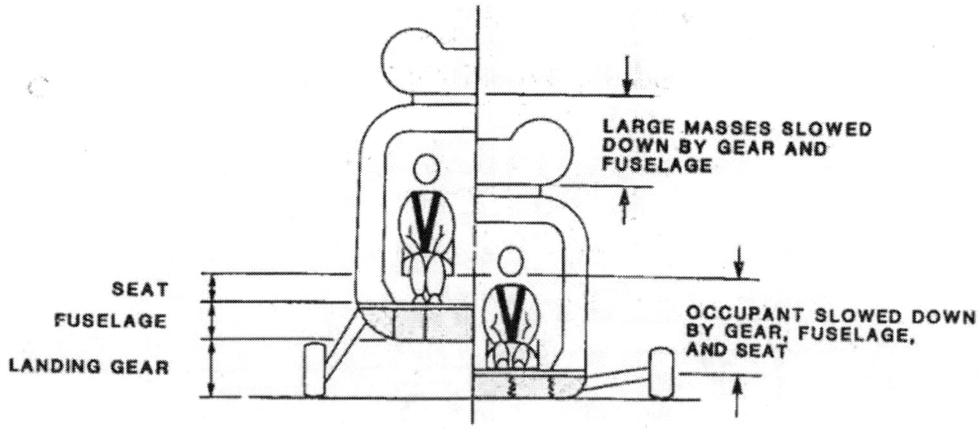

Schematic Version 2

AGATE-CP-443 Energy Absorption of Aircraft Structures as an Aspect of Crashworthiness, 1988

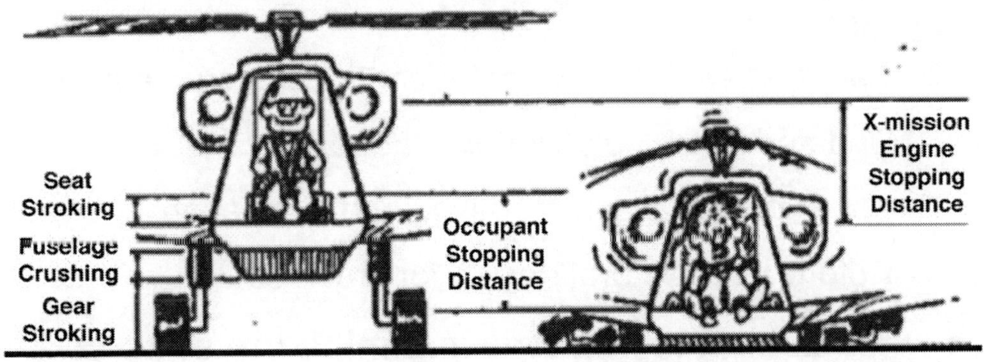

*** Formatting Issue - Best Available Paper/Graphic ***

Schematic Version 3

Small Airplane Crashworthiness Design Guide, attributed to Todd Hurley and Jill Vandenburg, 2002

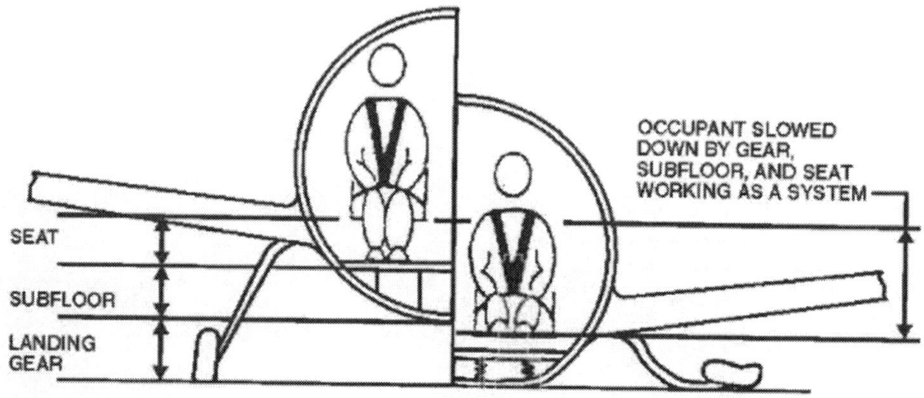

OBJECTIVES

- Examine the contributions to the current level of crash safety based on improvements in four areas:
  - Facilities and equipment for conducting crash testing
  - Updated crash certification requirements
  - Application of crash modeling and simulation
  - Rotorcraft structural design for crash safety
- Assess past technology, examine the state-of-the-art, and make recommendations for the future

*** Formatting Issue - Best Available Paper/Graphic ***

Facilities & Equipment for Crash Testing

- Full-Scale Crash Test Facilities
  - -Lunar Landing Research Facility (LLRF) at NASA Langley
  - -Impact Dynamics Research Facility (IDRF) at NASA Langley
  - -Landing and Impact Research (LandIR) Facility at NASA Langley
  - -LISA Aerospace Structures Impact Test Facility at Capua, Italy
  - -Water Test Facility at Yuma Proving Ground
- Vertical Drop Towers
  - -Vertical Test Apparatus (VTA) at NASA Langley
  - -Dynamics Drop Test Facility at the FAA Technical Center
- Equipment Supporting Full-Scale Crash Tests
  - - Data Acquisition Systems (DAS)
  - - Photogrammetry
  - -Anthropomorphic Test Devices (ATDs)

Facilities for Spacecraft Landing: LLRF

Lunar Landing Research Facility (LLRF)

- Steel A-frame structure
- Operational in 1963 to allow Apollo astronauts to practice the last 150-ft. of descent onto a simulated lunar surface at 1/6th gravitational conditions
- 400-ft. in length, 240-ft. tall, 265-ft. wide at base
- Awarded a National Historic Landmark for contributions to the Apollo Program

*** Formatting Issue - Best Available Paper/Graphic ***

Facilities for Spacecraft Landing: (LLRF)

Neil Armstrong at the LLRF

Lunar Excursion Module tethered from the LLRF

Facilities for Full-Scale Crash Testing: IDRF

At the end of the Apollo Program, the gantry facility was renamed Impact Dynamics Research Facility (IDRF) to investigate the crashworthiness of light aircraft/rotorcraft

*** Formatting Issue - Best Available Paper/Graphic ***

Facilities for Full-Scale Crash Testing: IDRF

Many full-scale crash tests of military helicopters have been performed at the IDRF

CH-47 Crash Test, 1976

UH-60 Crash Test, 1999

Facilities for Crash Testing: LandIR

In 2006, the IDRF was renamed Landing and Impact Research (LandIR) facility, since both spacecraft landing and aircraft crash testing were being performed

*** Formatting Issue - Best Available Paper/Graphic ***

Facilities for Crash Testing: LandIR Upgrade

Aerial View of LandIR Facility

- In 2011, construction of the Hydro Impact Basin (HIB) was completed
- HIB is 115-ft. long, 90-ft. wide, and 20-ft. deep
- It has been used to perform 47 full-scale impact tests since opening
- With the HIB, the LandIR facility now has full multi-terrain capability

Facilities for Crash Testing: LandIR Upgrade

Flight Path Determined By Swing Cable Length

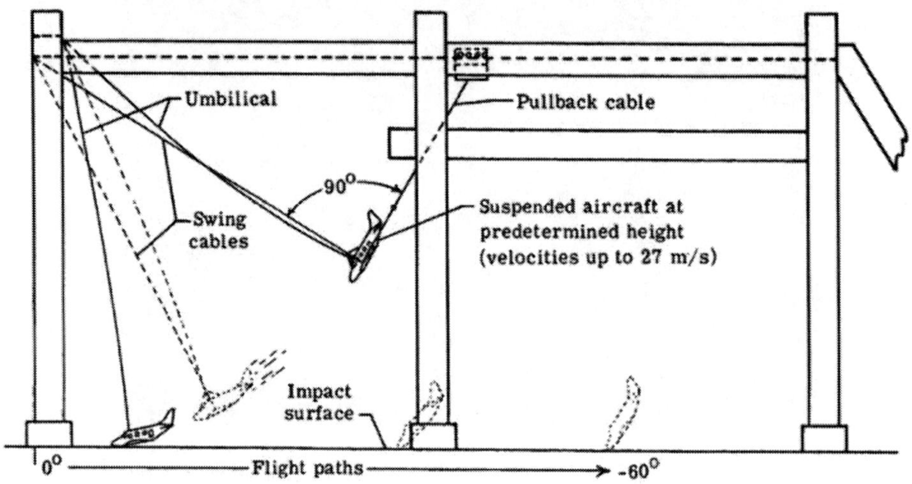

Test procedure changed to eliminate angular velocity, which is introduced using pendulum-swing

*** Formatting Issue - Best Available Paper/Graphic ***

Facilities for Crash Testing: LandIR Upgrade

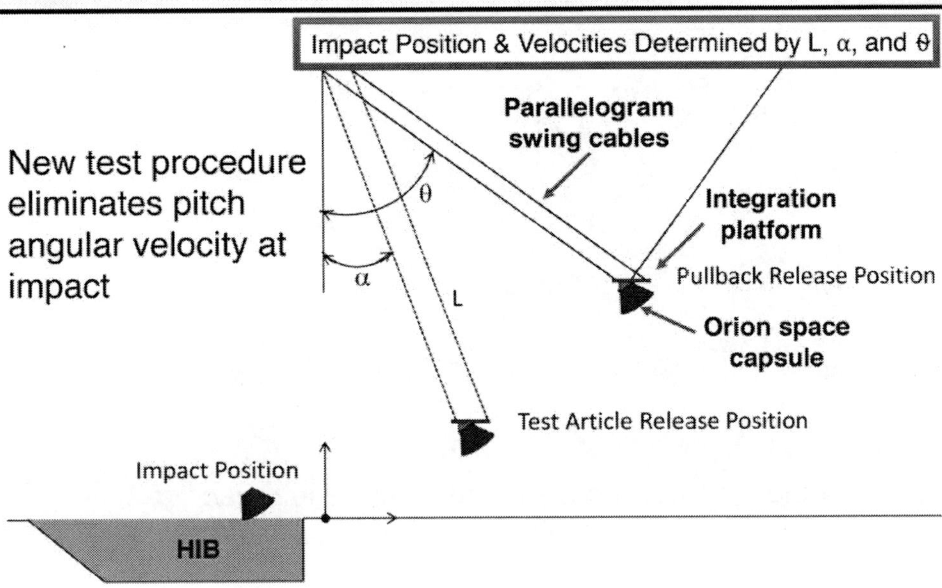

Facilities for Crash Testing: LandIR Upgrade

Orion space capsule impact into water

*** Formatting Issue - Best Available Paper/Graphic ***

Facilities for Crash Testing: LISA

- LISA Aerospace Structures Impact Test Facility, located in Capua, Italy, built in 2002
- Lift capacity of 44,000-lb.
- Has been dormant for the past 7 years

Facilities for Crash Testing: Water Impact Test Facility

- Water Test Facility at Yuma Proving Ground, 2012
- Impact Tests were conducted on UH-1 Huey
- Combined velocity impacts were performed using a trolley-and-guide-beam system

*** Formatting Issue - Best Available Paper/Graphic ***

Facilities for Crash Testing: Vertical Drop Towers

Vertical Test Apparatus (VTA) at LandIR

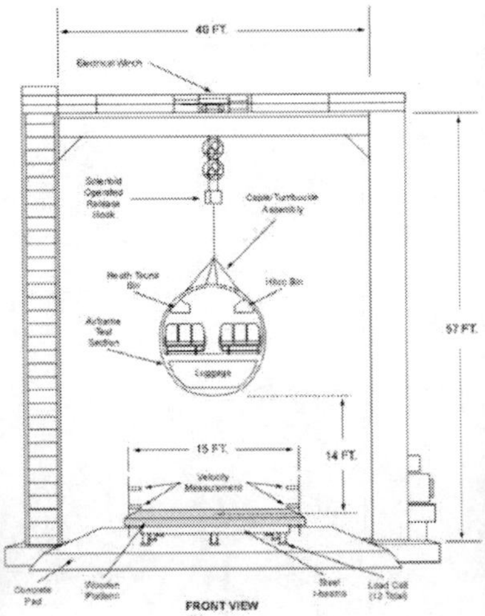

Vertical Drop Tower at FAA Technical Center, Atlantic City, NJ

Equipment for Crash Testing: Data Acquisition System (DAS)

DAS Properties	1980's Technology	Current Technology
Data collection	FM multiplexed analog tapes	Ruggedized onboard digital recorders
Turnaround time	several days, up to 3 weeks	Instantaneous
Channels recorded	96	512
Sampling rate per channel	4,000 samples/sec	100,000 samples/sec
Issues	Requires long umbilical cable to transmit data	No issues

*** Formatting Issue - Best Available Paper/Graphic ***

Equipment for Crash Testing: Photogrammetry

2013 Crash test of a CH-46E Sea Knight helicopter with 3-D photogrammetry

Animation of lateral displacement fringe plots based on photogrammetry

Equipment for Crash Testing: Anthropomorphic Test Devices (ATDs)

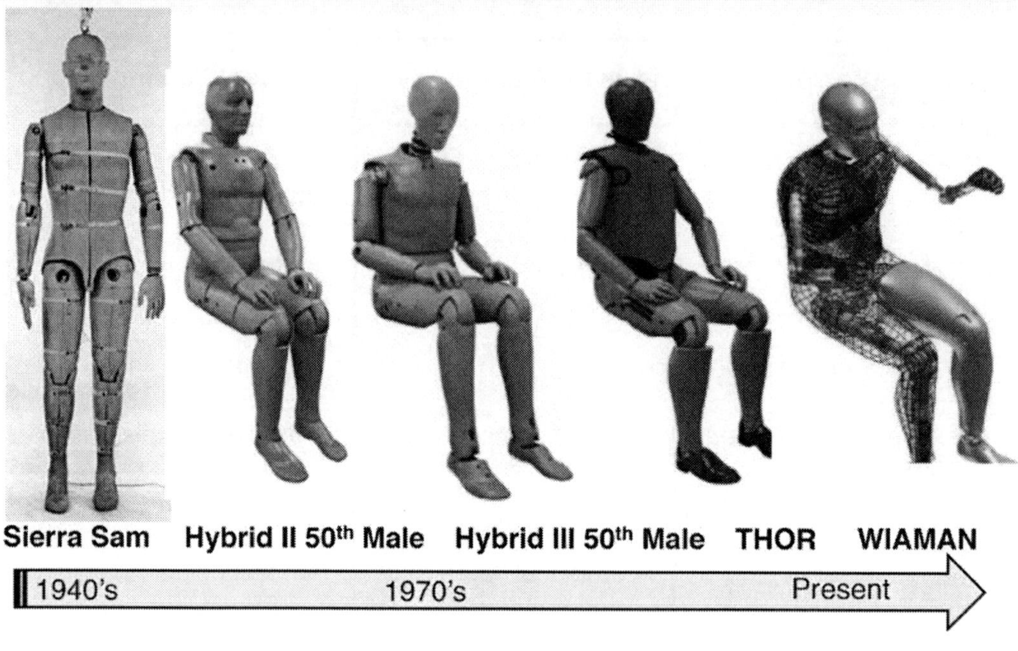

*** Formatting Issue - Best Available Paper/Graphic ***

Updated Crash Certification Requirements

- MIL-STD-1290(AV) A
- Full Spectrum Crashworthiness (FSC)
- FAA Seat Requirements for Civilian Aircraft/Rotorcraft

Updated Crash Certification Requirements

MIL-STD-1290(AV) A

- MIL-STD-1290(AV) A was initially released in 1974 and it established the minimum crash resistance criteria for implementation in the initial stages of aircraft design.
- The standard was revised and released in September 1988.
- In the late 1980's, an Aeronautical Design Standard (ADS-36) for Rotary-Wing Aircraft Crash Resistance was developed specifically for qualifying the RAH-66 Comanche.
- ADS-36 contains similar information found in the MIL-STD-1290A (AV). However, some of the criteria were relaxed so that the RAH-66 would have equivalent crash resistance as the UH-60 Black Hawk helicopter.
- The MIL-STD-1290(AV) A was cancelled by DOD in the mid-1990's to give defense contractors more freedom to develop creative & innovative designs without the constraints provided by the military standards.

*** Formatting Issue - Best Available Paper/Graphic ***

Updated Crash Certification Requirements

MIL-STD-1290(AV) A

Condition No.	Impact Direction (Aircraft Axes)	Velocity Change (ft/s)	Object/ Surface Impacted
1	Longitudinal (cockpit)	20	Rigid vertical barrier
2	Longitudinal (cabin)	40	
3	Vertical*	42	Rigid horizontal surface
4	Lateral, Type I	25	
5	Lateral, Type II	30	
6	Combined high angle* Vertical Longitudinal	 42 27	
7	Combined low angle Vertical Longitudinal	 14 100	Plowed soil

Updated Crash Certification Requirements

MIL-STD-1290(AV) A

Impact Condition	Performance Requirements for Crash Design Conditions
Combined	The designer shall ***analytically*** demonstrate the capability of the aircraft with 1 Standard Design Gross Weight (SDGW) rotor/wing lift and with landing gear extended to withstand the following combined impacts <u>without a reduction of the cockpit or cabin compartments that would seriously injure the occupants</u>: (1) a combined impact onto a rigid horizontal surface with vertical and longitudinal velocity changes of 42 and 27 ft/s, and (2) combined impact onto plowed soil with vertical and longitudinal velocity changes of 14 and 100 ft/s, respectively.

*** Formatting Issue - Best Available Paper/Graphic ***

Updated Crash Certification Requirements

MIL-STD-1290(AV) A

The MIL-STD also states that "instrumented full-scale crash test(s) are desirable to substantiate the capability of the aircraft system to prevent fatalities and minimize injuries during crashes. If the system testing is not conducted, then analysis shall be required to show that the individual components and subsystems function together effectively to achieve the specific overall level of crashworthiness."

- Thus, the military standard established *both* full-scale crash testing and analytical modeling as tools to achieve crash certification.

Updated Crash Certification Requirements

Full Spectrum Crashworthiness (FSC)

- US Army Aviation Applied Technology Directorate (AATD) sponsored the FSC, beginning in 2006(7) to develop comprehensive crash design requirements for a wide range of rotorcraft classes, types, configurations, and operating conditions that continue over the life cycle of the rotorcraft system

- Identify the key components that contribute to a system's crashworthiness

- Crashworthiness Index (CI) proposed as a new design standard to replace current specifications

	Basic Airframe Crashworthiness Rating Summary	Optimum Score	Accessed Score
1	**Crushing of Occupied Areas**		
1a	Crushing of Cockpit	15	16.52
1b	Crushing of Cabin	15	18.53
2	Absence of "Plowing" Tendency	10	10
3	Resistance to Longitudinal Impact Loads	10	10
4	**Resistance to Vertical Loads**		
4a	Vertical Impact, Gear Extended	80	61.5
4b	High Mass Retention	20	20
5	**Resistance to Lateral and Rollover Impact Loads**		
5a	Lateral Impact	15	13.5
5b	Static Rollover	15	15
6	**Landing Gear Vertical Force Attenuation**		
6a	Vertical Impact, Gear Retracted	40	38.8
6b	High Angle Vertical Impact	20	6.9
6c	Low Angle Vertical Impact	20	18.3
6d	Tail Boom Protection	5	5
7	Landing Gear Location	5	5
8	Effects of Blade Strike	20	20
9	Effect of Fuselage Separation	5	5
	TOTAL	**295**	**264.0**

Example from RDECOM TR 12-D-12

*** Formatting Issue - Best Available Paper/Graphic ***

Updated Crash Certification Requirements

FAA Seat Requirements for Code of Federal Regulations (CFR) Parts 23-29 Aircraft/Rotorcraft

- There are no airframe-level crashworthiness requirements for civilian rotorcraft, similar to MIL-STD-1290(A) AV or ADS-36.

- Instead, dynamic seat requirements are specified in CFR Parts 27 and 29 for normal and transport helicopters

- As with civilian helicopters, no aircraft-level crash requirements exist for General Aviation aircraft, regional jets, or transport aircraft. Seat requirements are specified in CFR Parts 23 and 25.

- In 2016. an Aviation Rulemaking Advisory Committee (ARAC) was formed to provide recommendations to the FAA for establishing airframe-level crash and ditching requirements for transport aircraft

Application of Crash Modeling and Simulation

- Description of Two Codes
  - -KRASH (1980's technology)
  - -LS-DYNA (Current technology)
- Illustrative Examples of Test-Analysis Comparisons
  - -KRASH simulation of the Sikorsky Advanced Composite Airframe (ACAP) crash test
  - -LS-DYNA simulation of a crash test of a MD-500 helicopter outfitted with a Deployable Energy Absorber (DEA)
- Modeling of Anthropomorphic Test Devices (ATDs)
- Modeling of Multi-Terrain Impact
- Expanded Capabilities of LS-DYNA

*** Formatting Issue - Best Available Paper/Graphic ***

Application of Crash Modeling and Simulation

Description of KRASH (1980's technology)

- Developed initially by the Lockheed-California Co. under US Army sponsorship in 1974. Later, the FAA supported further development of the code.
- Latest public domain version is KRASH85 and a commercial version, DRI-KRASH, was marketed by Dynamic Response Incorporated.
- KRASH models consist of massless interconnecting beams, concentrated rigid body masses, and spring elements.
- KRASH models typically rely on experimental data for input of spring properties, or a hybrid modeling approach is used in which finite element analyses are performed to supply data needed for input to the KRASH model.

Application of Crash Modeling and Simulation

Description of LS-DYNA (Current technology)

•LS-DYNA is a 3-dimensional transient dynamic finite element code for simulating problems involving high speed, nonlinear deformation of structures and solids, modal vibrations, and fluid-structure interactions

•The code is based on DYNA3D, a public domain code developed by Lawrence Livermore National Labs under Department of Energy sponsorship

•The code is developed and marketed by Livermore Software Technology Corporation (LSTC)

•LS-DYNA has a wide user base, and contains over 200 material property definitions, allows both implicit and explicit time integration, and contains a wide variety of contact algorithms

*** Formatting Issue - Best Available Paper/Graphic ***

Application of Crash Modeling and Simulation

Illustrative Example of KRASH (1980's technology)

Vertical Drop Test in 1987
(Velocity = 39 ft/s)

KRASH Model
- 53 discrete masses
- 23 springs
- 75 beam elements
- 44 nodes

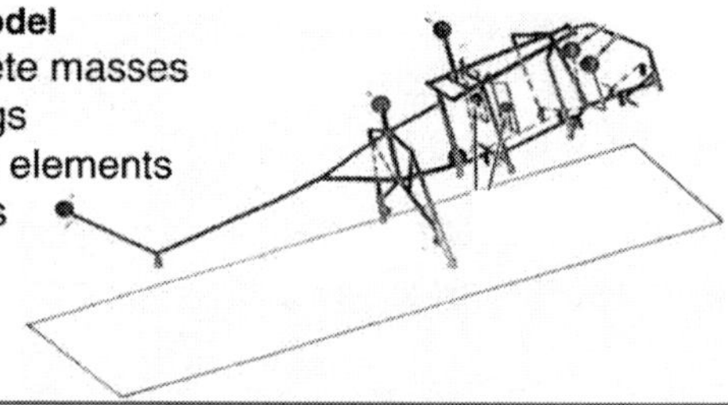

Test-Analysis Correlation of the Transmission Mass Acceleration

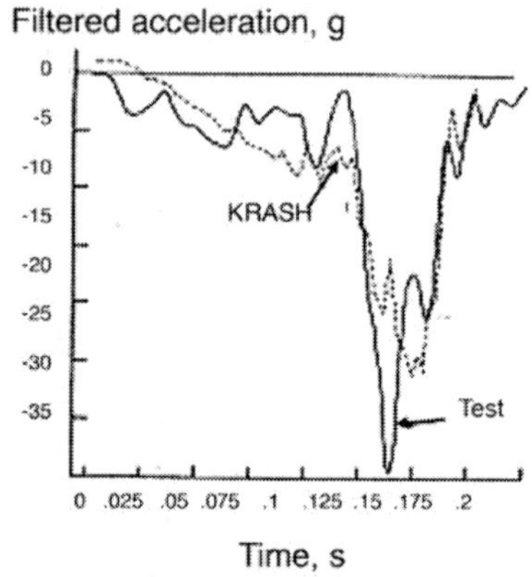

Figures reprinted from Perschbacher, J. P., Clarke, C., Furnes, K., and Carnell, B., "Advanced Composite Airframe Program (ACAP) Militarization Test and Evaluation (MT&E), Vol. V - Airframe Drop Test," USAATCOM TR 88-D-22E, March 1996.

Application of Crash Modeling and Simulation

Illustrative Example of LS-DYNA Simulation a Full-Scale Crash Test of an MD-500 Helicopter Retrofitted with a Deployable Energy Absorber, 2010

Crash test conducted at 25.6-ft/s vertical, and 38.8-ft/s forward velocity with -5.7° pitch, 7.0° roll, and 9.3° yaw onto concrete

*** Formatting Issue - Best Available Paper/Graphic ***

Application of Crash Modeling and Simulation

Illustrative Example of LS-DYNA

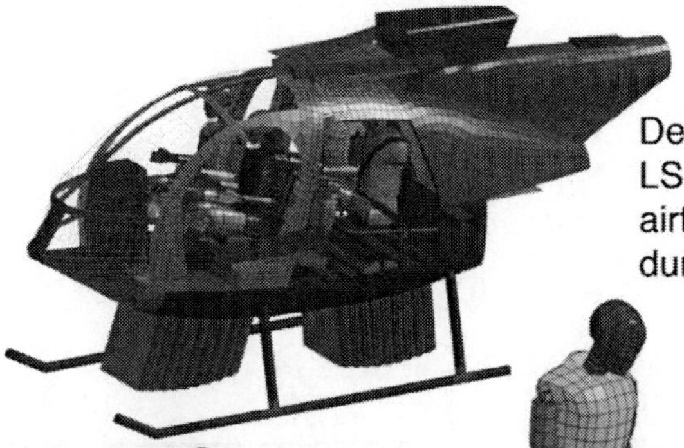

Developed a system-integrated LS-DYNA model, including the airframe, seats, restraints and dummies

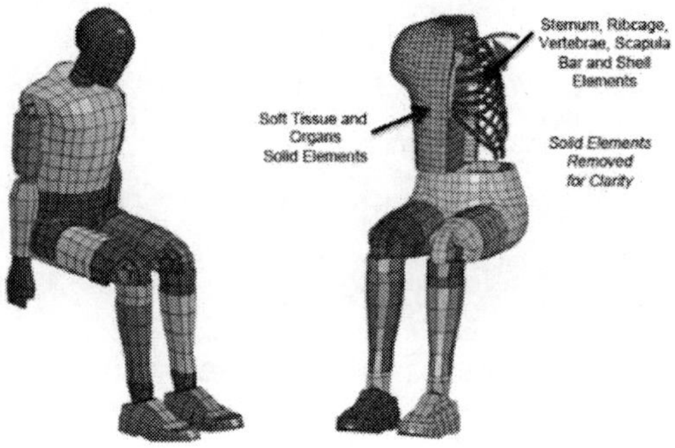

Application of Crash Modeling and Simulation

Illustrative Example of LS-DYNA: Test-Analysis Comparison

Initial impact Time of Max Acceleration

*** Formatting Issue - Best Available Paper/Graphic ***

Application of Crash Modeling and Simulation

Illustrative Example of LS-DYNA: Test-Analysis Comparison

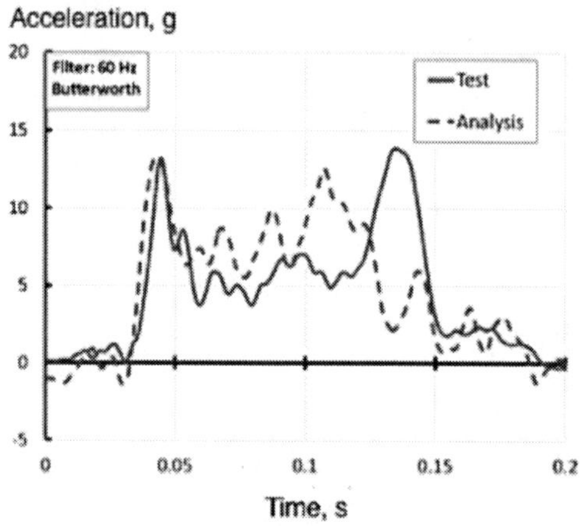

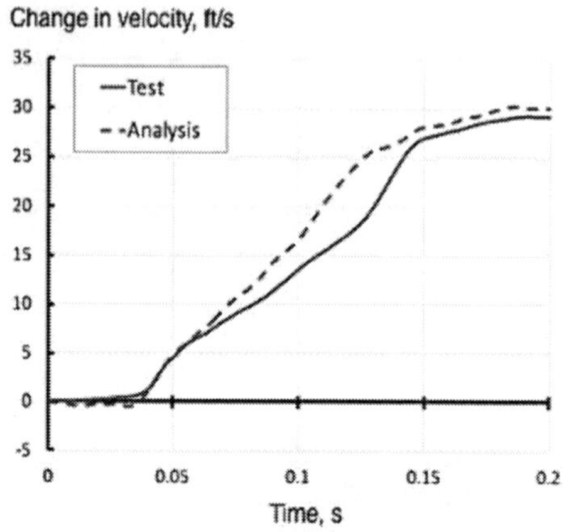

Comparisons of acceleration and change in velocity responses at the centerline of the floor beneath the passenger seat

Application of Crash Modeling and Simulation

Illustrative Example of LS-DYNA: Test-Analysis Comparison

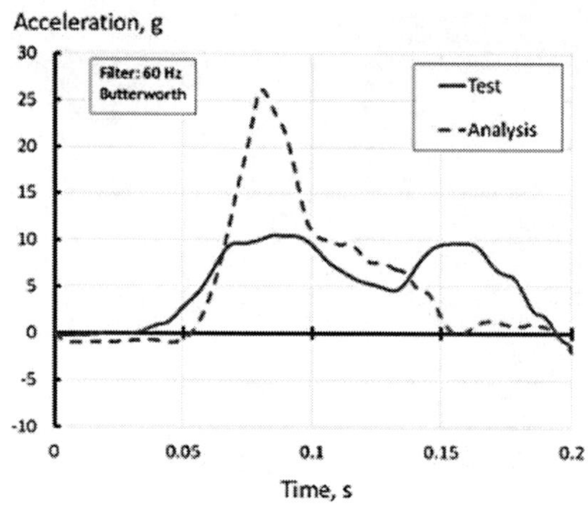

Comparison of test and analysis pelvic vertical acceleration response of the pilot ATD

*** Formatting Issue - Best Available Paper/Graphic ***

Application of Crash Modeling and Simulation

Illustrative Example of LS-DYNA: Test-Analysis Comparison

Results of this simulation effort led to two follow-on projects:

- Improved model correlation using a more rigorous computational approach to incorporate uncertainty propagation and quantification to assess model accuracy
- To further investigate the deficiencies of the LSTC-provided dummy models, a separate study was undertaken to examine Hybrid II, Hybrid III, and THOR-NT ATDs subjected to vertical loading, and two commercial ATD models

Application of Crash Modeling and Simulation

Illustrative Example of LS-DYNA: Test-Analysis Comparison

Follow-on study was undertaken to examine Hybrid II, Hybrid III, and THOR-NT dummies subjected to vertical loading and modeled using two commercial ATD models

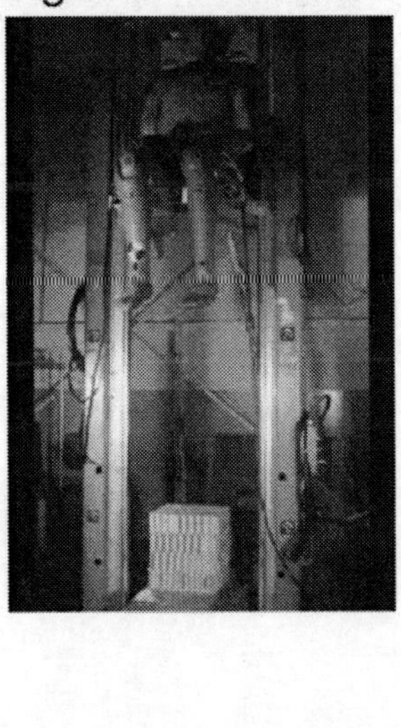

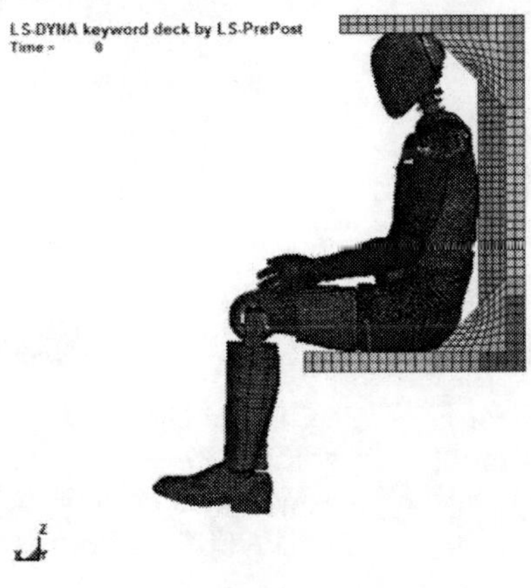

***** Formatting Issue - Best Available Paper/Graphic *****

Application of Crash Modeling and Simulation

Modeling of Anthropomorphic Test Devices (ATDs)

CAL3D

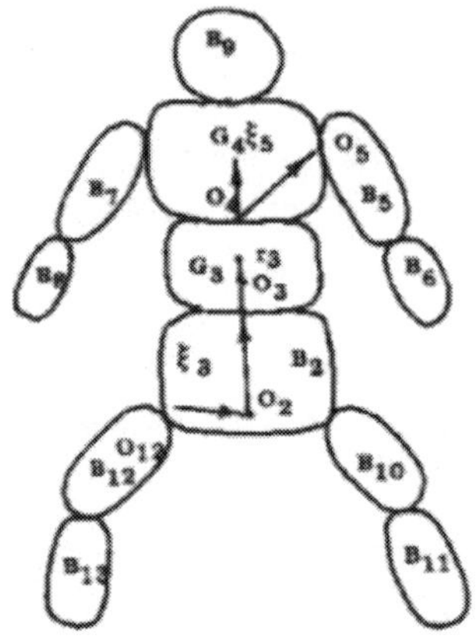

UCIN-3D

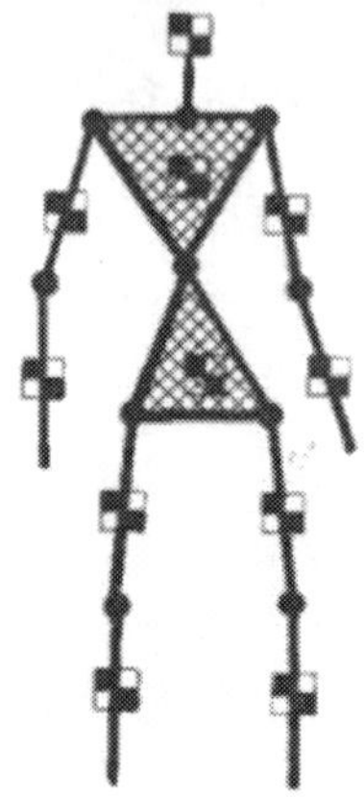

PROMETHEUS

- Early ATD models were constructed of rigid ellipsoids, up to 40 degrees of freedom, and joints modeled as simple hinges or ball-and-socket
- Some early dummy models had contact capability, as well as rudimentary injury prediction capability

Application of Crash Modeling and Simulation

Modeling of Anthropomorphic Test Devices (ATDs)
Examples of ATD models currently available in LS-DYNA

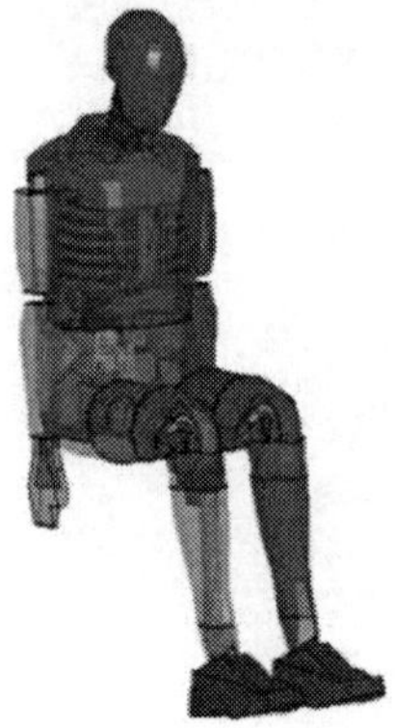

Hybrid III Fast
- 4000+ elements

Hybrid III Detailed
- 450,000+ elements
- Available as a 5th percentile female and a 50th percentile standing male

- Current ATD models are constructed using deformable finite elements
- Models are developed and calibrated to match ATD responses

*** Formatting Issue - Best Available Paper/Graphic ***

Application of Crash Modeling and Simulation

The Future: Modeling the Human Body

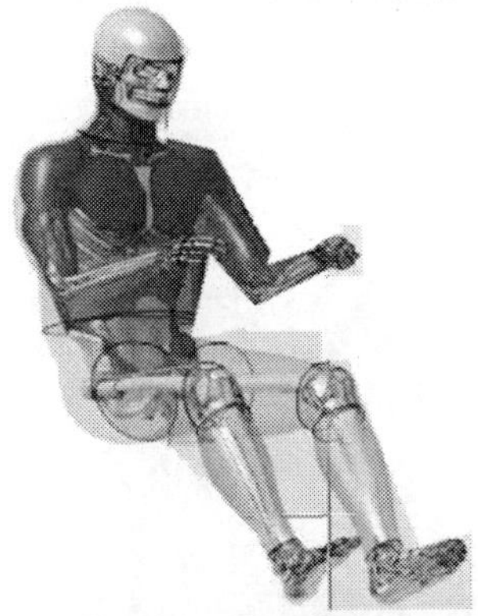

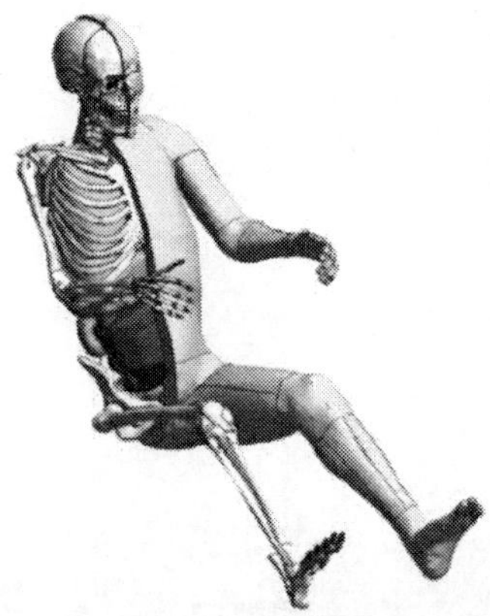

- Human body models are created to predict both hard and soft tissue injury
- Models are developed and calibrated to match human responses
- These models contain 1,000,000 nodes and 1,750,000 elements

Global Human Body Model

- Developed by consortium founded in 2006
- Consortium of automakers and suppliers working with institutes and universities

Total Human Model for Safety (THUMS)

- Developed by Toyota Motors and their Research Institute
- Intended for injury analysis

Application of Crash Modeling and Simulation

Modeling of Multi-Terrain Impact: What is the difference between these two crash tests?

TRACT 1 Full-Scale Crash Test of a CH-46E Sea Knight Helicopter, 2013

*** Formatting Issue - Best Available Paper/Graphic ***

Application of Crash Modeling and Simulation

Modeling of Multi-Terrain Impact: What is the difference between these two crash tests?

TRACT 2 Full-Scale Crash Test of a CH-46E Sea Knight Helicopter, 2014

Application of Crash Modeling and Simulation

Modeling of Multi-Terrain Impact

What is the difference between these two crash tests?
Answer: TRACT 2 exhibited half of the forward displacement of the TRACT 1 test, even though both airframes were tested under similar conditions
Why? Moist soil conditions.

LS-DYNA Model of TRACT2 Full-Scale Crash Test

*** Formatting Issue - Best Available Paper/Graphic ***

Application of Crash Modeling and Simulation

Modeling of Multi-Terrain Impact

Measuring the impact crater following TRACT2

LS-DYNA model of a penetrometer impacting soil

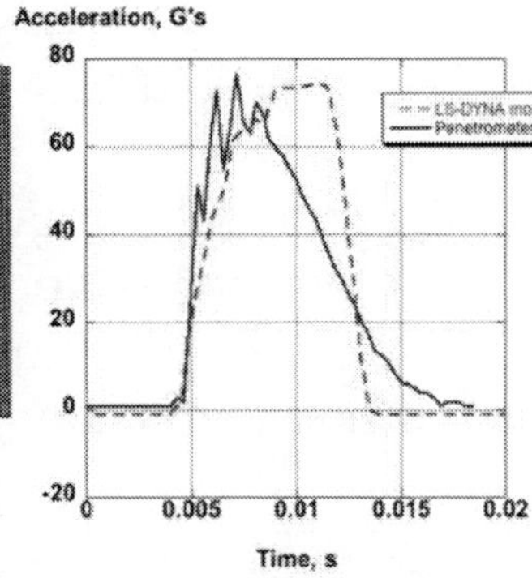

Test-analysis comparison

Modeling of Soils is Complicated

- Affected by moisture content
- Packing treatments
- Particle size
- Sand/clay mixture

Application of Crash Modeling and Simulation

Expanded Capabilities of LS-DYNA Today

Today, LS-DYNA has become more than a transient dynamic simulation code, but has expanded its capabilities to become a true multi-physics code. New capabilities include:

•Full 2- and 3-D finite elements
•Nonlinear and rigid body dynamics
•Quasi-static simulations
•Vibration modal analysis
•Linear static analysis
•Thermal analysis
•Underwater shock
•Failure and crack propagation
•Real-time acoustics
•Electromagnetism

•Fluid-structure interactions
-Eulerian fluids
-Arbitrary Lagrange-Euler
-Navier-Stokes fluids
-Compressible fluid solver
-Smooth Particle Hydrodynamics
-Element Free Galerkin
•Implicit spring back
•Radiation transport
•Multi-physics coupling

*** Formatting Issue - Best Available Paper/Graphic ***

Rotorcraft Structural Design for Crashworthiness

- Advanced Composite Airframe Program (ACAP)
  - -Bell ACAP Helicopter
  - -Sikorsky ACAP Helicopter
- Composite Energy Absorber Retrofit Program (TRACT 2)
  - -Corrugated web design
  - -Conusoid design
  - -Sinusoid design

Rotorcraft Structural Design for Crashworthiness

Advanced Composite Airframe Program (ACAP)

- ACAP sponsored by US Army Aviation Applied Technology Directorate (AATD)
- Began in 1979 and ended with full-scale crash tests of the Bell and Sikorsky helicopters in 1987
- Purpose: to demonstrate the potential of advanced composite materials to save weight and cost in airframe structures while achieving systems compatibility and meeting military requirements for vulnerability reduction, reliability, maintainability, and survivability

*** Formatting Issue - Best Available Paper/Graphic ***

Rotorcraft Structural Design for Crashworthiness

Advanced Composite Airframe Program (ACAP)

Crashworthy features of the Bell ACAP included:

-Main landing gear consisted of a two stage design
- •Oleo shock strut designed to limit velocity to 20-ft/s
- •Crushable graphite/epoxy tube for protection to 42-ft/s

-Tail gear contained a tube cutter

-Fuselage designed to be stiff and strong, surrounded underneath with composite energy absorbers fabricated of Kevlar face sheets and Nomex honeycomb

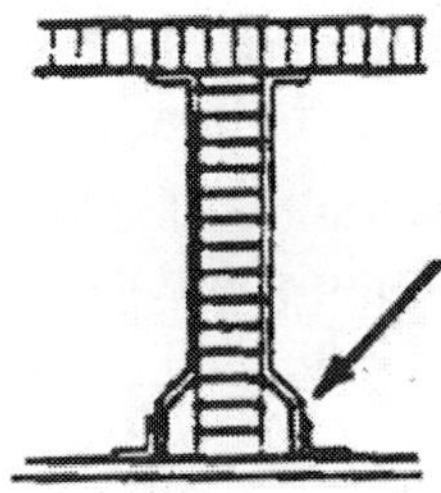

Rotorcraft Structural Design for Crashworthiness

Advanced Composite Airframe Program (ACAP)

- ▸Test data were lost due to premature failure of the umbilical cable
- ▸Test conducted at 41.4-ft/s vertical, 26.4-ft/s forward velocities, with 15° pitch and 14.5° roll
- ▸Findings:
  - •Rotor transmission mass was retained, no intrusion into the occupied cabin
  - •Protective shell maintained
  - •Right side energy absorbers fully crushed, less on the left
  - •Crew and troop seats stroked, but did not bottom out
  - •No fuel leakage during or after the test
- ▸ ***Systems approach for crashworthiness worked!***

Full-Scale Crash Test on August 27, 1987

*** Formatting Issue - Best Available Paper/Graphic ***

Rotorcraft Structural Design for Crashworthiness

Advanced Composite Airframe Program (ACAP)

Crashworthy features of the Sikorsky ACAP included:

-Main landing gear consisted of a two stage design
- •Air-oleo shock strut designed to limit velocity to 12.5-ft/s
- •Crushable 25-in.-long honeycomb canister for protection to 42-ft/s
- •Pins limit crushing of the honeycomb until the air-oleo strut has stroked 12-in.

-Landing gear attachment fittings designed for bearing failures

-Two stage beaded web crushable keel beams and transverse bulkheads

Two-stage beaded web energy absorber

Rotorcraft Structural Design for Crashworthiness

Advanced Composite Airframe Program (ACAP)

▸Test conducted at 39-ft/s vertical velocity, with 10° pitch and 10° roll

▸Findings:
- •Rotor transmission mass was retained, no intrusion into the occupied cabin
- •Protective shell maintained
- •Magnitudes of floor, seat, and occupant accelerations found to be survivable
- •Main landing gear failure of the forward side brace allowed rotation and premature bearing failure, that was compensated for by crushing of the subfloor

▸***Systems approach for crashworthiness worked!***

*** Formatting Issue - Best Available Paper/Graphic ***

Rotorcraft Structural Design for Crashworthiness

Advanced Composite Airframe Program (ACAP)

1999 Full-Scale Crash Test of the Sikorsky ACAP Flying Prototype
38-ft/s vertical and 32.5-ft/s forward velocities

Rotorcraft Structural Design for Crashworthiness

Advanced Composite Airframe Program (ACAP)

1999 Full-Scale Crash Test of the Sikorsky ACAP Flying Prototype
38-ft/s vertical and 32.5-ft/s forward velocities

*** Formatting Issue - Best Available Paper/Graphic ***

Rotorcraft Structural Design for Crashworthiness

Composite Energy Absorber Retrofit Program

- In 2012, NASA initiated the Transport Rotorcraft Airframe Crash Testbed (TRACT) program
- 2 CH-46E Sea Knight airframes were obtained
- The first helicopter was tested in a baseline configuration

- The second helicopter was retrofitted with three composite energy absorbers:
  - -Corrugated web design fabricated by the Australian Cooperative Research Centre for Advanced Composite Structures (CRC-ACS) and the German Aerospace Research Center (DLR)
  - -Conusoid design developed by NASA
  - -Sinusoid foam sandwich design developed by NASA

Rotorcraft Structural Design for Crashworthiness

Corrugated Web Energy Absorber

Trapezoidal crush element

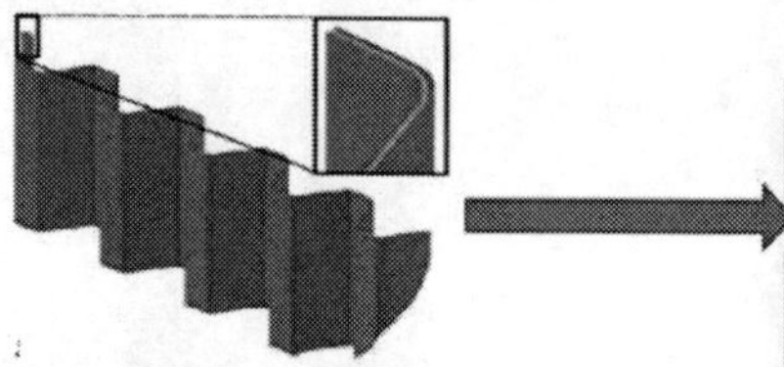

PAM-CRASH model of a corrugated web energy absorber

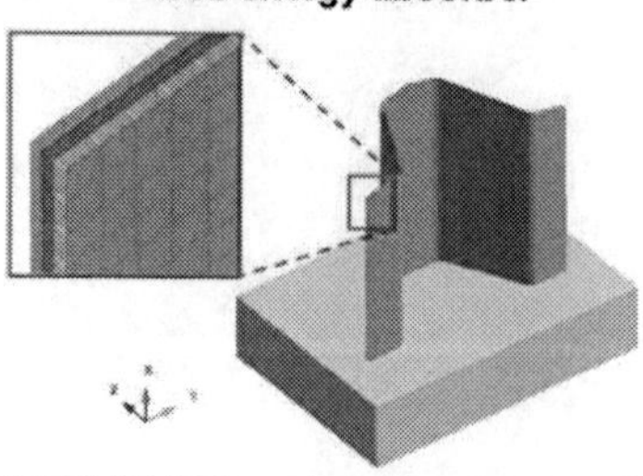

PAM-CRASH model of a corrugated web energy absorber fabricated of graphite-epoxy fabric

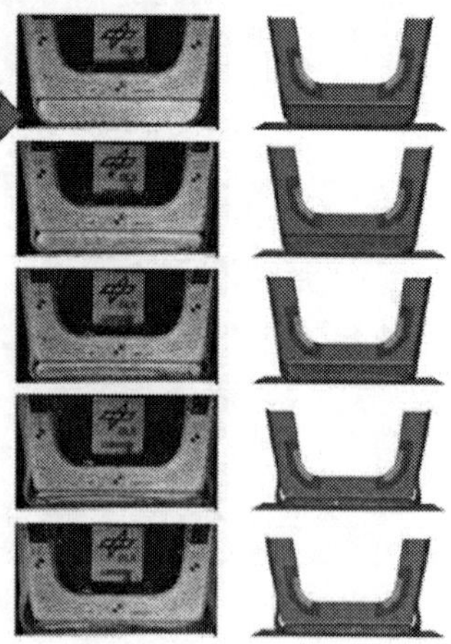

Corrugated web evaluated in a fuselage section drop test

*** Formatting Issue - Best Available Paper/Graphic ***

Rotorcraft Structural Design for Crashworthiness

Conusoid Energy Absorber (EA)

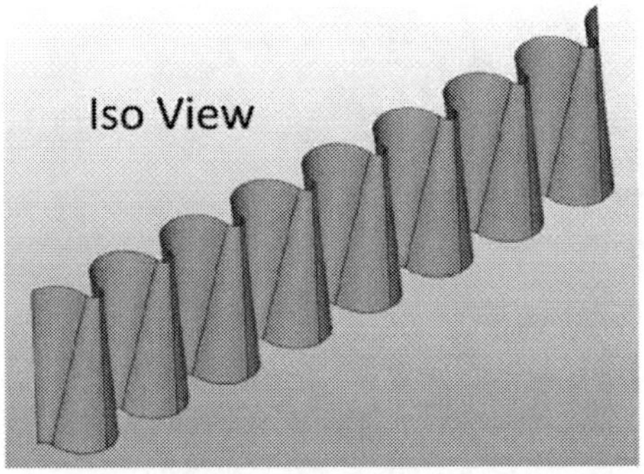

Photo of conusoid EA

- The conusoid is based on alternating right-side-up and up-side down half-cones placed in a repeating pattern.
- The conusoid combines a simple cone design with sinusoidal beam geometry to create a structure that utilizes the advantages of both configurations.
- The final design consisted of four layers of a hybrid carbon-Kevlar® plain weave fabric oriented at [+45°/-45°/-45°/+45°] with respect to the vertical direction.

Rotorcraft Structural Design for Crashworthiness

Conusoid Energy Absorber

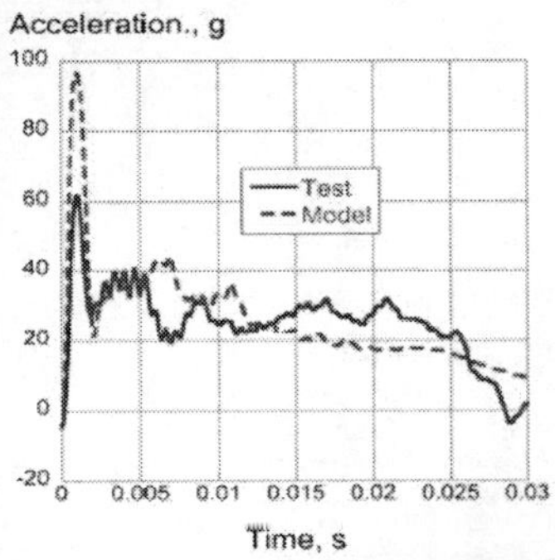

- Energy absorbers were designed to achieve 25- to 40-g average vertical acceleration
- The conusoid had an average vertical acceleration of 28.0-g for a pulse duration of 0.0-0.25-s
- Energy attenuated through plastic-like folding of the walls

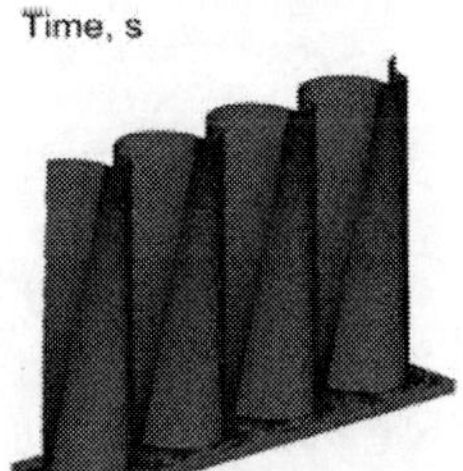

LS-DYNA model

*** Formatting Issue - Best Available Paper/Graphic ***

Rotorcraft Structural Design for Crashworthiness

Sinusoid Foam Sandwich Energy Absorber

- The sinusoid is based on a series of half circles with a diameter of 1.75-in.
- The final design made of hybrid carbon-Kevlar® plain weave fabric face sheets, two layers for each face sheet oriented at ±45° with respect to the vertical direction, and a closed-cell ELFOAM® P200 polyisocyanurate (2.0-lb/ft^3) foam core.
- Energy attenuation achieved through folding of the face sheets and foam crushing

Rotorcraft Structural Design for Crashworthiness

Sinusoid Foam Sandwich Energy Absorber

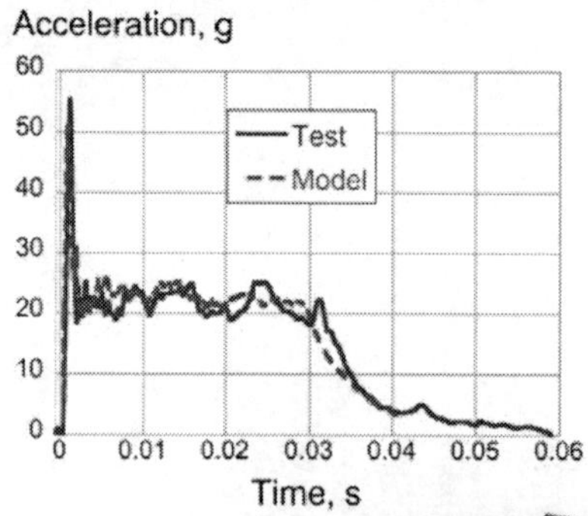

- The sinusoid had an average vertical acceleration of 21.8-g for a pulse duration of 0.0-0.03-s
- Energy attenuated through plastic-like folding of the walls and foam crushing
- Excellent prediction of the crush response

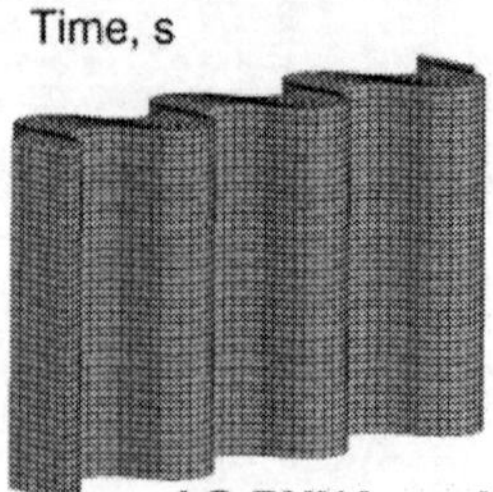

LS-DYNA model

*** Formatting Issue - Best Available Paper/Graphic ***

Rotorcraft Structural Design for Crashworthiness

- Systems Approach to Crashworthy Design Works!
  - -Bell ACAP Helicopter
  - -Sikorsky ACAP Helicopter
- Composite structures can be designed to crush in a stable manner and attenuate impact energy effectively
  - -Corrugated web design
  - -Conusoid design
  - -Sinusoid design

Recommendations for the Future

Facilities and Equipment for Conducting Crash Testing

- The US Government should take the lead in maintaining state-of-the-art facilities for crash testing and in upgrading equipment used to support crash testing
- ATD manufacturers should develop and calibrate ATDs designed for vertical or combined forward and vertical impact loading, accurately replicating human occupant responses

*** Formatting Issue - Best Available Paper/Graphic ***

Recommendations for the Future

Updated Crash Certification Requirements

- Decadal accident surveys should be routinely performed by the US Army, Navy, and Marine Corps for military helicopters and by the FAA/NTSB for commercial rotorcraft

- Whole-aircraft certification requirements should be developed for civilian/commercial rotorcraft

- Build confidence in crash certification by analysis

Recommendations for the Future

Crash Modeling and Simulation Techniques

- Develop accepted metrics for test-analysis correlation and establish the recommended approach as an industry-wide standard

- Develop an accurate ATD model to reliably predict the responses under vertical of combined forward and vertical impact loading

*** Formatting Issue - Best Available Paper/Graphic ***

Recommendations for the Future

Rotorcraft Structural Design for Improved Crashworthiness

- Develop, if possible, an energy absorbing subfloor concept optimized for multi-terrain impact loading
- Continue using the systems approach for crashworthiness in rotorcraft design – it works!
- The rotorcraft community is urged to continue using composite materials, where they make sense
- As novel rotorcraft designs are developed, their crash performance may require new certification standards and rethinking of the current crash design approaches - please keep crash safety as a priority in the early design stages

CONCLUSIONS

- **Documented advances in rotorcraft crashworthiness that have occurred over the past 40 years in 4 categories**
  - Facilities and Equipment for Conducting Crash Testing
  - Updated Crash Certification Requirements
  - Crash Modeling and Simulation
  - Rotorcraft Structural Design for Crashworthiness
- **Provided recommendations such that the progress in rotorcraft crashworthiness made to date can be continued into the future**

*** Formatting Issue - Best Available Paper/Graphic ***

ACKNOWLEDGEMENTS

Thanks and deepest appreciation to:

- Susan Gorton, Revolutionary Vertical Lift Technology Project (RVLT) Manager at NASA Langley
- The AHSI Awards Committee
- Colleagues and coworkers at NASA Langley
- Members of the AHSI Crash Safety Technical Committee
- My family and friends

The theme of the 2½ day conference is *"Innovative Structures and Survivability Solutions for Vertical Lift."* The meeting will consist of unclassified, unrestricted presentations to be held at the Hampton Roads Convention Center in Virginia. The meeting will highlight research and development efforts, both current and planned, related to manned and unmanned rotorcraft structures, crash safety, and vulnerability reduction. Lodging is available next door at the Embassy Suites.

If you are interested in presenting, please submit an abstract **NLT than Monday, July 24, 2017**. Abstracts should include presentation title, presenter(s) name and affiliation(s), contact number and email address. Abstracts should be submitted to **structures2017@vtol.org**. **Final presentations will be due by October 2, 2017.**

General Chair: Ms. Susan Gorton, NASA Langley Research Center
Technical Chairs: Mr. Nate Bordick, US Army ADD-AATD; Dr. Karen Jackson, NASA Langley, and Dr. Mark Robeson, US Army ADD-AATD

www.vtol.org/structures

*** Formatting Issue - Best Available Paper/Graphic ***

Questions???

UNCLASSIFIED//FOUO

PROGRAM EXECUTIVE OFFICE

AVIATION

Version #3
As of 19APR17

UNCLASSIFIED//FOUO

U.S.ARMY

Program Executive Office, Aviation

PROGRAM EXECUTIVE OFFICE AVIATION

The Mission:

PEO Aviation, in collaboration with stakeholders, designs, develops and delivers capabilities, focused on Army materiel requirements and priorities

The Vision:

Effective and efficient execution of the Aviation Portfolio to provide enhanced reach & reliability to combat commanders.

The Magnitude:

- *Eight Project Offices, 97 Separate Programs*
- *Managing ~$5.6B Total FY17 Resources*
- *$25.7B over POM years FY 17-21*
- *Total FMS value $35.8B over POM years FY 16-20*
- *204 FMS Cases — 61 Countries*
- *~1,800 Government & ~1,300 Support Contractors*

The Operation:

Centralized Management for All Assigned Army Aviation Programs of Record

Enhance Reliability, Deliver Capabilities, Safety, and Interoperability

Maintain Combat Overmatch through Recapitalization and Modernization

America's Force of Decisive Action 2

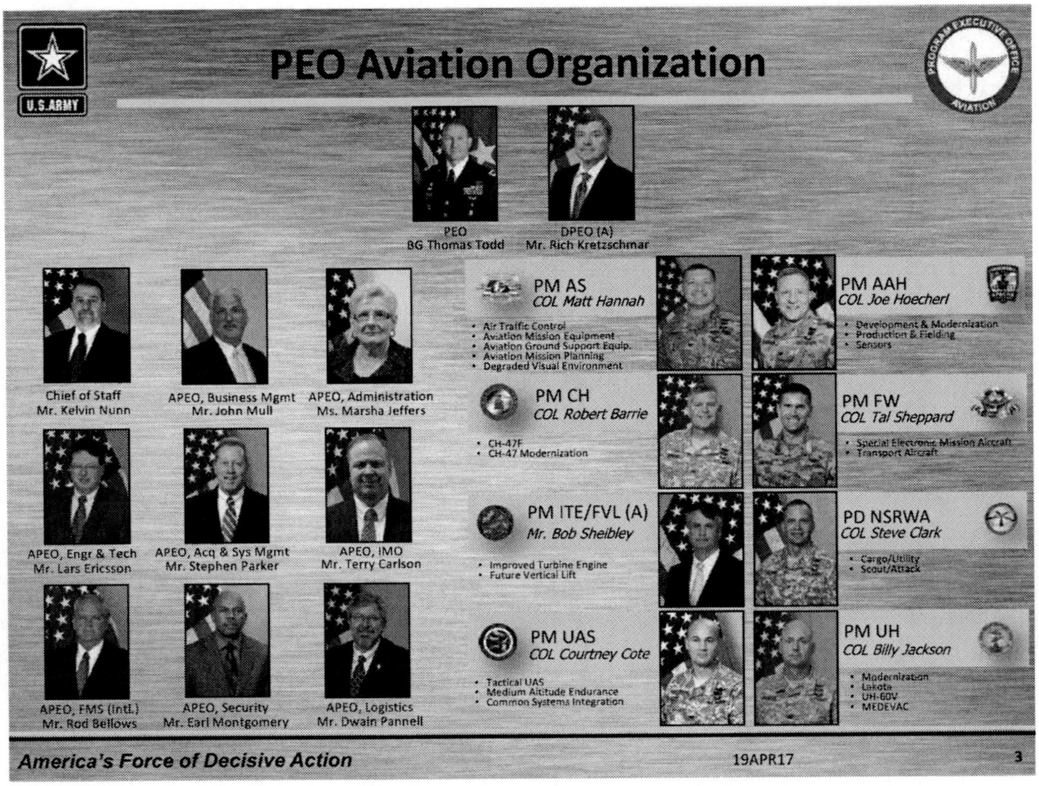
U.S.ARMY
PEO Aviation Organization
PROGRAM EXECUTIVE OFFICE AVIATION
PEO
BG Thomas Todd
DPEO (A)
Mr. Rich Kretzschmar
Chief of Staff
Mr. Kelvin Nunn
APEO, Business Mgmt
Mr. John Mull
APEO, Administration
Ms. Marsha Jeffers
APEO, Engr & Tech
Mr. Lars Ericsson
APEO, Acq & Sys Mgmt
Mr. Stephen Parker
APEO, IMO
Mr. Terry Carlson
APEO, FMS (Intl.)
Mr. Rod Bellows
APEO, Security
Mr. Earl Montgomery
APEO, Logistics
Mr. Dwain Pannell
PM AS
COL Matt Hannah
• Air Traffic Control
• Aviation Mission Equipment
• Aviation Ground Support Equip.
• Aviation Mission Planning
• Degraded Visual Environment
PM AAH
COL Joe Hoecherl
• Development & Modernization
• Production & Fielding
• Sensors
PM CH
COL Robert Barrie
• CH-47F
• CH-47 Modernization
PM FW
COL Tal Sheppard
• Special Electronic Mission Aircraft
• Transport Aircraft
PM ITE/FVL (A)
Mr. Bob Sheibley
• Improved Turbine Engine
• Future Vertical Lift
PD NSRWA
COL Steve Clark
• Cargo/Utility
• Scout/Attack
PM UAS
COL Courtney Cote
• Tactical UAS
• Medium Altitude Endurance
• Common Systems Integration
PM UH
COL Billy Jackson
• Modernization
• Lakota
• UH-60V
• MEDEVAC
America's Force of Decisive Action
19APR17
3

U.S.ARMY
Aviation Materiel Enterprise
PROGRAM EXECUTIVE OFFICE AVIATION
PEO IEW&S
PEO C3T
SOCOM
AMCOM
HQDA
USASAC
PEO Aviation
Aviation Enterprise
PEO Missiles & Space
PEO Soldier
U.S. ARMY RDECOM
RDECOM
PEO EIS
DCMA
ARMATI COMPARATOR CUSTOS
USAACE
INSCOM
ACC
Strategic Capabilities Office
America's Force of Decisive Action
4

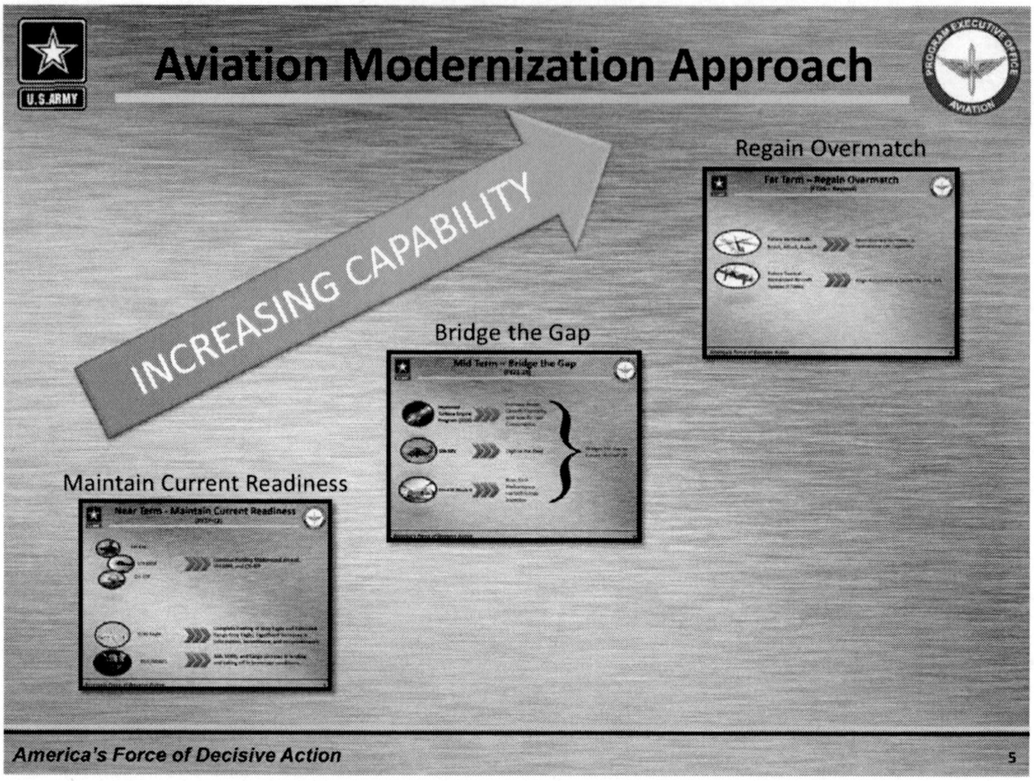
U.S.ARMY
Aviation Modernization Approach
PROGRAM EXECUTIVE OFFICE AVIATION
INCREASING CAPABILITY
Regain Overmatch
Bridge the Gap
Maintain Current Readiness
America's Force of Decisive Action
5

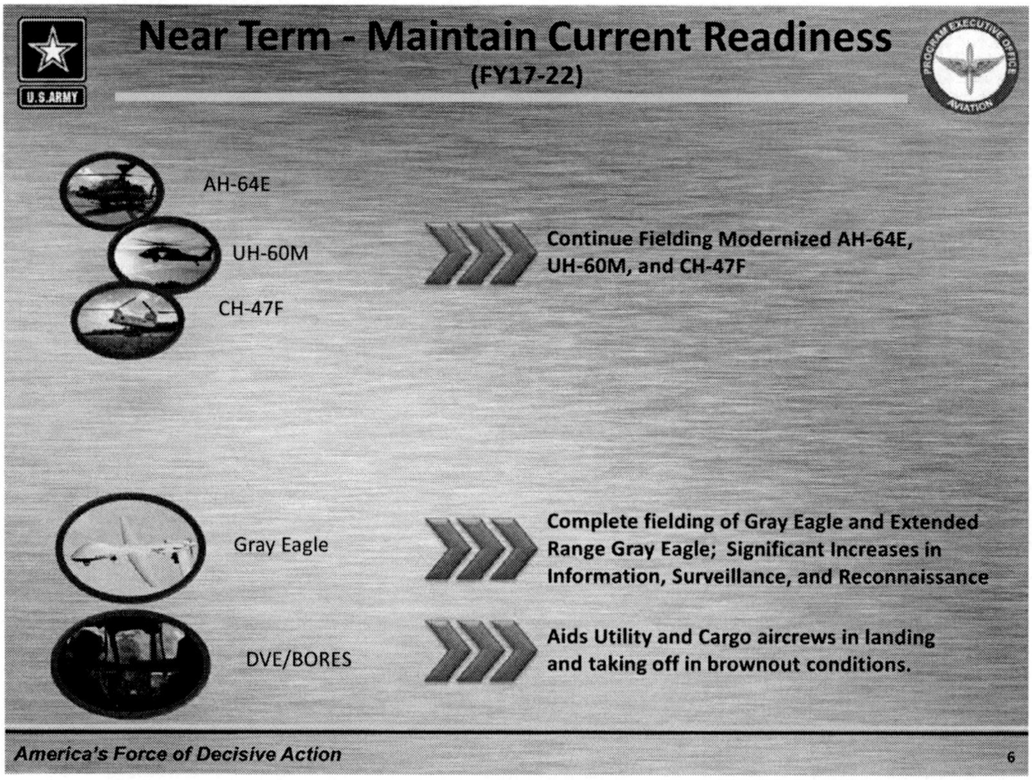
U.S.ARMY
Near Term - Maintain Current Readiness
(FY17-22)
PROGRAM EXECUTIVE OFFICE AVIATION
AH-64E
UH-60M
CH-47F
Continue Fielding Modernized AH-64E, UH-60M, and CH-47F
Gray Eagle
Complete fielding of Gray Eagle and Extended Range Gray Eagle; Significant Increases in Information, Surveillance, and Reconnaissance
DVE/BORES
Aids Utility and Cargo aircrews in landing and taking off in brownout conditions.
America's Force of Decisive Action
6

Multiyear Contracts

- ACC and PEO Aviation successfully teaming to award multiyear contracts that provide significant savings to the Government over single year contracting
  - Chinook
    - FYs: 13-17
    - Savings: 19.18%
    - Quantities: Base - 155; Options – 60; Total – 215
      - 18 options remain available
  - Apache
    - FYs: 17-21
    - Estimated Savings: 10.9%
    - Quantities: Min 240 (48 Reman per year), Max 450 (Reman + New Build)
  - Black Hawk
    - FYs 17-21
    - Estimated Savings 10.7%
    - Quantities: Base – 257; Min – 220; Max - 360
      - (Base and Min Qtys include 40 FMS a/c)

ACC/PEO Aviation Teaming

America's Force of Decisive Action 7

UNCLASSIFIED

Gray Eagle / Extended Range Gray Eagle

	MQ-1C Baseline Gray Eagle	MQ-1C Extended Range Gray Eagle (ER)
Aircraft Configuration / Endurance		
Electro-Optical (EP)/Infrared (IR) & Synthetic Aperture Radar (SAR)	23 hrs	34 hrs*
EO/IR, SAR, Tactical SIGNT Payload, 2 HELLFIREs	15 hrs	27.5 hrs*
Gross Take off Weight (GTOW)	3,600 GTOW	4,200 GTOW
Space, Weight, & Power	400 lb internal 500 lb each wing 9.5 kW	540 lb internal 500 lb each wing 500 lb centerline hard point 9.5 kW
Max Fuel Capacity	600 lbs	860 lbs
Heavy Fuel Engine (JP-8/F-24)	Yes (OEM 160) – 160 hp	Yes (OEM 180) – 180 hp
Supportability	Soldier Maintained	Soldier Maintained (50% more payload/avionics volume, greater access to avionics, improved landing gear, longer life servos)

Much Greater Endurance

Much Larger Payload

America's Force of Decisive Action 8

Enhanced Scheduled Maintenance - ESM2

- The ESM2 program is an initiative to optimize the utilization of time constrained components and extend the current Phase Maintenance cycle.
- Major Changes in the Scheduled Maintenance program are:
  - Extension of the Phase Maintenance Inspection (PMI) cycle from 360/720 to 480/960
  - Provides an additional 240 FH per maintenance cycle
  - Creates a savings of 0.47 Maintenance Man Hours per Flight Hour (MMH/FH)
  - All maintenance tasks and intervals jointly approved by Sikorsky Engineering and Aviation Engineering Directorate (AED)

AMCOM & PEO AVN Teaming

A typical Assault Battalion saves ***2,510 MMH*** *of scheduled maintenance annually.*

Mid Term – Bridge the Gap
(FY21-25)

Improved Turbine Engine Program (ITEP) → Increases Power, Growth Capability, and Specific Fuel Consumption

UH-60V → Digitizes the Fleet

CH-47F Block II → Buys Back Performance via technology insertion

} Bridges the Gap to Future Vertical Lift

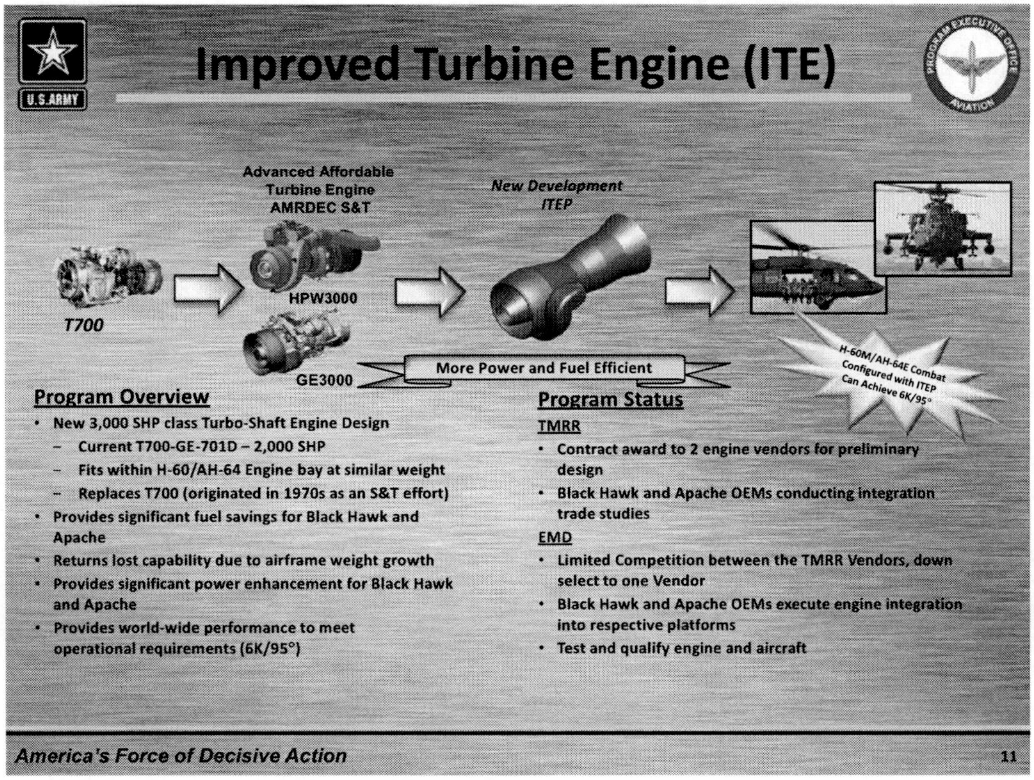
U.S.ARMY
Improved Turbine Engine (ITE)
PROGRAM EXECUTIVE OFFICE AVIATION
Advanced Affordable Turbine Engine AMRDEC S&T
New Development ITEP
HPW3000
T700
GE3000
More Power and Fuel Efficient
H-60M/AH-64E Combat Configured with ITEP Can Achieve 6K/95°
Program Overview
• New 3,000 SHP class Turbo-Shaft Engine Design
– Current T700-GE-701D – 2,000 SHP
– Fits within H-60/AH-64 Engine bay at similar weight
– Replaces T700 (originated in 1970s as an S&T effort)
• Provides significant fuel savings for Black Hawk and Apache
• Returns lost capability due to airframe weight growth
• Provides significant power enhancement for Black Hawk and Apache
• Provides world-wide performance to meet operational requirements (6K/95°)
Program Status
TMRR
• Contract award to 2 engine vendors for preliminary design
• Black Hawk and Apache OEMs conducting integration trade studies
EMD
• Limited Competition between the TMRR Vendors, down select to one Vendor
• Black Hawk and Apache OEMs execute engine integration into respective platforms
• Test and qualify engine and aircraft
America's Force of Decisive Action
11

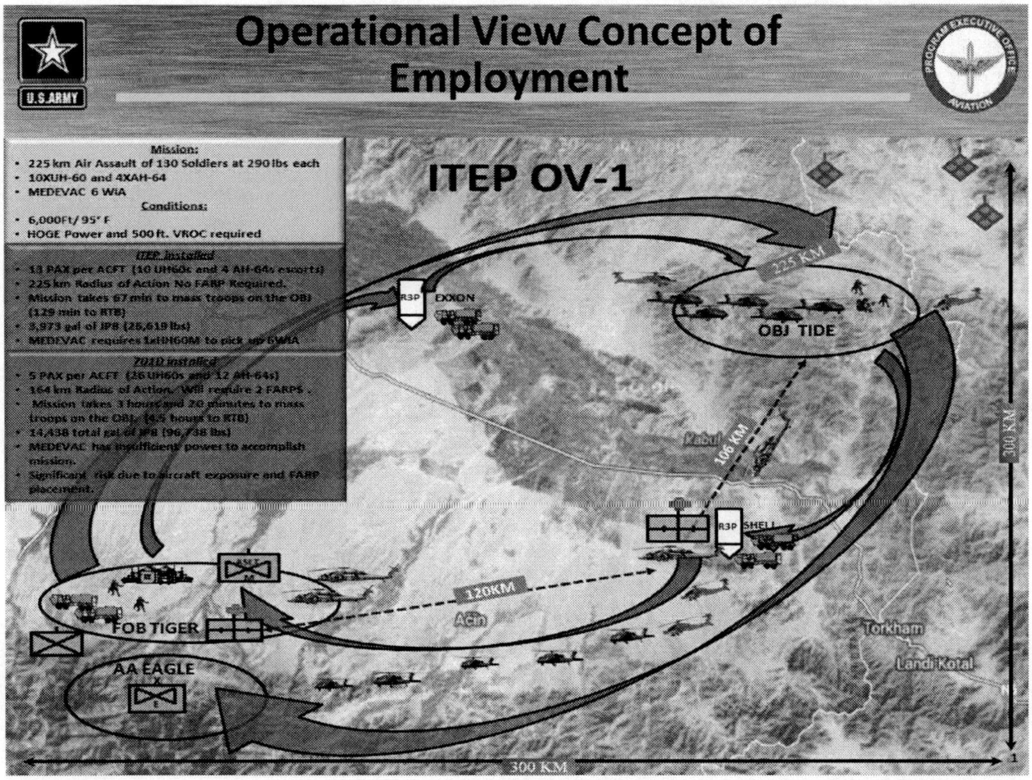
U.S.ARMY
Operational View Concept of Employment
PROGRAM EXECUTIVE OFFICE AVIATION
Mission:
• 225 km Air Assault of 130 Soldiers at 290 lbs each
• 10XUH-60 and 4XAH-64
• MEDEVAC 6 WIA
Conditions:
• 6,000Ft/95° F
• HOGE Power and 500 ft. VROC required
ITEP Installed
• 13 PAX per ACFT (10 UH60s and 4 AH-64s escorts)
• 225 km Radius of Action No FARP Required.
• Mission takes 67 min to mass troops on the OBJ (129 min to RTB)
• 3,973 gal of JP8 (26,619 lbs)
• MEDEVAC requires 1xUH60M to pick up 6WIA
701D installed
• 5 PAX per ACFT (26 UH60s and 12 AH-64s)
• 164 km Radius of Action. Will require 2 FARPS.
• Mission takes 3 hours and 20 minutes to mass troops on the OBJ. (4.5 hours to RTB)
• 14,438 total gal of JP8 (96,738 lbs)
• MEDEVAC has insufficient power to accomplish mission.
• Significant risk due to aircraft exposure and FARP placement.
ITEP OV-1
R3P EXXON
225 KM
OBJ TIDE
Kabul
106 KM
R3P SHELL
120KM
FOB TIGER
Acin
Torkham
Landi Kotal
AA EAGLE
300 KM
300 KM

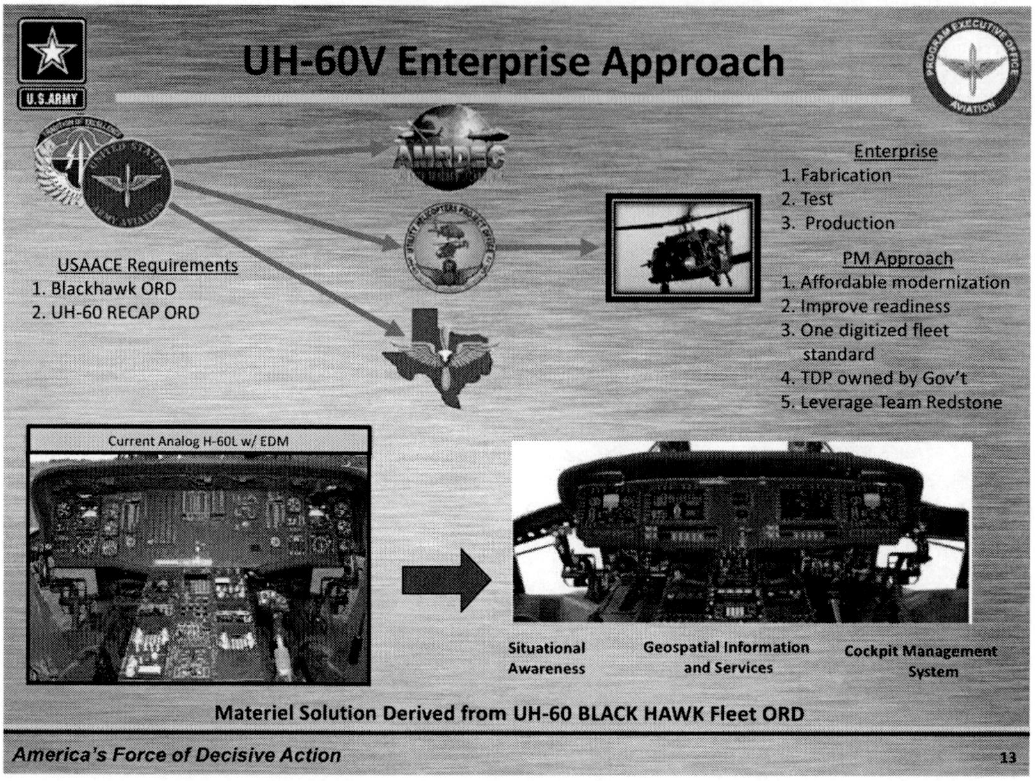
U.S.ARMY
UH-60V Enterprise Approach
AMRDEC
Enterprise
1. Fabrication
2. Test
3. Production
PM Approach
1. Affordable modernization
2. Improve readiness
3. One digitized fleet standard
4. TDP owned by Gov't
5. Leverage Team Redstone
USAACE Requirements
1. Blackhawk ORD
2. UH-60 RECAP ORD
Current Analog H-60L w/ EDM
Situational Awareness
Geospatial Information and Services
Cockpit Management System
Materiel Solution Derived from UH-60 BLACK HAWK Fleet ORD
America's Force of Decisive Action
13

U.S.ARMY
H-47 Block II Changes
Product Improvement Plan
ARC-231 integration
New data loader
Common fill ports
Improved fuel sys info
Electromagnetic hardening
On-Going Engineering Change Proposals
Rotor System
Advanced Chinook Rotor Blade
Upgraded hub
Improved Drive Train
Upgraded Sync/Rotor Shafts
Uprated forward,
combiner transmissions
Modified/Upgraded Aft transmission
Block II Airframe Mods
Electrical System
3rd Gen. for emer. power
Regulated Transformer Rectifiers
Aft Section and Pylon
Strengthened for fatigue & growth (G baseline)
Nose Section
Strengthened for fatigue & growth (G baseline)
Common Avionics Architecture System Update
Flight Controls
Redesigned trim actuators
Digital Advanced Flight Control System update
Fuel System
New single tank,
Externally replaceable pumps,
New composite pods
America's Force of Decisive Action
14

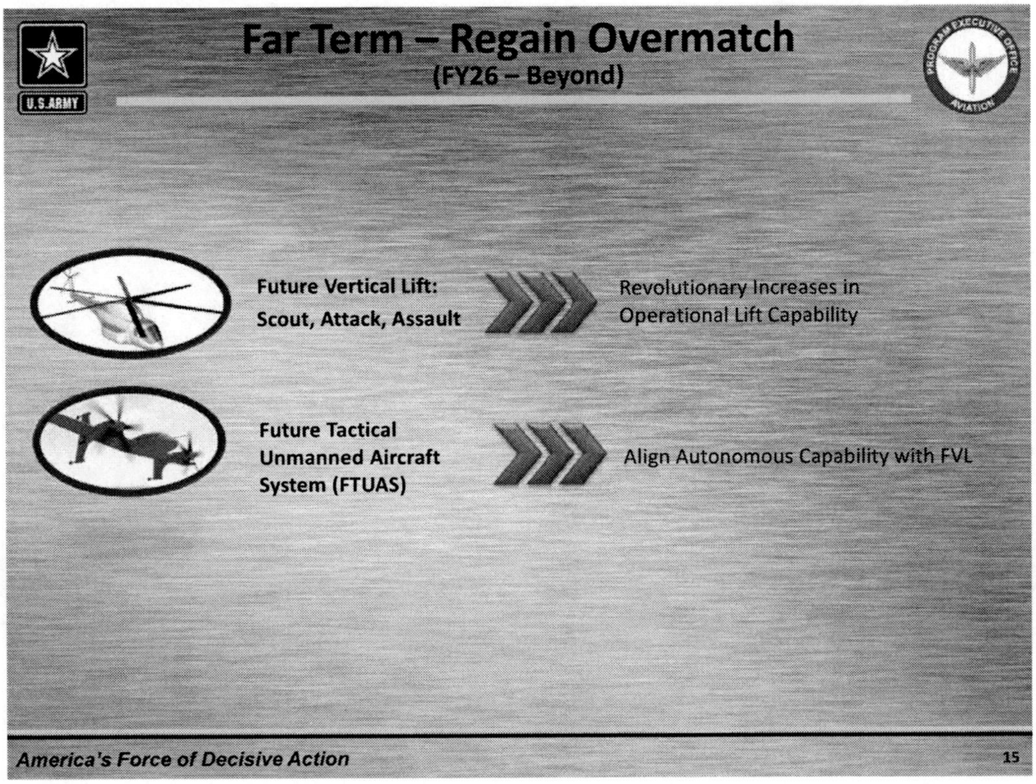
U.S. ARMY
Far Term – Regain Overmatch
(FY26 – Beyond)
PROGRAM EXECUTIVE OFFICE AVIATION
Future Vertical Lift:
Scout, Attack, Assault
Revolutionary Increases in
Operational Lift Capability
Future Tactical
Unmanned Aircraft
System (FTUAS)
Align Autonomous Capability with FVL
America's Force of Decisive Action
15

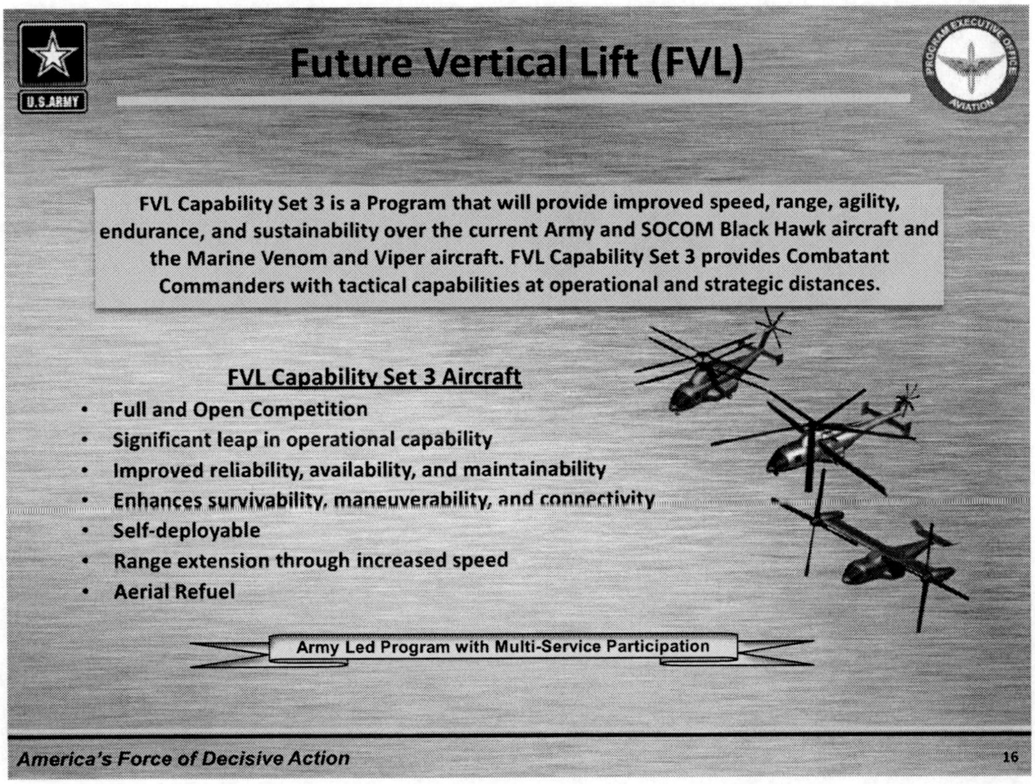
U.S. ARMY
Future Vertical Lift (FVL)
PROGRAM EXECUTIVE OFFICE AVIATION
FVL Capability Set 3 is a Program that will provide improved speed, range, agility, endurance, and sustainability over the current Army and SOCOM Black Hawk aircraft and the Marine Venom and Viper aircraft. FVL Capability Set 3 provides Combatant Commanders with tactical capabilities at operational and strategic distances.
FVL Capability Set 3 Aircraft
• Full and Open Competition
• Significant leap in operational capability
• Improved reliability, availability, and maintainability
• Enhances survivability, maneuverability, and connectivity
• Self-deployable
• Range extension through increased speed
• Aerial Refuel
Army Led Program with Multi-Service Participation
America's Force of Decisive Action
16

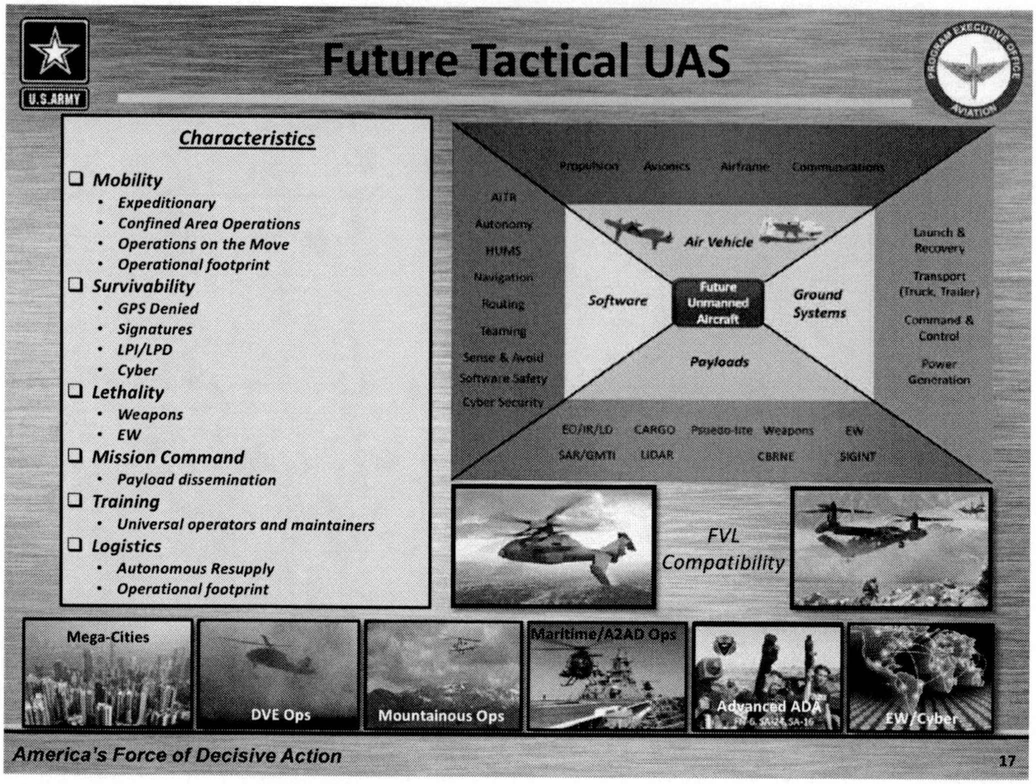
U.S.ARMY
Future Tactical UAS
PROGRAM EXECUTIVE OFFICE AVIATION
Characteristics
Mobility
Expeditionary
Confined Area Operations
Operations on the Move
Operational footprint
Survivability
GPS Denied
Signatures
LPI/LPD
Cyber
Lethality
Weapons
EW
Mission Command
Payload dissemination
Training
Universal operators and maintainers
Logistics
Autonomous Resupply
Operational footprint
Propulsion
Avionics
Airframe
Communications
AITR
Autonomy
HUMS
Navigation
Routing
Teaming
Sense & Avoid
Software Safety
Cyber Security
Air Vehicle
Software
Future Unmanned Aircraft
Ground Systems
Payloads
Launch & Recovery
Transport (Truck, Trailer)
Command & Control
Power Generation
EO/IR/LD
CARGO
Pseudo-lite
Weapons
EW
SAR/GMTI
LIDAR
CBRNE
SIGINT
FVL Compatibility
Mega-Cities
DVE Ops
Mountainous Ops
Maritime/A2AD Ops
Advanced ADA
EW/Cyber
America's Force of Decisive Action
17

PROGRAM EXECUTIVE OFFICE
AVIATION

H-47 Configuration Roadmap

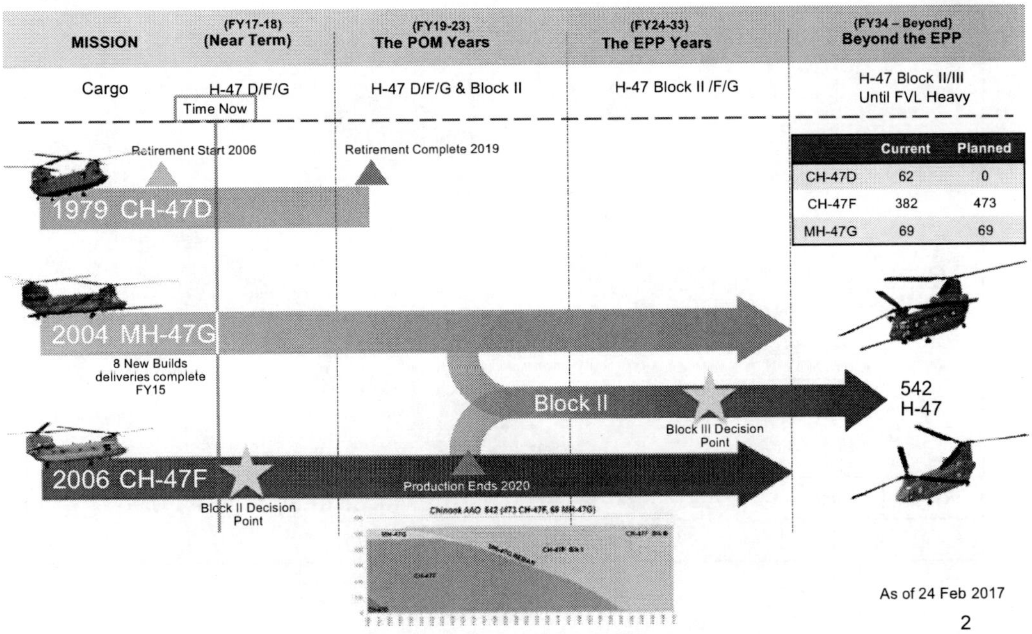

	Current	Planned
CH-47D	62	0
CH-47F	382	473
MH-47G	69	69

As of 24 Feb 2017

2

CH-47F Block II Product Overview

3

H-47 ACRB Overview

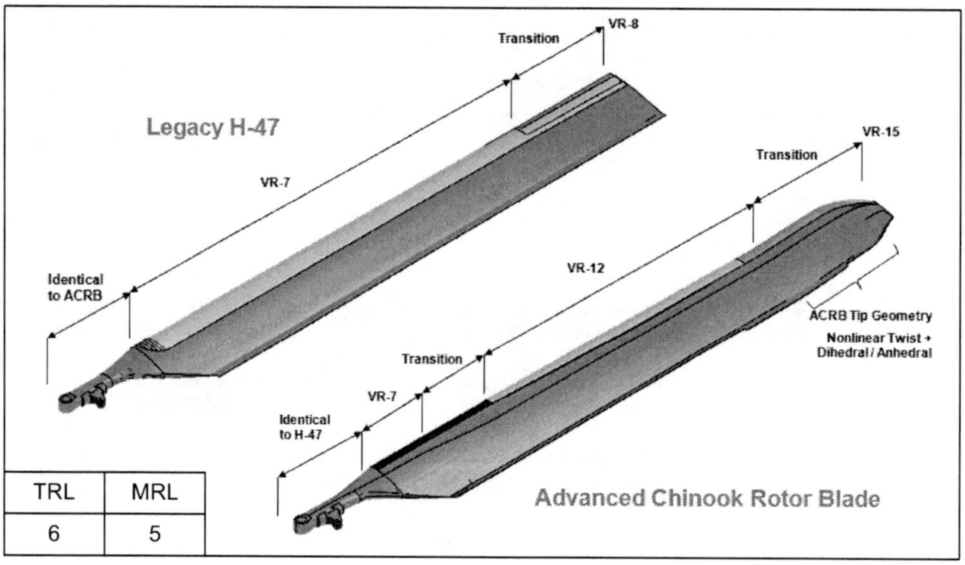

TRL	MRL
6	5

4

ACRB Flight Test

First Flight Dec 6
Completed LGW Hover Points

First Sling Load Dec 9
Completed MGW Hover Points

First Tethered Hover Dec 11
Completed Initial Performance Assessment

First Forward Flight Jan 5

<u>Test Points Complete:</u>
Ground Runs
Hover Test Points
LGW & MGW Forward Flight BU
Handling Qualities
HGW Forward Flight BU
Air Resonance Evaluation
Load Survey
MRD Configuration Testing

<u>OBJECTIVES</u>

- ✓ **+1500 lbs. @ 4K95**
- ❑ **Mitigate risk to EMD**

5

Block II Rotor System

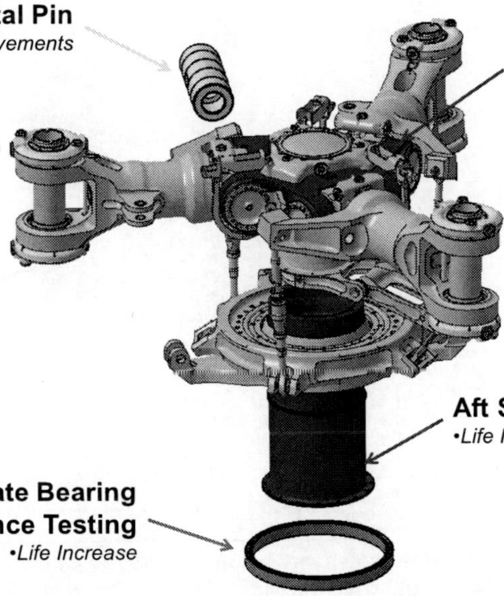

6

Improved Drive Train

Benefits:
- Posture aircraft for Future Affordable Turbine Engine
- Commonality with 47G and reduce closed loop parts

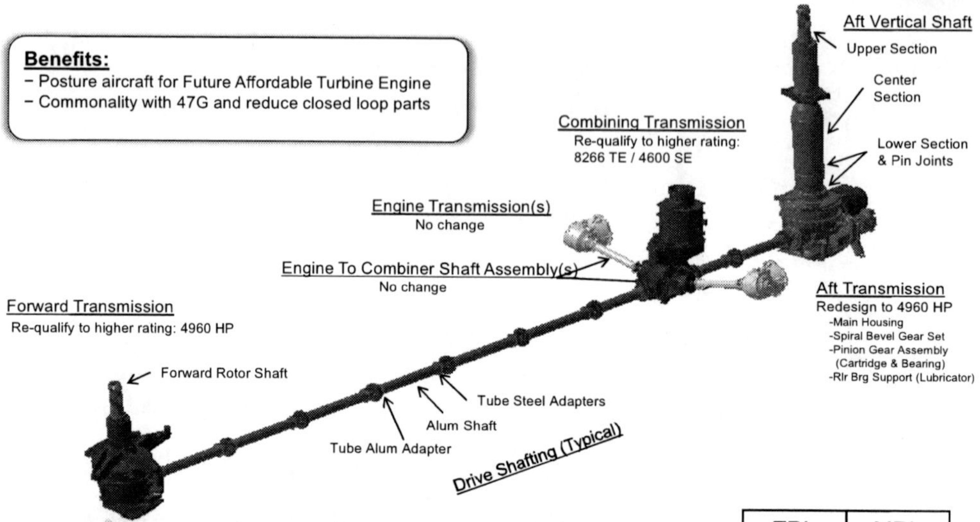

- No Change (MYII Configuration)
- Change required for Block II
- No redesign - Requalified to higher rating

TRL	MRL
6	5

7

Block II Configuration

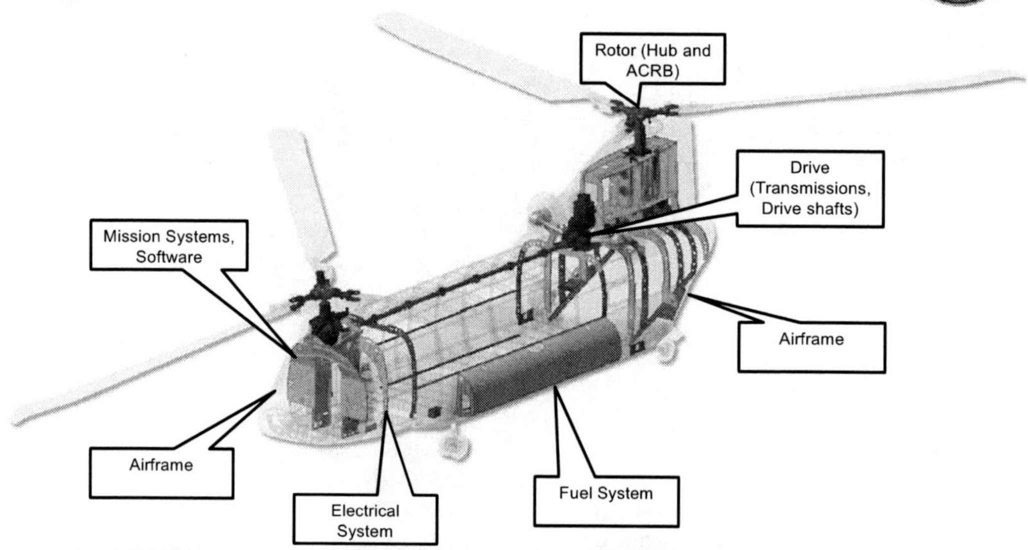

Block II Configuration Provides Necessary Lift and Improved Capability to the CAB Commander in the Near Term While Providing Growth Capability for the Future

8

Block II – Airframe

SEC 41	New Build
SEC 43	Salvage and Mod
SEC 46	New Build
Aft Pylon	New Build
Ramp	Refurbished

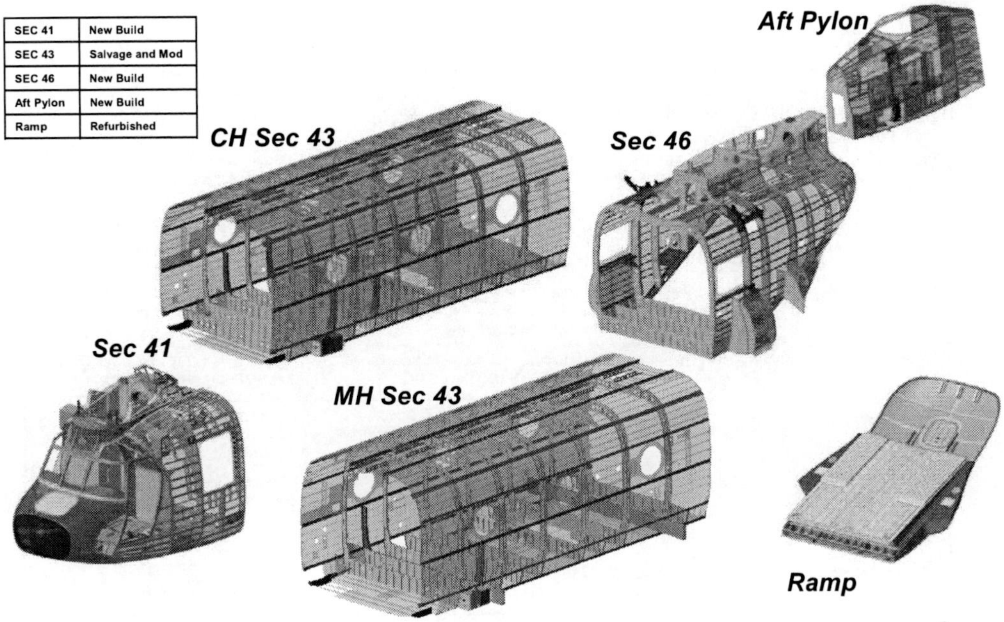

Fuel System

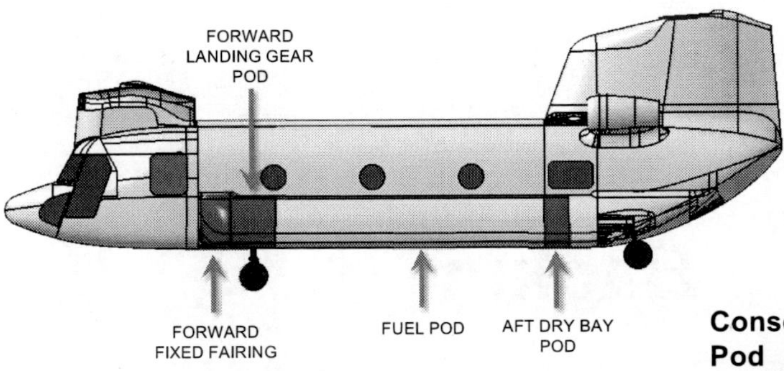

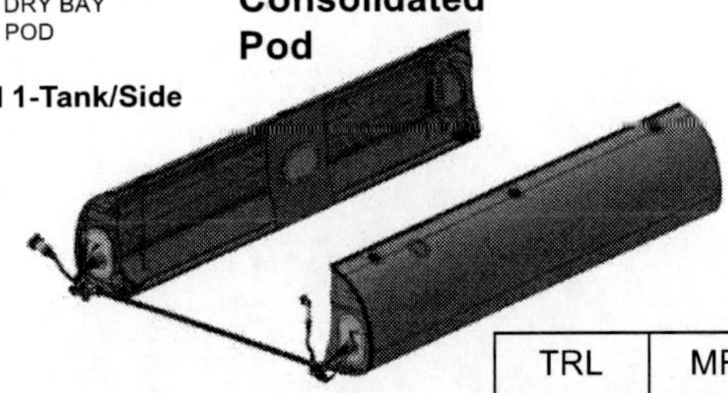

Legacy 3-Tank/Side Arrangement to Consolidated 1-Tank/Side

- Reduction in part count
- Forward landing gear shifts to G-Model location
- 1110 gals (increase of 80 gals)
- Reduced maintenance burden/O&S costs

TRL	MRL
7	5

Configuration Summary

Element	Block II Approach (BTA)	Rationale
Airframe	"Block II" Approach – Long Nose (G-based), Retain –F Cabin (modified for internal sys. & fuel), , G-Aft (with mounting provisions for "Block II+ " engine (alignment with longer term plan)	Lower NRE, common maintenance procedures, DVE provisions, larger gunner window/emergency escape
Ramp, Doors, Landing Gear	Retained –F Ramp and Doors, Retained LG, but relocated forward (common w/MH-47G) New single tank per side fuel sys.	Weight savings, common LG location for maintenance
Fuel	New Single Tank (1110 Gal total), New Structure	Weight savings, reduced parts count, increased fuel available, reduced O&S costs
Transmission	Re-rated Combiner and Forward, leading to 8266(TE)/4600(SE), Redesigned Aft XMSN (4960 SHP), Retain Engine xmsn, New Sync Shafts	Sea Level/ Std payload increases, reduced O&S costs, postured for increased power requirements
Rotor	Updated Hub (lead spline), Slider, Updated Aft Shaft, New Forward Shaft. High Hot: Nr=103% (& Trim/LCTA)	Allows 54K MGW with lower O&S cost across MH/CH fleets
Electrical	Modified current – cleaned up DC, enhanced DC power, current APU, New 10-15 KVA Backup Gen, LED lighting	Upgraded and cleaner power for modern electronics, load capability for emergency situations
Flight Controls	Redesigned LCTA, DAFCS Update	Reduced maintenance, improved handling/DVE
Engine	HW 714 Engine – O&S improvements	Reduced O&S costs
Mission Systems	Updated CAAS & DAFCS (as driven by above changes), New Rotor Torque Sensor	Improved human interface and aircraft torque reliability

11

Backup

12

ACRB to Legacy Blade Comparison

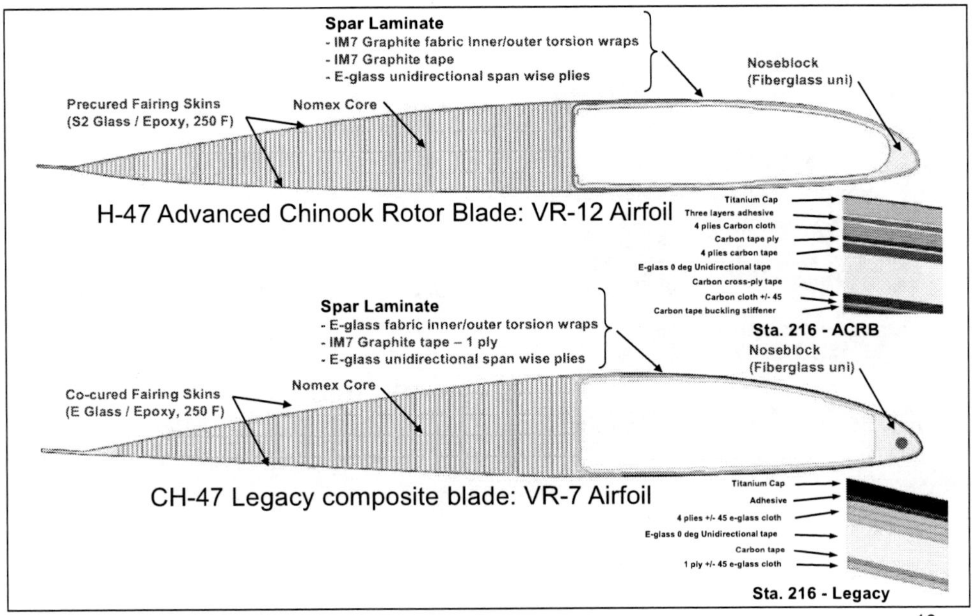

13

Chinook CH-47 Block II Schedule

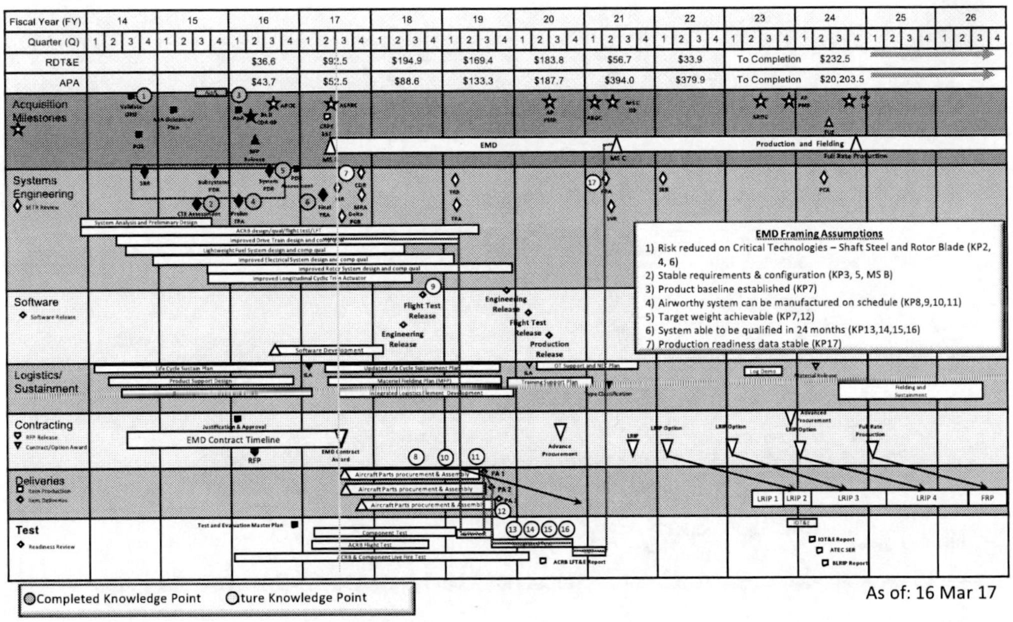

As of: 16 Mar 17

14

Apache Project Office

Apache Update
AHS

Mr. Alan McClendon
Apache PM
Technical Management Division Chief

Distribution A: Public Release

AH-64 Configuration Roadmap

Apache Project Office

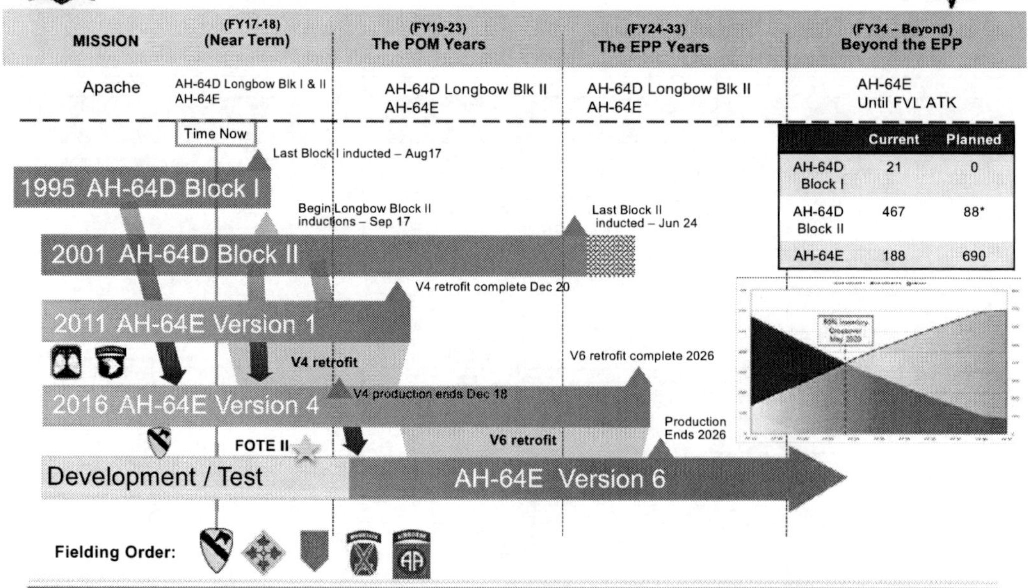

	Current	Planned
AH-64D Block I	21	0
AH-64D Block II	467	88*
AH-64E	188	690

Distribution A: Public Release

Apache Life-Cycle Plan

Apache Project Office

Best-Value Sustainment and Modernization Program

- Maintains Apache Readiness Today
- Ensures Apache Preparedness for Tomorrow

AH-64A (1984-1994) 821

- FLIR sensors
  - Targeting & Pilotage
- Improved fire control
- HELLFIRE Missile
  - Laser Seeker
- 30 MM gun
- Improved survivability
  - Redundant design
  - Ballistic tolerance
  - Backup Control System
  - Crashworthy fuel system

Analog cockpit

AH-64D - (Block I) (1995-2002) 284

- Longbow Fire Control Radar
- Radar Frequency Interferometer (RFI)
- HELLFIRE missile
- Radio Frequency Seeker
- Glass cockpit

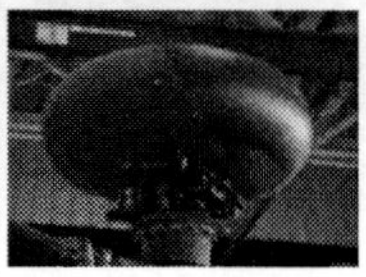

AH-64D (Block II) (2003-2013) 429, 68 (new)

- Digital Interoperability
- Digital Map
- Integrated Communications Suite
- MSPU
- M-TADS/PNVS
- CMWS
- ASPI
- VNSight
- Manned Unmanned Teaming – Level 2
- Ground Fire Acquisition System

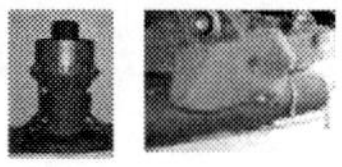

AH-64E (2011-2025) 634, 56 (new)

- Improved Flight Performance
  - Improved Drive System (IDS)
  - Engine: 701D with EDECU
  - Composite Main Rotor Blades
- Increased Computer Processing/OSA
  - Mission Processor, Aircraft Interface Units, Solid State Recording Device, Gun Subsystem Controller
  - IHADSS with Magnetic Head Tracker
- Improved Situational Awareness
  - Cockpit Decision Aiding
  - Color TV
  - Level IV UAS Control
- Joint Interoperability
  - ARC-231/SATCOM
  - Link-16/SRW
- Improved Sustainability
  - New Fuselage
  - System Level Embedded Diagnostics
- Improved Survivability
- Improved Targeting
  - Radar Range Extension
  - Maritime Targeting Mode
  - RFI Passive Ranging

Distribution A: Public Release

3

AH-64E Tech Insertions Funding Based Strategy

Apache Project Office

FY 2012 – 2015: LRIP / Lot 2	2016 – 2018: Version 4 (v4)	2019 - 2026: Version 6 (v6)
• Level IV Unmanned Aircraft System • Improved Communications Suite • Improved Electronics / Modular Open Systems Approach • Improved Drive System • Composite Main Rotor Blades • New Fuselages • Integrated Aircraft Survivability Equipment • Modernized Signal Processor Unit – Integrated (MSPU-I) • Training Device Concurrency • Radar Electronics Unit (REU) • IMC/IFR Hardware / Software • Pure 701D Engines with Enhanced Electronic Controls	• Link-16 • System Level Embedded Diagnostics (SLED) • Smart Tool for Apache Maintenance Picture (STAMP) • External Fuel System (Full Provisions) • Air to Air to Ground video relay (AAG)	• Radar Frequency Interferometer (RFI) Passive Ranging/Range Extension • Fire Control Radar Enhancements • Longbow Integrated Maintenance Support System Ground Analysis Software Improvement • Interactive Electronic Technical Manual Upgrade • Maritime Targeting Mode • Multi-Mode Laser • Image Blending • Soldier Radio Waveform (SRW) • Embedded Diagnostics • Common Configuration Fielding 2019

- 690 AH-64E aircraft is the planned end state (634 Remanufacture / 56 New Build)
- Capabilities are time-phased based on funding only
- All Version 4-6 capabilities can be applied as field retrofits
- Capability insertion strategy applies to both AH-64E Reman and New Build aircraft

Distribution A: Public Release

Apache Modernization

Apache Project Office

- **Apache will remain the Army attack helicopter until at least 2048 per US Army Equipment Modernization Plan**
  - AH-64E production complete ~2026
  - Future Vertical Lift (FVL) planned IOC in 2034
- **Modernization necessary to keep Apaches relevant and dominant.**

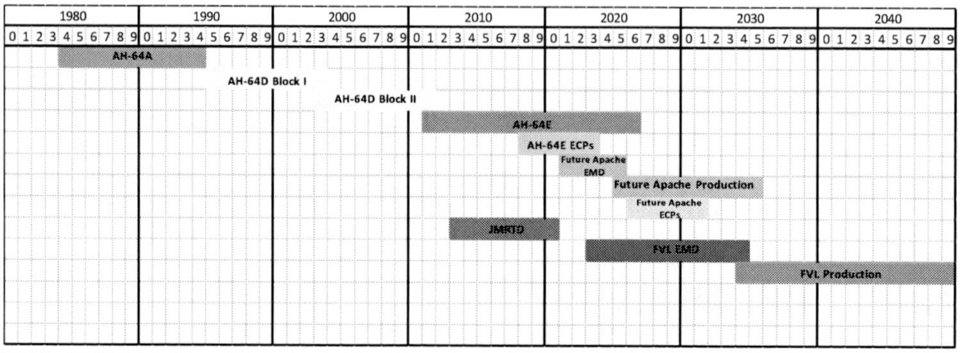

JMRTD – Joint Multi-Role Rotorcraft Technology Development

5

Distribution A: Public Release

Potential Capability Improvements

Apache Project Office

AH-64E ECPs

- Platform and Aircraft Performance
  - Composite Tail Rotor Blade (Phase I)
  - Composite Structures (Composite Tail Boom, Composite Vertical Stabilizer)
  - Improved IR Suppressor
  - Improved Main Rotor Hub
- Sensors and Situational Awareness
  - Modernized Radar Above Rotor
  - Sensor Improvements
  - MUMT- X/IOP
  - Advanced Cockpit (Head/Helmet Mounted Display)
  - Advanced Air Data Sensor
  - Next Gen Targeting Sensor
  - HR Turret (TADS/PNVS)
  - DVE – TAWS
- Avionics and Crew Awareness
  - Mission Processor
  - Improved Crew Ballistic Protection
  - Helmet Mounted Display
  - Advanced CDAS
  - Adaptive Crew Seats
  - LRU SWAP Improvements
  - Improved CBM Capabilities
  - Modernized Pylon & Weapon Interface
- Weapons
  - Small Guided Munitions
  - Modernized Rocket Launcher (Inc 1, Inc 2)
  - Next Generation Air-to-Ground Missile
  - Airburst 30mm
  - Improved Gun Turret
  - ASE Upgrades
  - Directed Energy Weapons (Phase I, Phase II)

AH-64E ECPs

- Platform and Aircraft Performance
  - ITEP
  - ITEP IR Suppressor
  - 3850 SHP Transmission
  - Composite Structures
  - Fuselage Center Section Upgrade
  - Improved Drive System
  - Propulsor
  - Compounding Wing
  - Improved Tail Rotor Drive System
- Sensors and Situational Awareness
  - Mission Processor Refresh
  - Modernized Radar
  - EO/IR Modular Sensor
  - Sensor Improvement (Conformal Radar, 360° Sensor)
  - DVE - Synthetic Vision
  - Integrated RF / Sensor Fusion
- Avionics and Crew Awareness
  - Advanced ASE Suite
  - Adaptive Crew Seat
  - Avionics Re-Architecture (Mission Processor Refresher, IDM Elimination)
  - Electrical Power
  - Advanced Cockpit (see list for sub-projects)
  - Advanced PVI and Displays
  - Advanced Flight Controls/Grip/Audio Interface
- Weapons
  - Counter Precision Fires

Distribution A: Public Release

AH-64E Multi-Year Summary

Apache Project Office

Apache Program Summary:

- AH-64E consists of two MDAPs: (AAO – 690)
  - AH-64E Remanufacture program (AAO – 634)
  - AH-64E New Build program (AAO – 56)
- **Milestone B:** June 2006
- **Milestone C:** September 2010
- **IOT&E:** April 2012
- **FRP:** August 2012
- **FOT&E I:** August 2014
- **FOT&E II:** April 2018

This will be the first Multiyear Procurement for the AH-64E-Model

Apache Multiyear Contract Parameters

- Five year contract covering FY 17-21
- Contract Need Date: 30 April 2017
- PoP: 60 Months
- Type Contract: Firm Fixed Price
- Contract Baseline Quantities: 244 remanufacture aircraft
- Options for up to 450 total aircraft includes USG Reman, New Build, and FMS

Multi-Year Contract Milestones:

- **MYP J&A HQDA:** 23DEC14
- **Boeing accepts RFP:** 16APR15
- **Boeing Proposal:** 02OCT15
- **MY Exhibits Submitted:** 28SEP15
- **CAPE Analysis:** SEP 15 – FEB 16
- **AP Award:** 22MAR16
- **POM Approval:** SEP 16
- **Negotiations:** OCT – DEC 16
- **SECDEF Approval:** 23JAN17
- **MY Award:** 15MAR17

Status:

- Multi-Year contract was awarded on 15MAR17
- First delivery under Multi-Year planned for March 2018
- Plan to cut-in Version 6 configuration to production line in 2019
- FMS cases will be included in Multi-Year

Green **color indicates action complete**

Distribution A: Public Release

7

International Apache Fleet

Apache Project Office

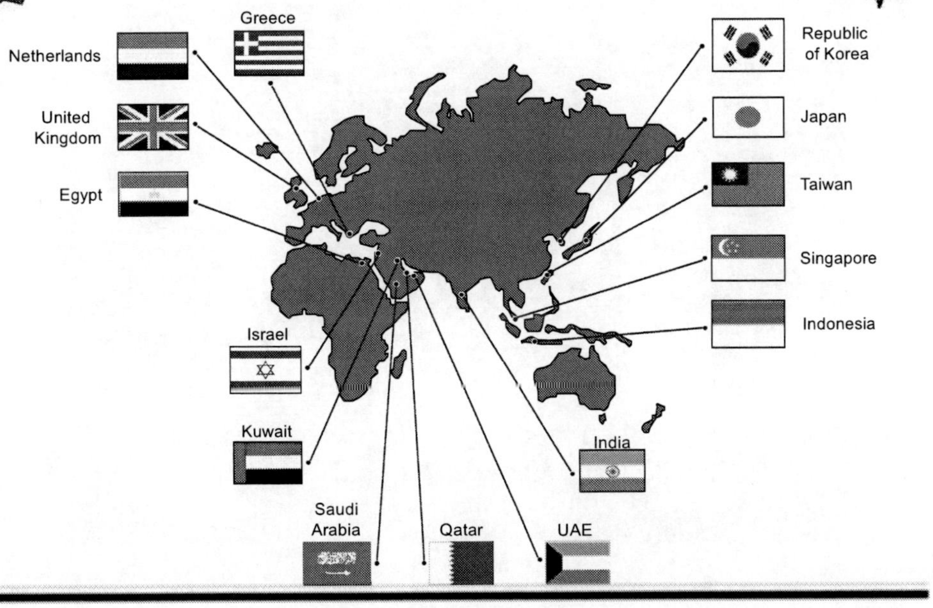

Distribution A: Public Release

As of 5/23/2017

Common Configuration

Apache Project Office

Purpose:

- **Better align FMS and USG configuration for improved interoperability and supportability**
- **Mutual benefit from unique development programs**
- **Reduce cost and time to export USG upgrades**
- **Lower recurring production costs**
- **Lower software development costs**
- **Reduce life cycle costs through common software baseline**
- **Position to support MY**

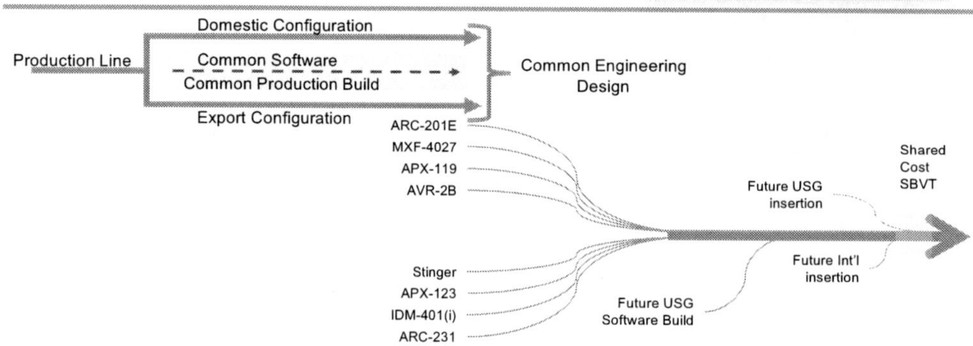

Distribution A: Public Release

Questions

Apache Project Office

Backup

Apache Project Office

AH-64E Program Schedule

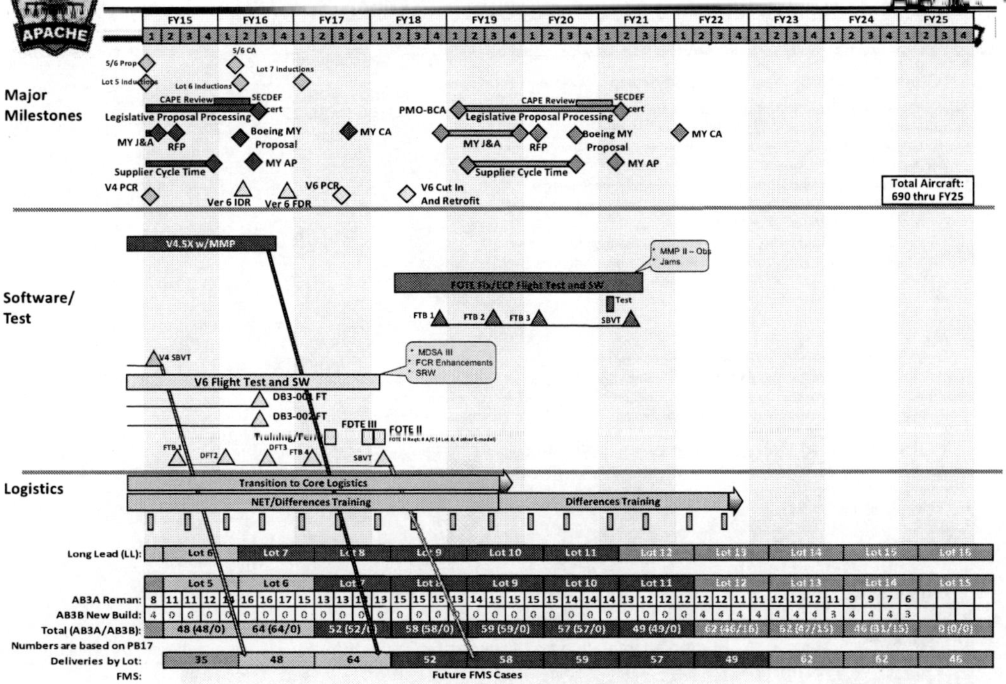

Distribution A: Public Release

UNCLASSIFIED

PM UAS

Mr. Dennis Sparks
Technical Management Chief,
PM Unmanned Aircraft Systems
AHS | May 2017

Distribution Statement A – Approved for Public Release: Distribution is Unlimited

9 May 2016

UNCLASSIFIED

Army Unmanned Aircraft Systems (PM Portfolio)

INSCOM and Special Operations (Echelon Above Division)

Echelons Above Division (EAD)/Strategic Reconnaissance, Surveillance, and Target Acquisition (RSTA) supporting the ISR process. Full Motion Video (FMV), Signals Intelligence (SIGINT), Synthetic Aperture Radar, Ground Moving Target Indicator Radar (SAR/GMTI), comms, and weapons. Assigned to Aerial Intelligence Brigades and the 160th Special Operations Aviation Regiment.

Division Level

Provides dedicated RSTA supporting the Commander's Common Operational Picture (COP), and the ISR Process. Mission-configured support including FMV, SAR/GMTI, SIGINT, comms, and weapons. Assigned to Active Component Divisions and the National Training Center.

Brigade Level

Provides organic RSTA at Brigade level and Below supporting the Commander's COP. Mission Configured support including FMV and comms. Assigned to the Brigade Combat Team, Combat Aviation Brigades, and Special Forces Group/Ranger Regiment.

Battalion and Company Level

Provides the small unit the organic capability to perform Beyond Visual Line-Of-Sight (BLOS) RSTA with payloads such as FMV.

Command & Control and Dissemination

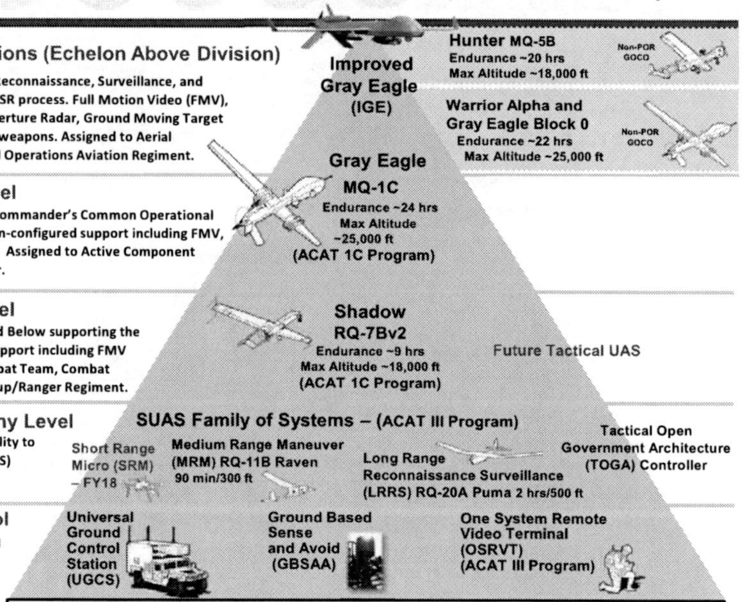

FOCUS OF ARMY UAS:
- Strategic/National
- Operational
- Tactical

Rapid Integration and Acceptance Center (RIAC) | International Programs (FMS)

2

UNCLASSIFIED

UAS: Intersection of the Army

UAS interacts with all aspects of operations...mixing stakeholders, missions & capabilities

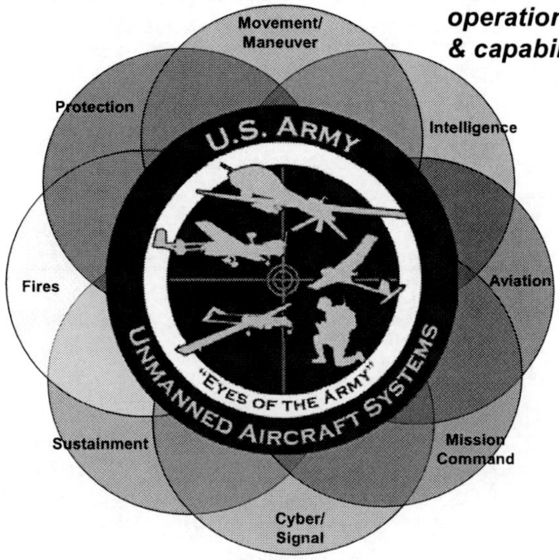

- **Manned/Unmanned Teaming**
- **Indirect Fire**
- **Data Dissemination**
- **Recon/Surveillance/BDA**
- **Electronic Warfare**
- **Special Operations Forces**
- **Communication/Data Relay**
- **Attack**
- **Cargo**

UAS has revolutionized how the Army fights...it will continue to do so

UAS...a Combined Arms Asset

3

Unmanned Aircraft Systems Portfolio

MC-1C Gray Eagle

- ACAT 1C Program -- In Full Rate Production Phase
- Primary missions: Provide Division and Echelons Above Divisions Reconnaissance, Surveillance, Target Acquisition (RSTA), extended Command & Control (C2), Attack, Manned Unmanned Teaming (MUM-T)
- Deployed in support of Overseas Contingency Operations (OCO)
- A/C OEM: General Atomics, Inc., San Diego, CA
- Key subcontractors: L3, Textron Systems, Sierra Nevada Corporation (SNC)

RQ-7B v1 /v2 Shadow TUAS

- ACAT 1C Program -- Shadow v1 In Operations and Sustainment Phase and Shadow v2 In Production & Deployment phase
- Primary missions: Provides brigade commanders with organic tactical level RSTA, and battle damage assessment (BDA) capabilities
- Features Electro Optical and Infrared (EO/IR) sensors
- Deployed in support of Overseas Contingency Operations
- Aircraft OEM: Textron Systems, Hunt Valley, MD
- Key subcontractors: L3, Sierra Nevada Corporation, Lockheed Martin, Stark Aerospace, Raytheon

Family of Small Unmanned Aircraft Systems

- ACAT III Program --- In Operations and Support Phase
- Provides echelons below battalion operators enhanced reconnaissance, surveillance, and battlefield awareness
- Provides Short Range Micro (SRM), Medium Range Maneuver (MRM), and Long Range Recon and Surveillance (LRRS) capabilities
- RQ-11B Raven performs Medium Range role; RQ-20A Puma performs LRRS role; and SRM procurement FY18
- AAO: MRM 1213; LRRS 3604; SRM 1580
- Contractors: (*Products*) Aerovironment, Altavian, Elbit Systems of America, Innovative Automation Technologies, and Lockheed Martin. (*Services*) Bosh

One System Remote Video Terminal (OSRVT)

- ACAT III Program -- In Production and Deployment Phase
- OSRVT is a kit integrated with the Remote Operations Video Enhanced Receiver (ROVER) 6 System that provides full motion video and telemetry data from multiple unmanned platforms: Gray Eagle, Shadow, Hunter, Raven, Predator, other UAS and manned platforms (Apache, Kiowa, Litening/Sniper Pods (Air Force F-15))
- Prime Contractor: Textron Systems
- Hardware Contractor: L3 Communications

America's Force of Decisive Action 4 ARMY STRONG.

MQ-1C Gray Eagle (Extended Range)

Provides extended range, multi-purpose UAS capable of executing reconnaissance, security, attack, and intelligence collection across the range of military operations. Supports Echelons Above Division.

Characteristics / Description

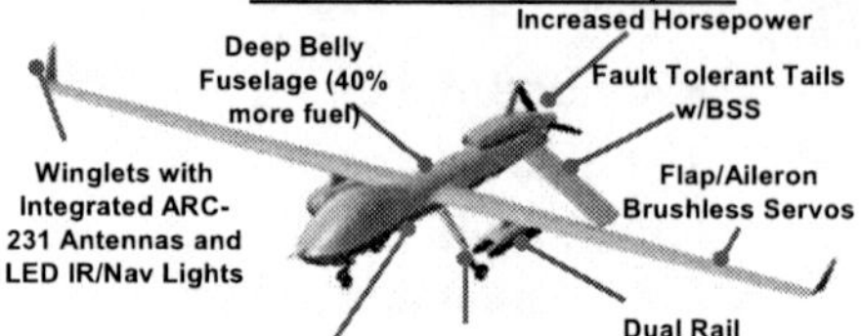

Attribute	Capability
EO/IR & SAR	34+ hrs
EO/IR, SAR, TSP, 2 Hellfires	27+ hrs
Gross Take off Weight	4,200 GTOW
Space, Weight, & Power	540 lb internal; 500 lb each wing; 500 lb centerline hard point; 9.5 kW
Max Fuel Capacity	900 lbs
Aircraft Communications Radios	(3) ARC-231s
Heavy Fuel Engine (JP-8/F-24)	Yes (OEM 180) – 180 hp
Supportability	Soldier Maintained: (50% more payload/avionics volume, greater access to avionics, improved landing gear, longer life servos)

Mission Requirements

- Next-generation derivative of baseline Gray Eagle intended for use in RSTA and Attack missions fully integrated into Army unified land operations as follows:
  - RSTA configuration is a clean wing with no pylons, a Common Sensor Payload (CSP) High Definition (HD), and STARLite Extended Range system
  - INSCOM/SOCOM configuration consists of dual rail Hellfire pylon with 2 Hellfire missiles, Tactical Sign Payload (TSP) system, CSP HD, and STARlite SAR/GMTI
  - Attack configuration is comprised of two dual rail Hellfire pylons with 4 Hellfire missiles, CSP HD, and STARLite SAR/GMTI
- GA-ASI developed with Internal Research and Development (IRAD) funds

Current Status

- In Production
- E3, Environmental, First Article Testing: Apr–Nov 2017
- Log Demo: Jan 2018
- FOTE II: Apr 2018
- Trans Mobility: May 2018

Contractors

- General Atomics - San Diego, CA (Aircraft)
  - L3 Communications - Salt Lake City, UT (Datalinks)
  - Textron Corporation - Hunt Valley, MD (Ground Control Station (GCS))
  - Raytheon – McKinney, TX (Common Sensor Payload)
  - Northrop Grumman – Baltimore, MD (STARlite SAR Payload)
  - BAE – Nashua, NH (Tactical SIGINT Payload)

America's Force of Decisive Action 5 ARMY STRONG.

RQ-7Bv2 Improvements

Block III Engine
- More horsepower
- Improved Reliability
- Lower acoustic signature
- Requires AV structural mods

Small Mission Computer (SMC)
- Enable autonomy and AiTR
- 50x greater computer processing power
- FACE Compliant: Increased interoperability

Communications Relay
- Updated Comms Relay Interface Box on the AV to convert VOIP to analog
- Replace 152 with 152A (JTRS) radio
- New multi-band antennas

Weatherization
- Allows flights in 2" of rain

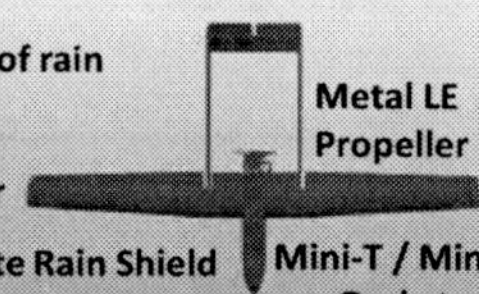

Assured Position Navigation and Timing
- Enables Shadow missions in a GPS denied threat environment.
- Vision Based Navigation, Anti-Jam GPS Antenna, GDT RF tracking, more tightly coupled GPS/INS.

Second Source Laser Designation Payload
- Allows for integration of multiple LD payloads

America's Force of Decisive Action 6 ARMY STRONG.

Future Tactical Unmanned Aircraft System

Characteristics

- ***Mobility***
  - ***Expeditionary***
  - ***Confined Area Operations***
  - ***Operations on the Move***
  - ***Operational footprint***
- ***Survivability***
  - ***GPS Denied***
  - ***Signatures***
  - ***LPI/LPD***
  - ***Cyber***
- ***Lethality***
  - ***Weapons***
  - ***EW***
- ***Mission Command***
  - ***Payload dissemination***
- ***Training***
  - ***Universal operators and maintainers***
- ***Logistics***
  - ***Autonomous Resupply***
  - ***Operational footprint***

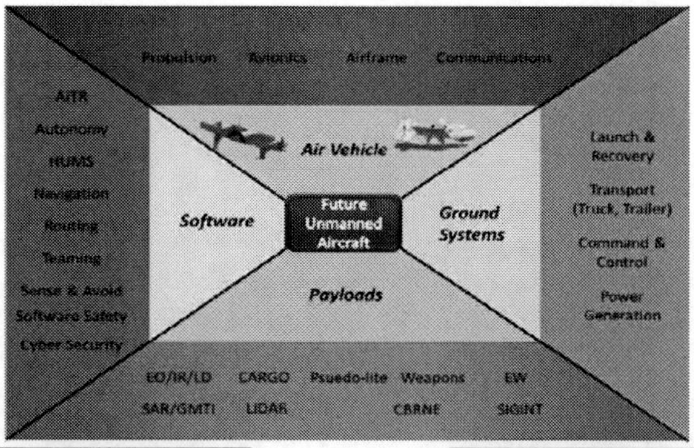

FVL Compatibility

Software Obsolescence Overview

Description

- **Has the same basic functionality as the current UGCS Software**
- **Provide re-architected GCS Software that:**
  - **Addresses software safety**
  - **Meets cyber security requirements**
  - **Is FACE™ conformant**
  - **Reduces Hardware dependencies**
  - **Supports Scalability**
- **Provide a new, common Human Machine Interface (HMI), enabling the Universal Operator concept**

Why Do Software Obsolescence

- **Existing UGCS software is:**
  - **Old and "brittle" → Costly to maintain**
  - **Expensive to add new capabilities**
  - **Was not designed to address current Cyber, Information Assurance, Safety, and Airworthiness requirements**
- **Need to comply with Army Common Operating Environment in the Real Time Safety Critical Computing Environment → Future Airborne Capabilities Environment (FACE™)**

Top Level Technical Approach

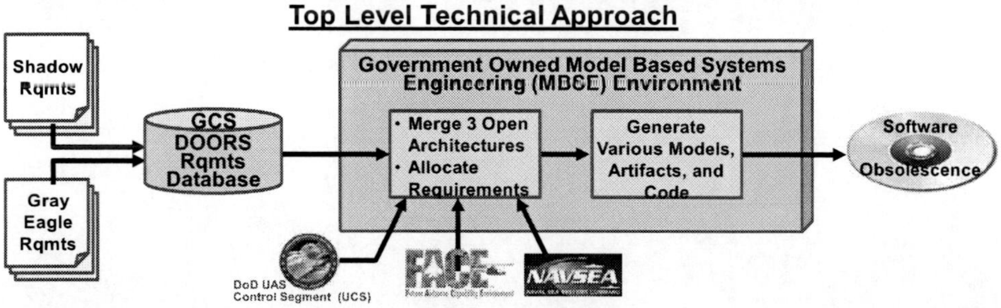

Top Level Technical Approach

Merging Multiple Architectures

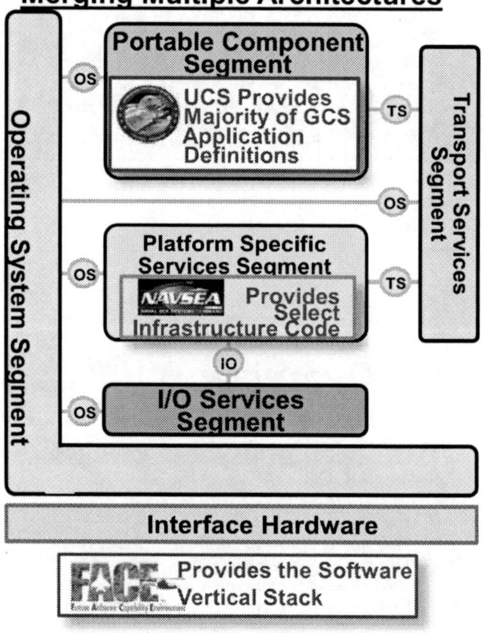

Key Technical Enablers

- Each architecture focused on different aspects of the overall software
- Data Models:
  - Facilitates connection of differing architectures
  - FACE will allow us to define the syntactic and semantic aspects of the Data Model
  - Data Model tools will help document and develop Data Models
- NAVSEA is providing some of our Infrastructure software:
  - Software Containers
  - Standardized Service Management
  - Track Manager
- NAVSEA Infrastructure software facilitates connecting non-FACE based applications into the overall architecture via software containers

America's Force of Decisive Action 9 ARMY STRONG.

UNCLASSIFIED
H-60 Black Hawk Update
MAY 2017
Mr. TRAVIS S. SINCLAIR
H-60M CHIEF ENGINEER
Commitment to Soldiers & Acquisition Excellence
UNCLASSIFIED

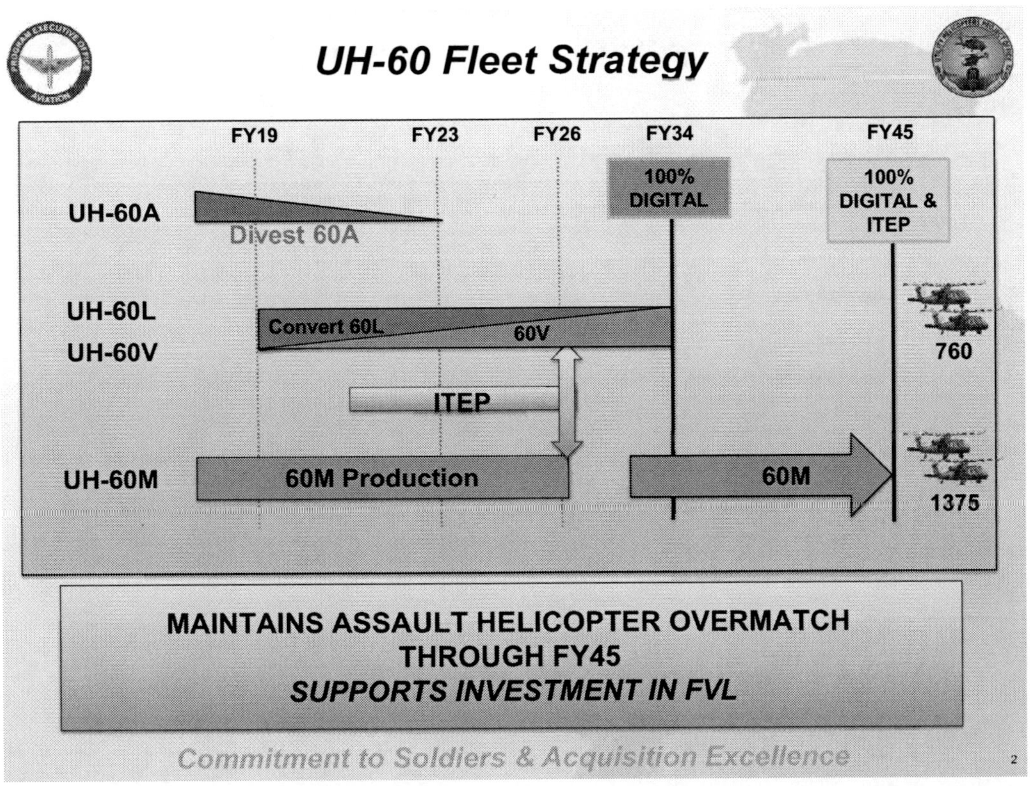
UH-60 Fleet Strategy
FY19
FY23
FY26
FY34
FY45
100% DIGITAL
100% DIGITAL & ITEP
UH-60A
Divest 60A
UH-60L
UH-60V
Convert 60L
60V
760
ITEP
UH-60M
60M Production
60M
1375
MAINTAINS ASSAULT HELICOPTER OVERMATCH THROUGH FY45
SUPPORTS INVESTMENT IN FVL
Commitment to Soldiers & Acquisition Excellence
2

H-60M Program Overview

(H-60 AAO = 2135)

- Digital cockpit, Fully-coupled flight director, Wide-chord blades & 701D engine
- H-60M currently in Full Rate Production
  - ➢ Produced the 1,000th M-model on October 13, 2016
  - ➢ Began 4th CAB NETT at Fort Carson, CO on February 27, 2017

Commitment to Soldiers & Acquisition Excellence 3

UH-60V Program Overview

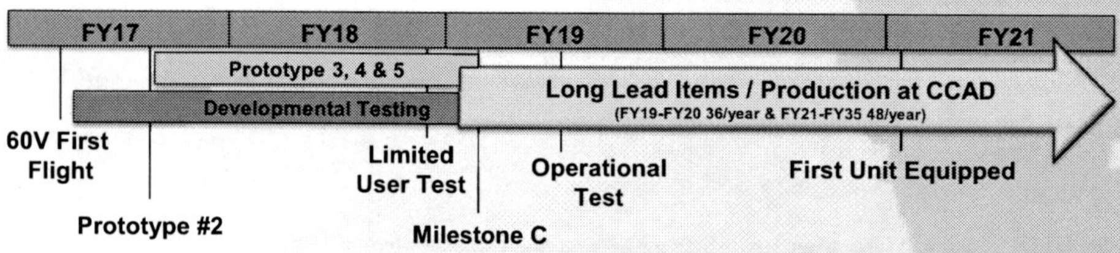

- Reuse existing H-60L airframes
- Upgrade of avionics to meet mission requirements
  - Digital Data Bus; Upgraded Communications and Navigation Suite
  - Digital Moving Map; Global Air Traffic Management (GATM)
- Cockpit modification during RECAP at CCAD
- 60M-like Pilot Vehicle Interface
- Government Rights Technical Data Package

60V Benefits

- Saves $B++
- Promotes Organic Industrial Base
- Open Architecture
- GATM
- Full and Open Competition
- Achieve Tech Data Overmatch
- Rapid Cyber Security Reaction
- Fully Leverage 60L Investment

Commitment to Soldiers & Acquisition Excellence 4

Handling Qualities Improvement Project (HQIP)

Description

- Improve H-60 handling qualities to reduce pilot workload in a Degraded Visual Environment (DVE)
- Modernized Control Laws (MCLAWS)
  - Attitude Command Attitude Hold (ACAH)
  - Automatically engages in low airspeed below 35 kts
  - Reduces flight control inputs to perform maneuvers in low airspeed
- Coupled Approach to Point (APTP)
  - Stable hover at selected point/altitude

Schedule

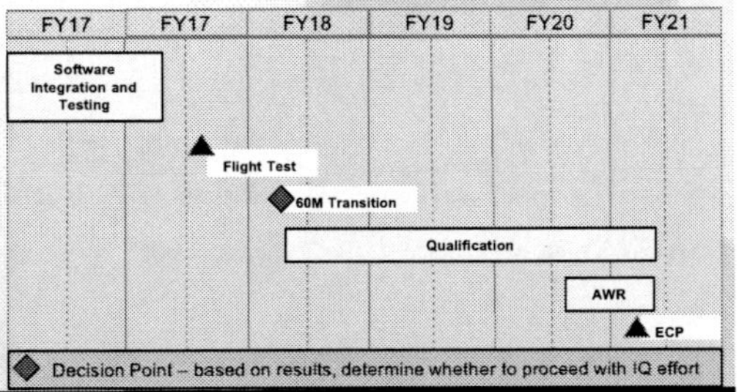

Strategy

- Optimize the current platform capability with improved control system software and symbology
- Demonstrate in flight on a production representative H-60M
- **U.S. Army Aero Flight Dynamics Directorate (AFDD)**
  - Revises modern control laws (MCLAWS) software
  - Performs EH-60L prototype flight testing
- **Sikorsky Aircraft Corporation**
  - Modifies UH-60M flight control computer (FCC) and flight management system (FMS) software
- **U.S. Army Aviation Applied Technology Directorate (AATD)**
  - Performs UH-60M production demonstration flights

Overview

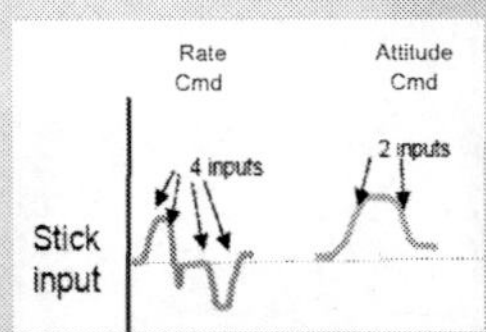

ACAH reduces flight control inputs to perform maneuvers in low airspeed. APTP is a coupled approach to a hover.

Commitment to Soldiers & Acquisition Excellence

5

Full Authority Flight Controls

Description

- Develop a solution for the integration of full authority flight controls into the H-60
- Achieve Level I Handling Qualities in GVE and DVE
- Work with Airworthiness Engineering Directorate to determine airworthiness requirements for a full authority flight control system
- Integrate into a production H-60M and perform a flight demonstration

Schedule

✓ Signed AATD/Sikorsky Agreement	MAR 2016
✓ Development of 1st Flight Requirements	JAN 2017
• Delivery of Kit to AATD	NOV 2017
• Flight Test AWR Signed	FEB 2018
• AATD Flight Demonstration	MAY 2018

Strategy

- Cooperative Development Agreement established between Sikorsky and U.S. Army Aviation Applied Technology Directorate (AATD)
- Sikorsky will provide full authority flight control components
- UH PO will provide HH-60M aircraft for full authority conversion
- AATD will perform integration of the full authority system and perform flight testing
- Modified aircraft available to support future autonomous testing activities

Overview

Commitment to Soldiers & Acquisition Excellence

6

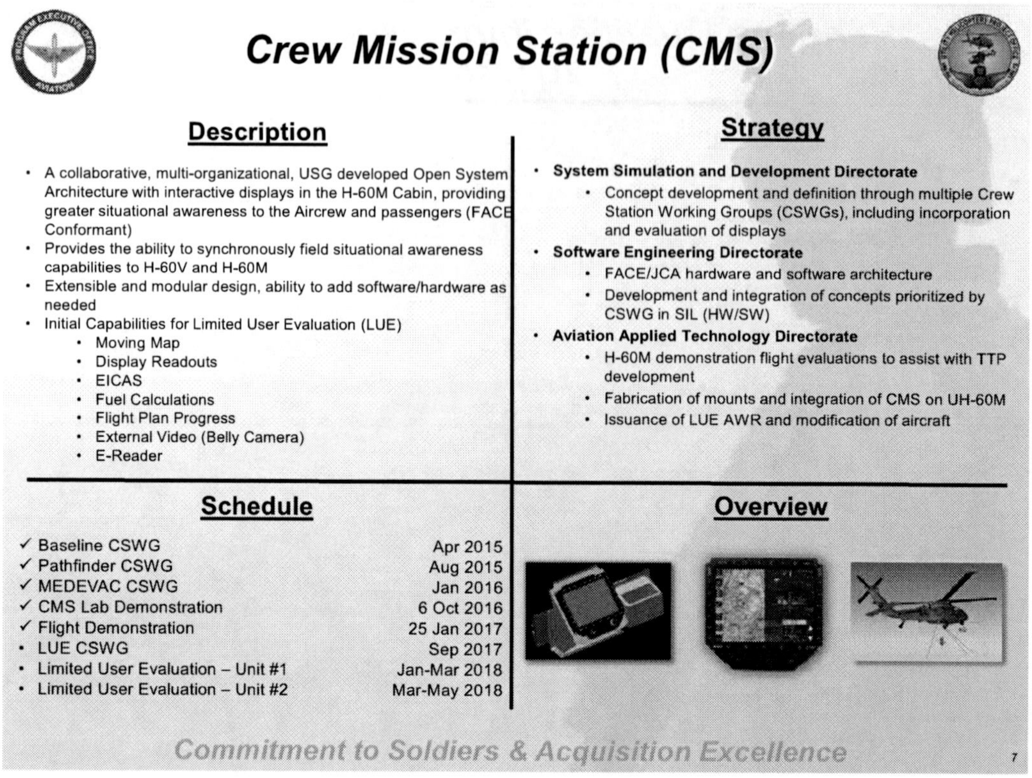
Crew Mission Station (CMS)
Description
• A collaborative, multi-organizational, USG developed Open System Architecture with interactive displays in the H-60M Cabin, providing greater situational awareness to the Aircrew and passengers (FACE Conformant)
• Provides the ability to synchronously field situational awareness capabilities to H-60V and H-60M
• Extensible and modular design, ability to add software/hardware as needed
• Initial Capabilities for Limited User Evaluation (LUE)
• Moving Map
• Display Readouts
• EICAS
• Fuel Calculations
• Flight Plan Progress
• External Video (Belly Camera)
• E-Reader
Strategy
• System Simulation and Development Directorate
• Concept development and definition through multiple Crew Station Working Groups (CSWGs), including incorporation and evaluation of displays
• Software Engineering Directorate
• FACE/JCA hardware and software architecture
• Development and integration of concepts prioritized by CSWG in SIL (HW/SW)
• Aviation Applied Technology Directorate
• H-60M demonstration flight evaluations to assist with TTP development
• Fabrication of mounts and integration of CMS on UH-60M
• Issuance of LUE AWR and modification of aircraft
Schedule
✓ Baseline CSWG Apr 2015
✓ Pathfinder CSWG Aug 2015
✓ MEDEVAC CSWG Jan 2016
✓ CMS Lab Demonstration 6 Oct 2016
✓ Flight Demonstration 25 Jan 2017
• LUE CSWG Sep 2017
• Limited User Evaluation – Unit #1 Jan-Mar 2018
• Limited User Evaluation – Unit #2 Mar-May 2018
Overview
Commitment to Soldiers & Acquisition Excellence
7

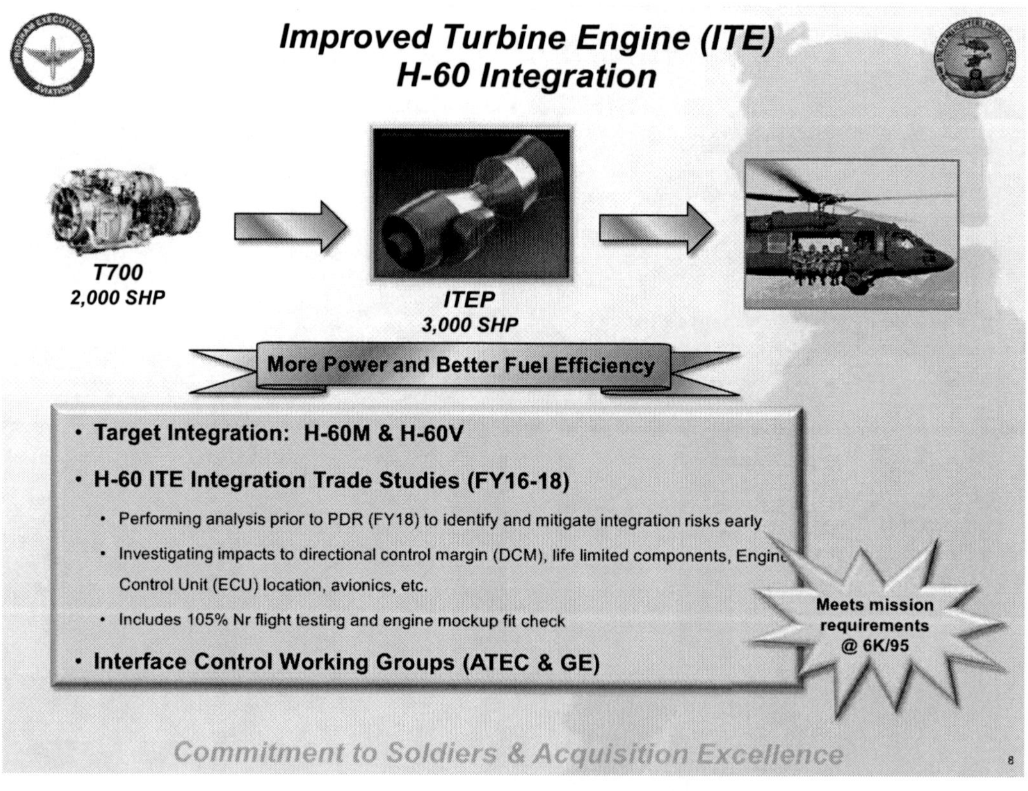
Improved Turbine Engine (ITE)
H-60 Integration
T700
2,000 SHP
ITEP
3,000 SHP
More Power and Better Fuel Efficiency
• Target Integration: H-60M & H-60V
• H-60 ITE Integration Trade Studies (FY16-18)
• Performing analysis prior to PDR (FY18) to identify and mitigate integration risks early
• Investigating impacts to directional control margin (DCM), life limited components, Engine Control Unit (ECU) location, avionics, etc.
• Includes 105% Nr flight testing and engine mockup fit check
• Interface Control Working Groups (ATEC & GE)
Meets mission requirements @ 6K/95
Commitment to Soldiers & Acquisition Excellence
8

H-60A/L Divestment
BLACK HAWK Exchange & Sales Team (BEST)

- Army Acquisition Executive and Department of the Army G4 Approved Exchange & Sale Program
- Eliminates Program Management Office Divestment
- Provides H-60A/L Fair Market Value toward replacement UH-60M aircraft via Reimbursable Transfer or Credit to Multi-Year Contract
- Customers are Federal & State Agencies for Surplus Property (SASPs); the OEM, Sikorsky Aircraft Corporation; and US Contractors via GSA auction

Commitment to Soldiers & Acquisition Excellence 9

Our Commitment: Provide Soldiers the best platforms Aviation has to offer.

- **THE WARFIGHTER IS OUR PRIORITY**
  Full Spectrum – Disaster Relief – Homeland Defense, MEDEVAC, Air Assault, OCO, Low – High Intensity
- **POSTURE FOR THE NEAR/MID-TERM**
  Digitization of UH-60V
  Keep Products Affordable
  Remain Relevant for the Future
- **SET CONDITIONS FOR FUTURE SUCCESS**
  ITEP – One stepping stone
  - Improves current fleets
  - Enable Future Technology

KEY POINTS:
- Broad and Constant Mission Sets
- Employed Globally
- Provides our Army Enduring Overmatch
- Ensure our Soldiers can achieve the Maximum Effect at home and abroad.

Commitment to Soldiers & Acquisition Excellence 10

The Strength of Our Nation...
...At Home and Abroad.

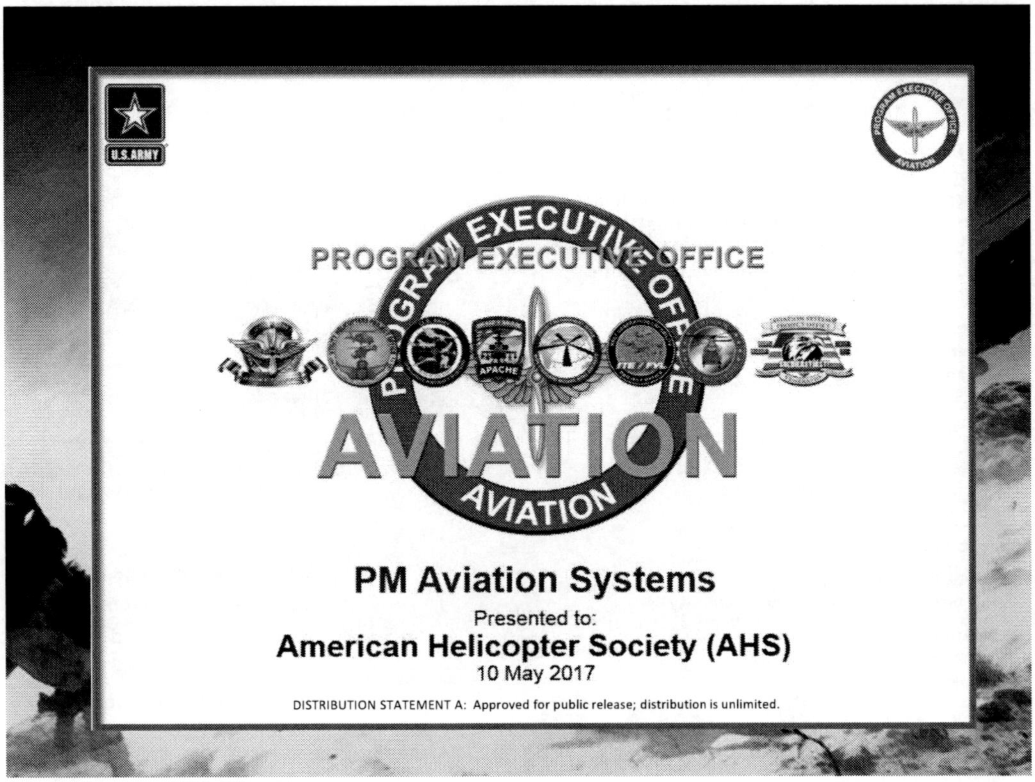
U.S. ARMY
PROGRAM EXECUTIVE OFFICE
AVIATION
PM Aviation Systems
Presented to:
American Helicopter Society (AHS)
10 May 2017
DISTRIBUTION STATEMENT A: Approved for public release; distribution is unlimited.

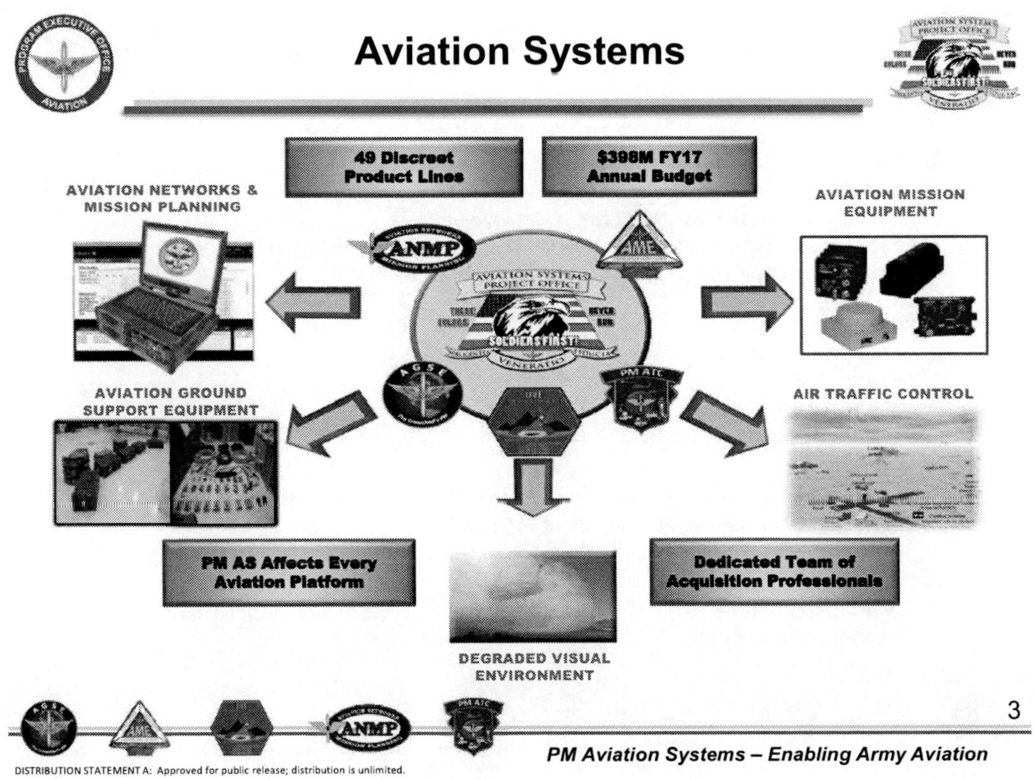
Aviation Systems
49 Discreet Product Lines
$398M FY17 Annual Budget
AVIATION NETWORKS & MISSION PLANNING
AVIATION MISSION EQUIPMENT
ANMP
AME
AVIATION SYSTEMS PROJECT OFFICE
AVIATION GROUND SUPPORT EQUIPMENT
AIR TRAFFIC CONTROL
PM ATC
PM AS Affects Every Aviation Platform
Dedicated Team of Acquisition Professionals
DEGRADED VISUAL ENVIRONMENT
3
PM Aviation Systems – Enabling Army Aviation
DISTRIBUTION STATEMENT A: Approved for public release; distribution is unlimited.

Degraded Visual Environment (DVE)
Brownout Rotorcraft Enhancement System (BORES)

Program Description ACAT II - Pre-MS B

The Degraded Visual Environment/Brownout Rotorcraft Enhancement System (DVE/BORES) is an Army aviation program providing critical operational capabilities for the Combat Aviation Brigades (CAB). Initially BORES will be fielded on the UH and CH aircraft to enable landing and takeoff in a brownout environment.

Capabilities:
- **Multi sensor, fused image**
- **Hazard avoidance**
- **Forward looking**
- **Single aircraft**

DVE/BORES – Increment 1

- **Enable Brownout operations**
- **Establish the foundation for growth**
- **Synchronize programs to enable future combat capability**
- **Capability Development Document (CDD) is out for staffing**

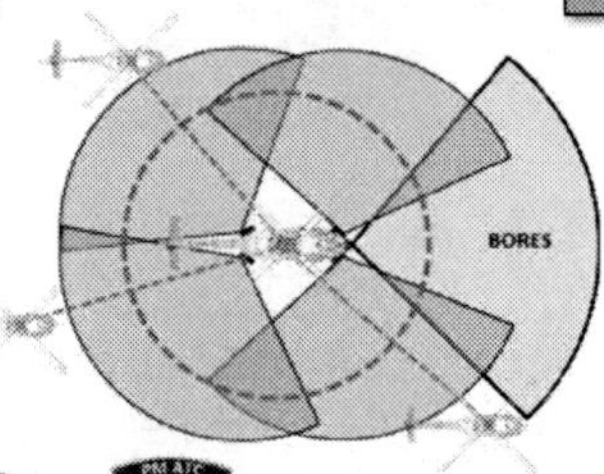

DVE Full Capability – Increment 2

- **Enable capabilities beyond Brownout "Own the Environment"**
- **Enable multiple flight regimes**
- **Technology Maturation:**
  - **360 degree field of view**
  - **Interoperability**
  - **Component efficiency**

4

PM Aviation Systems – Enabling Army Aviation

Leveraging Future Capabilities

- **Advanced Ground Support Equipment to Support:**
  - Enhanced Aircraft Capabilities
  - Greater Mobility and Sustainability in Deployment
  - Tools, Test Equipment, Material Handling Equip, and Information In Forward Operations
  - Adaptable to Split Base and Asymmetrical Operations
- **Navigation**
  - Enhanced Security
  - Without GPS
  - Improving Anti-Jam
  - Denied/Obscured/Contested Environments
- **Future Airborne Capability Environment (FACE) as Open Systems Solution**
  - Accepted Common Data Standards at Boundaries
  - Retains OEM Proprietary Secret Sauce
  - Increases Competition
  - Improves Portability
  - Speeds Development

5

PM Aviation Systems – Enabling Army Aviation

Leveraging Future Capabilities

- **Networking**
  - Aviation's Increasing Capability As User and Provider of Network Services
  - Information Assurance of the Platform and Mission Command Systems
  - Team Member in Defining Future Vertical Network
  - Ease of Use (Self-forming, Self-healing, etc.)
  - Denied/Obscured/Contested Environments

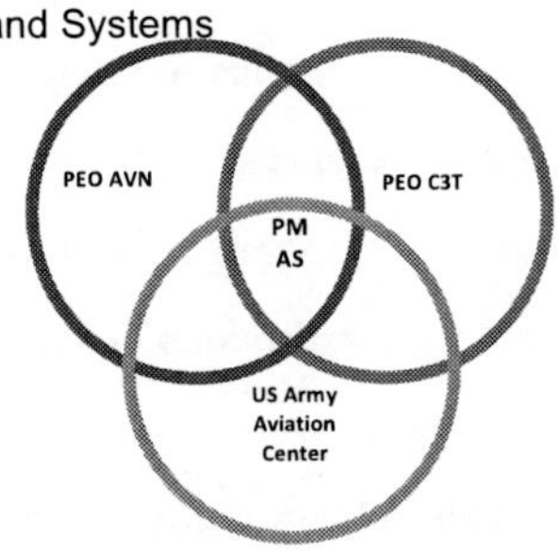

- **Contributions to Space, Weight, and Power Reductions**
  - Materials
  - Multifunctional
  - Green
  - Software instead of Hardware
- **Cyber Security**
  - Increasing Threats and Increasing Risks to Soldiers, Units, and Platforms
  - An Environment of Shared Risk and Responsibility Among All Network User
  - DoD Risk Management Frameworks (RMF) Invokes Significant Measures to Lessen Risk
  - Brings a New Rule Set to the Complex Environment of Aviation Platforms

6

PM Aviation Systems – Enabling Army Aviation

DISTRIBUTION STATEMENT A: Approved for public release; distribution is unlimited.

Network: Data Exchange

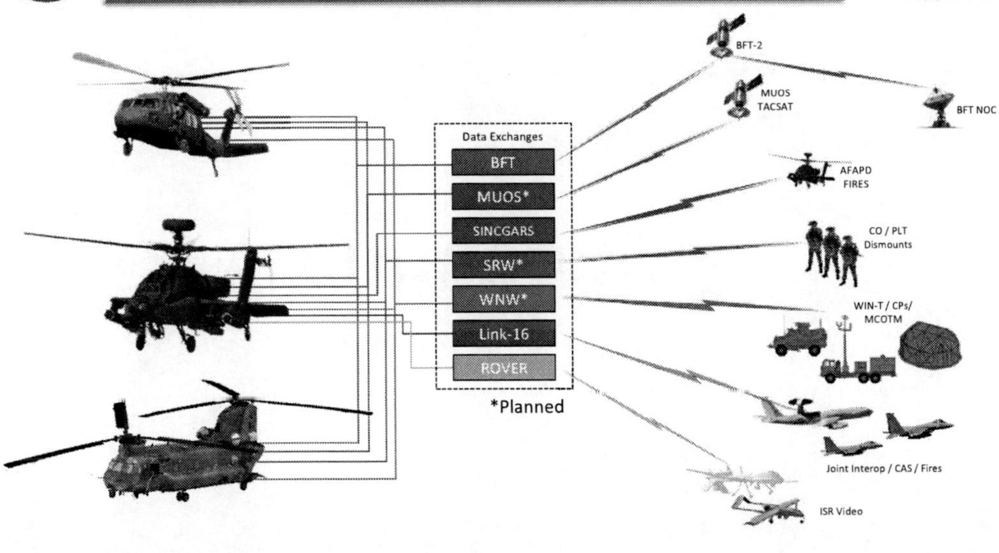

7

PM Aviation Systems – Enabling Army Aviation

DISTRIBUTION STATEMENT A: Approved for public release; distribution is unlimited.

Future Capability

- Product common across all platforms
- Common radio control
- Common message parser
- SWAP Savings
- FACE compliant/open architecture
- Multi-core process certification
- Build hardware agnostic software
- Cyber: Maintain Information Assurance
- Increases interoperability capability while minimizing OFP updates
- Landing platform for 3rd Party applications

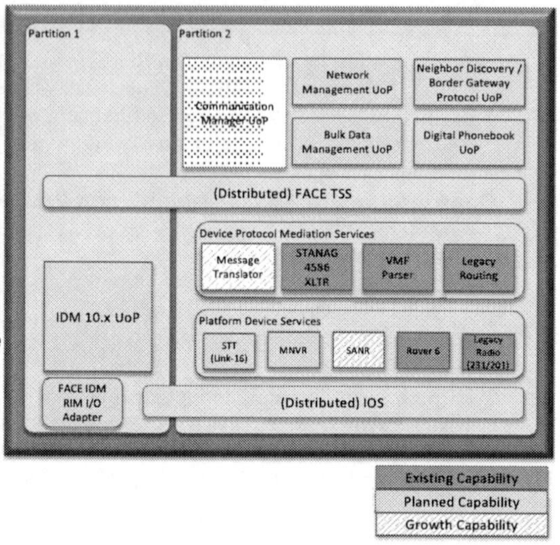

PM Aviation Systems – Enabling Army Aviation

DISTRIBUTION STATEMENT A: Approved for public release; distribution is unlimited.

Questions

PM Aviation Systems – Enabling Army Aviation

Improved Turbine Engine/Future Vertical Lift (ITE/FVL) Project Office Update

Mr. Bob Sheibley
Acting Project Manager
Improved Turbine Engine / Future Vertical Lift Project Office

10 May 2017

ITE/FVL Organization

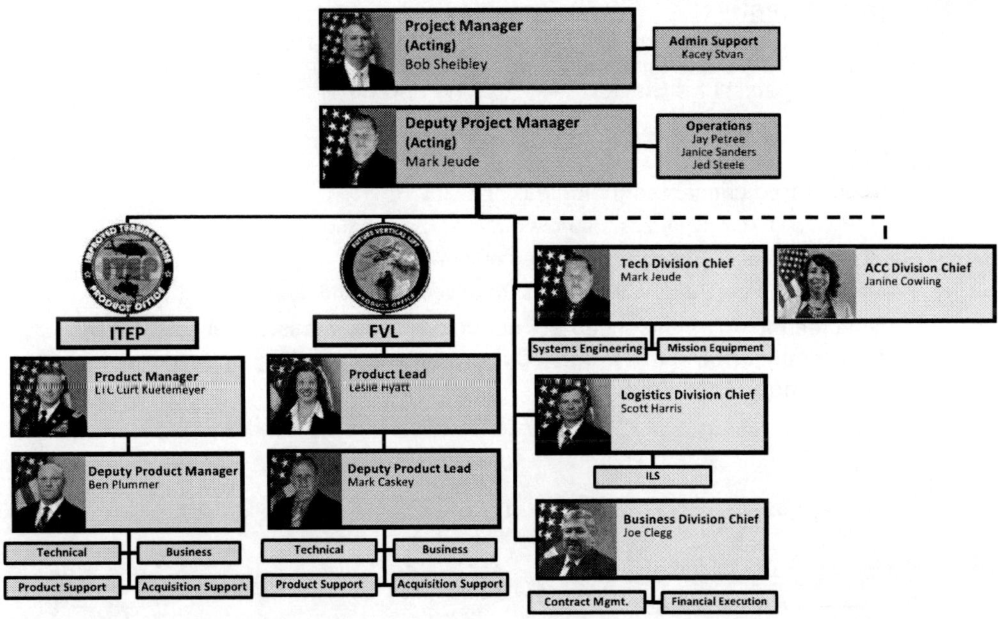

Improved Turbine Engine Program (ITEP)

Legacy T700 Engine:

1. 2000 Shaft Horse Power Engine Class
2. Originally designed with 1970's technology
3. 30 years of evolutionary growth
4. Engine performance has been maximized

Improved Turbine Engine (ITE)

1. 3000 Shaft Horse Power Class Engine
2. Will fit within the current Black Hawk and Apache engine bay at similar weight
3. Provides world wide performance to meet operational requirements at 6K/95
4. More power with greater fuel efficiency
5. Equal or better sustainability and reliability with growth margin

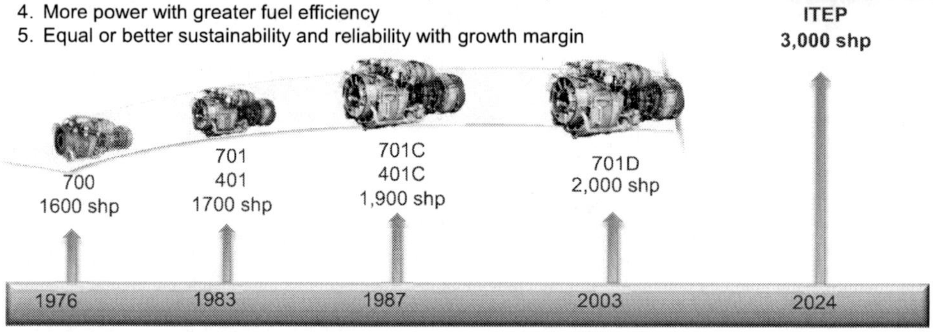

S&T Contribution to ITEP

Program Objective:

Advanced Affordable Turbine Engine (AATE) was an S&T program to develop improved, 3000 hp class turboshaft engine technology providing improved operational capability for Black Hawk, Apache and other Future Vertical Lift aircraft.

Results:

- Demonstrated critical technologies:
  - Advanced Aerodynamics: Improved fuel efficiency and power
  - Hybrid Ceramic Bearings: Improved reliability
  - Advanced Inlet Particle Separator: Removed sand and dust at inlet
- Critical technologies assessed at Technology Readiness Level (TRL) 6 by independent review team from Deputy Assistant Secretary of the Army for Research and Technology

Compressor/ Advance Aerodynamics

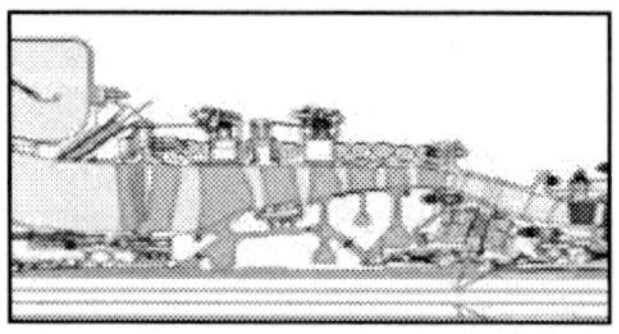

Hybrid Ceramic Bearing

Inlet Particle Separator

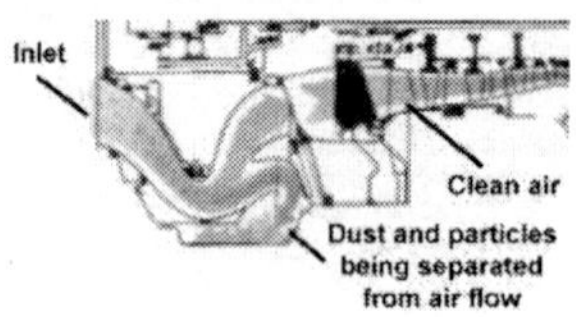

What is left to do

- Design and build an engine by integrating critical technologies
- Integrate engine onto the platforms (AH and UH)
- Test and qualify engines and aircraft

TMRR

- Preliminary Design and Technical Efforts
  - Conduct System Level, Component Level, and Software Requirements Reviews
  - Establish system level requirement verification methodologies
  - Develop engine and control system Interface Control Documents (ICDs)
  - Finalize preliminary design configuration and specification(s) to validate readiness to proceed to CDR
- Integration Efforts
  - Development of airframe ICDs
  - Identify platform configuration changes, determine integration risk mitigation strategies
  - More than 15 analysis and trade studies for each airframe including the following focus areas:
    - Engine Control Unit Location
    - Aircraft Software and Electrical Interface
    - Drive System Impact
    - Flight Technology/Performance Analysis

EMD

- Critical Design and Technical Efforts
  - Mature successful PDR designs via analysis, engineering, and rig testing prior to CDR
  - Conduct a CDR to finalize engine design
  - Begin building engine components and engines for testing
- Testing Efforts
  - Approve all official test plans/analysis
  - Witness testing and evaluate results to validate the engine meets requirements
  - Complete component, software, and engine level tests to grant a preliminary flight rating for platform tests
  - Complete remaining component, software, and engine-level testing/qualification required to complete EMD and grant a military qualification rating
  - Generate AH-64E and UH-60M test airworthiness release to begin preliminary flight tests
  - Demonstrate TRL of 8 and MRL of 8 to enter LRIP

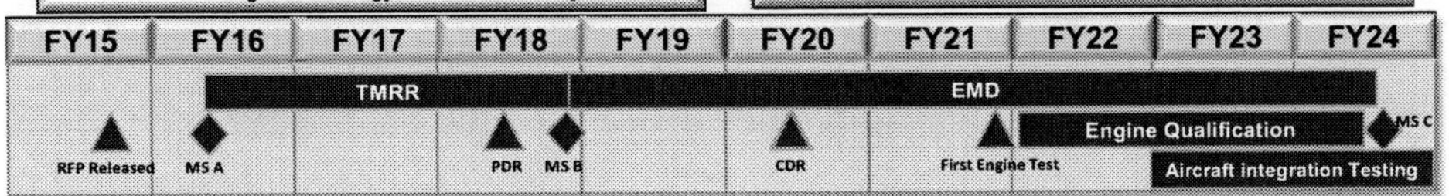

ITEP Interoperability

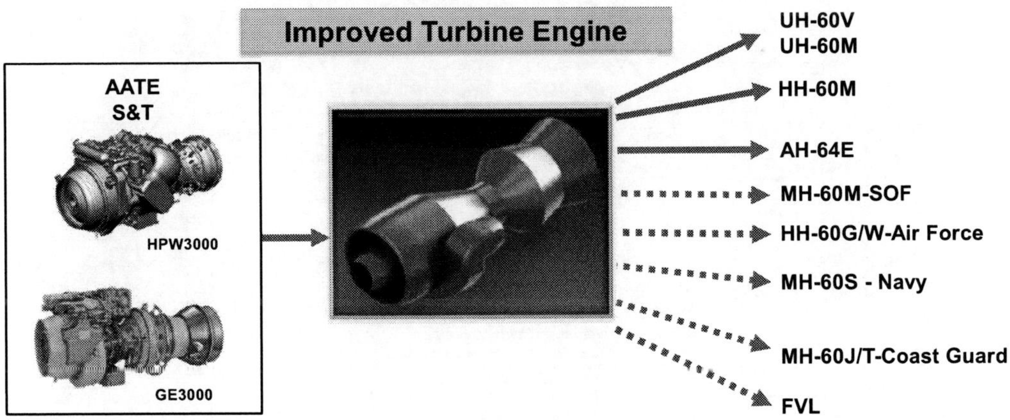

——	Solid denotes planned system
·······	Dash denotes potential system

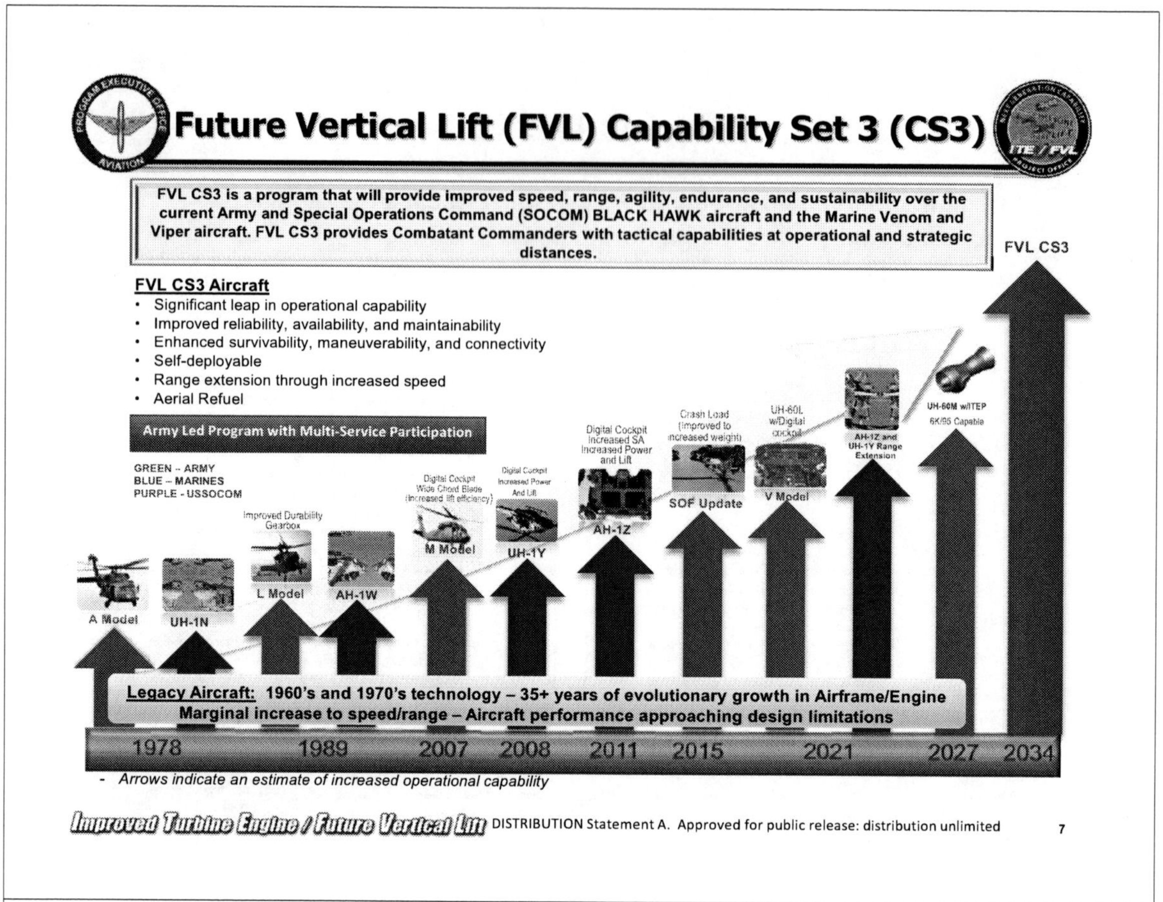
Future Vertical Lift (FVL) Capability Set 3 (CS3)
FVL CS3 is a program that will provide improved speed, range, agility, endurance, and sustainability over the current Army and Special Operations Command (SOCOM) BLACK HAWK aircraft and the Marine Venom and Viper aircraft. FVL CS3 provides Combatant Commanders with tactical capabilities at operational and strategic distances.
FVL CS3 Aircraft
• Significant leap in operational capability
• Improved reliability, availability, and maintainability
• Enhanced survivability, maneuverability, and connectivity
• Self-deployable
• Range extension through increased speed
• Aerial Refuel
Army Led Program with Multi-Service Participation
GREEN – ARMY
BLUE – MARINES
PURPLE - USSOCOM
A Model
UH-1N
Improved Durability Gearbox
L Model
AH-1W
Digital Cockpit Wide Chord Blade (Increased lift efficiency)
M Model
UH-1Y
Digital Cockpit Increased SA Increased Power and Lift
AH-1Z
Crash Load (improved to increased weight)
SOF Update
UH-60L w/Digital cockpit
V Model
AH-1Z and UH-1Y Range Extension
UH-60M w/ITEP
6K/95 Capable
FVL CS3
Legacy Aircraft: 1960's and 1970's technology – 35+ years of evolutionary growth in Airframe/Engine
Marginal increase to speed/range – Aircraft performance approaching design limitations
1978
1989
2007
2008
2011
2015
2021
2027
2034
- Arrows indicate an estimate of increased operational capability
Improved Turbine Engine / Future Vertical Lift DISTRIBUTION Statement A. Approved for public release: distribution unlimited
7

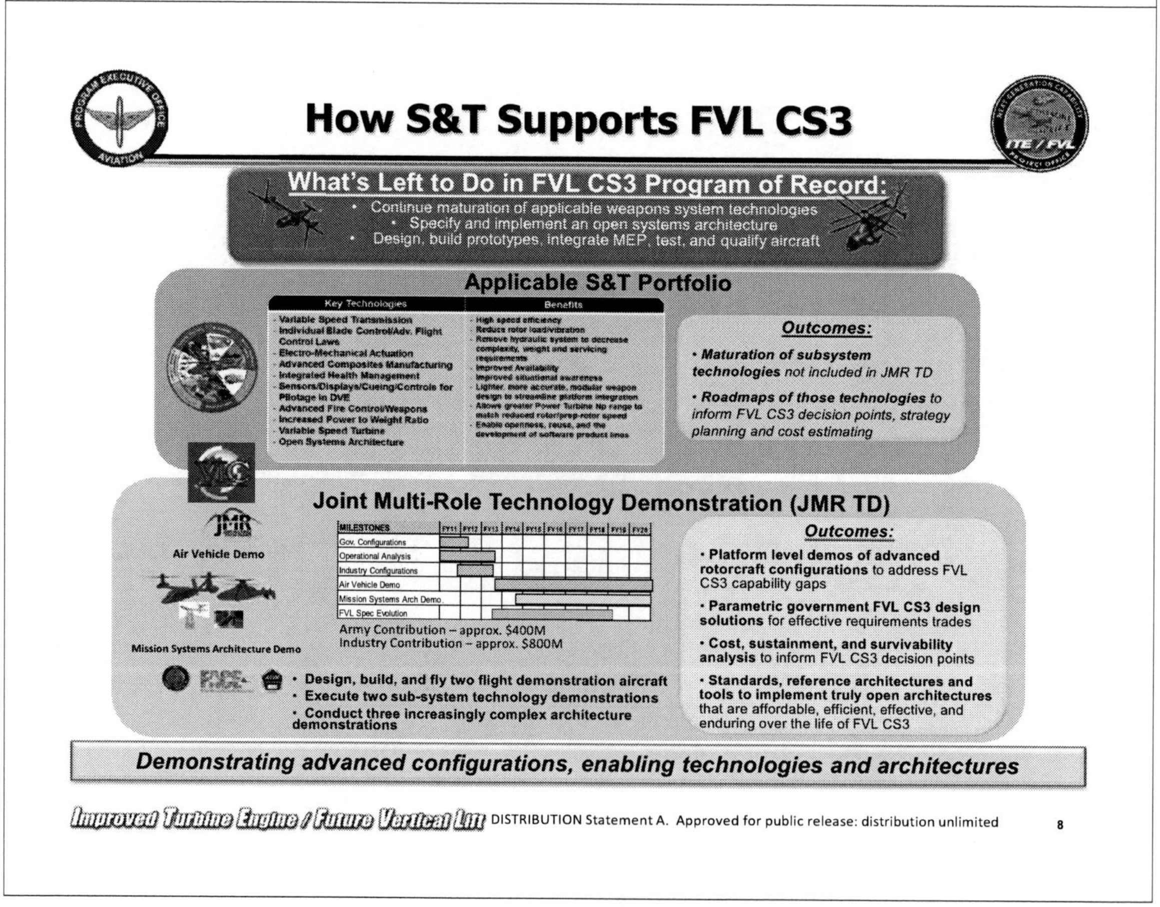
How S&T Supports FVL CS3
What's Left to Do in FVL CS3 Program of Record:
• Continue maturation of applicable weapons system technologies
• Specify and implement an open systems architecture
• Design, build prototypes, integrate MEP, test, and qualify aircraft
Applicable S&T Portfolio
Key Technologies
- Variable Speed Transmission
- Individual Blade Control/Adv. Flight Control Laws
- Electro-Mechanical Actuation
- Advanced Composites Manufacturing
- Integrated Health Management
- Sensors/Displays/Cueing/Controls for Pilotage in DVE
- Advanced Fire Control/Weapons
- Increased Power to Weight Ratio
- Variable Speed Turbine
- Open Systems Architecture
Benefits
- High speed efficiency
- Reduce rotor load/vibration
- Remove hydraulic system to decrease complexity, weight and servicing requirements
- Improved Availability
- Improved situational awareness
- Lighter, more accurate, modular weapon design to streamline platform integration
- Allows greater Power Turbine Np range to match reduced rotor/prop-rotor speed
- Enable openness, reuse, and the development of software product lines
Outcomes:
• Maturation of subsystem technologies not included in JMR TD
• Roadmaps of those technologies to inform FVL CS3 decision points, strategy planning and cost estimating
Joint Multi-Role Technology Demonstration (JMR TD)
Air Vehicle Demo
Mission Systems Architecture Demo
MILESTONES
FY11 FY12 FY13 FY14 FY15 FY16 FY17 FY18 FY19 FY20
Gov. Configurations
Operational Analysis
Industry Configurations
Air Vehicle Demo
Mission Systems Arch Demo
FVL Spec Evolution
Army Contribution – approx. $400M
Industry Contribution – approx. $800M
• Design, build, and fly two flight demonstration aircraft
• Execute two sub-system technology demonstrations
• Conduct three increasingly complex architecture demonstrations
Outcomes:
• Platform level demos of advanced rotorcraft configurations to address FVL CS3 capability gaps
• Parametric government FVL CS3 design solutions for effective requirements trades
• Cost, sustainment, and survivability analysis to inform FVL CS3 decision points
• Standards, reference architectures and tools to implement truly open architectures that are affordable, efficient, effective, and enduring over the life of FVL CS3
Demonstrating advanced configurations, enabling technologies and architectures
Improved Turbine Engine / Future Vertical Lift DISTRIBUTION Statement A. Approved for public release: distribution unlimited
8

FVL CS3 Key Technologies

Component	Key Technology	Benefit
Aircraft Systems 	- Variable Speed Transmission - Individual Blade Control/Adv. Flight Control Laws - Electro-Mechanical Actuation - Advanced Composites Manufacturing - Integrated Health Management - Sensors/Displays/Cueing/Controls for Pilotage in DVE - Advanced Fire Control/Weapons	- High speed efficiency - Reduce rotor load/vibration - Remove hydraulic system to decrease complexity, weight and servicing requirements - Improved Availability - Improved situational awareness - Lighter, more accurate, modular weapon design to streamline platform integration Capabilities/Attributes: Speed, Range, Endurance, Agility, Survivability, Sustainability, Affordability
Powerplant Systems 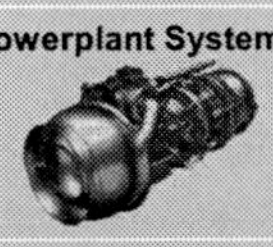	- Increased Shaft Horsepower - Variable Speed Turbine	Allows greater Power Turbine Np range to match reduced rotor/prop-rotor speed Capabilities/Attributes: Range, Endurance, Affordability
Mission Systems Architectures (MSA) 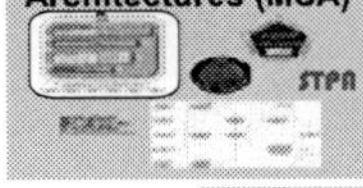	- Open Systems Architecture	- Enable openness, reuse, and the development of software product lines Capabilities/Attributes: Agility, Interoperability, Survivability, Sustainability, Affordability

Key Technologies Support Achievement of Key Capabilities/Attributes

FVL CS3 Program Status

MDD 1QFY17 → AoA 1QFY19 → Milestone A → Milestone B → Milestone C FY29 → Full Rate Production FY33

MDD (28 Oct 16)

- Approved entry into the Materiel Solution Analysis phase
- Designated Army as Lead Service, with multi-Service participation (USMC and SOCOM)
- Initiate AoA

Materiel Solution Analysis Phase

- OSD Initial AoA SAG on 14 Apr 2017 approved the AoA Study Plan
- Execute AoA
- Develop draft Capabilities Development Document
- Finalize Statement of Work and Performance Specification
- Conduct Milestone A

Questions?

Improved Turbine Engine / Future Vertical Lift DISTRIBUTION Statement A. Approved for public release: distribution unlimited 11

US Army Aviation Program Manager Briefings

0800-0820	PEO Overview	Mr. Lars Ericsson, APEO for Eng & Tech
0820-0850	Cargo	Mr. Tom Neupert, Tech Chief
0850-0920	Apache Attack	Mr. Alan McClendon, Tech Chief
0920-0950	Utility	Mr. Travis Sinclair, Chief Engineer, H-60M
0950-1020	Break	
1020-1050	UAS	Mr. Dennis Sparks, Tech Chief
1050-1120	Aviation Systems	Mr. Art Medellin, Tech Chief
1120-1150	ITE/FVL	Mr. Bob Sheibley, PM
1150-1200	Wrap-Up	

AHS Igor I Sikorsky Human Powered Helicopter Competition

History and Lesson Learned

Presented at the 73rd Annual Forum and Technology Display
May 10, 2017
Fort Worth, Texas, USA

SIKORSKY
A LOCKHEED MARTIN COMPANY

Benjamin R Hein
AHS I. I. Sikorsky HPHC Chair - 2013

Winning Flight on YouTube

You've seen this video, right? It has 3.67 million views (Jul 11, 2013)

Do you know how we got here?

https://www.youtube.com/watch?v=syJq10EQkog

Original Conception Picture

4.4 Flight Requirements

4.4.1 The flight requirements shall consist of hovering for one minute while maintaining flight within a 10-meter square. During this time the lowest part of the machine shall exceed momentarily 3 meters above the ground.

5. THE AWARD. It has been recognized from the beginning that the monetary award offered has been relatively small ($10,000 originally, now $20,000). Efforts are under way to try to increase the amount. Yet if the last few years are any indication, it is clear that there is much more to our contest than just a monetary award. There is not only the excitement of possibly winning the AHS HPH competition, but also the incentive of setting a new world record. There is the considerable technical and athletic challenge, the satisfaction of conquering new territory, and the enjoyment of making new things happen. ALL THIS combines to entice people to participate in our Human-Powered Helicopter Competition.

October 30, 1992 ---------

3

Rough History of HPH

1959	Kremer prize	Henry Kramer
1977 Aug 23	Gossamer Condor	Figure 8 0.5mi 10ft alt, £50K, Mc Cready
1979 Jun 12	Gossamer Albatross	English Channel 22.2mi 2hr49m, £100K, Mc Cready
1980 May	Igor I. Sikorsky Prize	Prize Issued for $10K, increased to $20K later EC Sonny Darling ton, first coordinator
1981 – 1989	CalPoly SL Obispo	Leonardo I&II, III
1984 – 1986	A Day Fly	Nihon Unieversity, Akira Naito
1989 Dec	L. DaVinci III	50ft radius, tip propellers ~8s 8in
1988-1992	Papillion A, B, C	didn't fly, coaxial, coaxial, tip-driven 2-blade
1993	Yuri I	Quad-rotor, First flight, 80 lbs structure, 5m radius, 20m/s
1994	Yuri I	0.2m 19.46s, unofficially 24s 0.7m
2009 May	Sikorsky	Increase prize to $250K USD
2011 May 12	UMD Gamera I	Judy Wexler 4.2s
2011 July 13	UMD Gamera I	Judy Wexler 11.4s
2012 Jun 12	UMD Gamera II	50s
2012 Aug 28	UMD Gamera II	65s
2012 Aug 29	AeroVelo Atlas	First Flight
2013 Feb 27	UMD Gamera II	Baltimore conference flights
2013 Feb 28	AeroVelo Atlas	Second Flight
2013 Apr 18	AeroVelo Atlas	Third Flight, without tip flaps
2013 April 27	UMD Gamera II	Flights at Athletic Center
2013 June 14	AeroVelo Atlas	Wins the Prize
2013 Sep 25	UMD Gamera II	1min 37.5s flight (world record?)

4

Original Analysis

HUMAN POWERED HELICOPTER

FEASIBILITY STUDY
FOR AHS BOARD MEETING
26 March 1979

PREDESIGN STUDY
MAN POWERED HELICOPTER

GROSS WEIGHT : 250 LB 2 BLADED ROTOR SYSTEM
HOVER HEIGHT : 10 FT $C_L = 0.85$ $C_D = 0.018$

MAIN ROTOR ONLY

$$HP = \frac{1.3(K)(GW)}{550}\sqrt{\frac{GW}{2\pi(R^2)(\rho)}} + \frac{.78(GW)(\Omega R)}{550}\frac{C_D}{C_L}$$

WEIGHT ANALYSIS

TOTAL HELICOPTER WEIGHT	=250	LB
− ENGINE WEIGHT	=150	LB
AIRFRAME+ROTOR WEIGHT	100	LB
− AIRFRAME WEIGHT	20	LB
BLADE WEIGHT	40	LB/BLADE

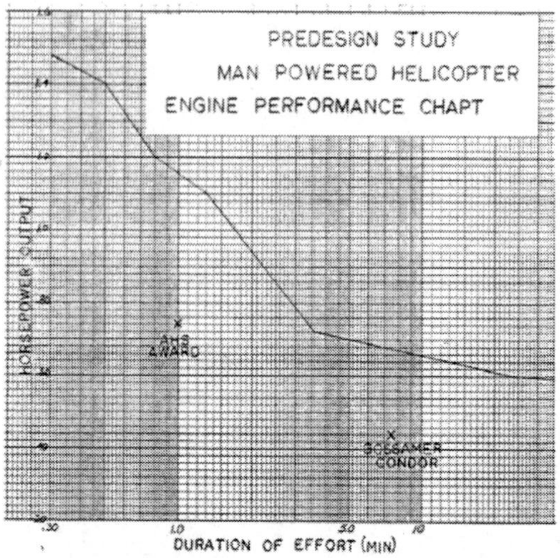

AS A DESIGN POINT CONSIDER A TIP SPEED OF VT=65 FT/SEC
THIS GIVES AS THE OPTIMUM RADIUS R= 40.4 FEET
AND WILL REQUIRE 0.70 HORSEPOWER

Courtesy of Dr. Gordon Leishman via AHS

5

Human Power Plant

Power Available:

~600 W for 60s

~1000 W for 10 s

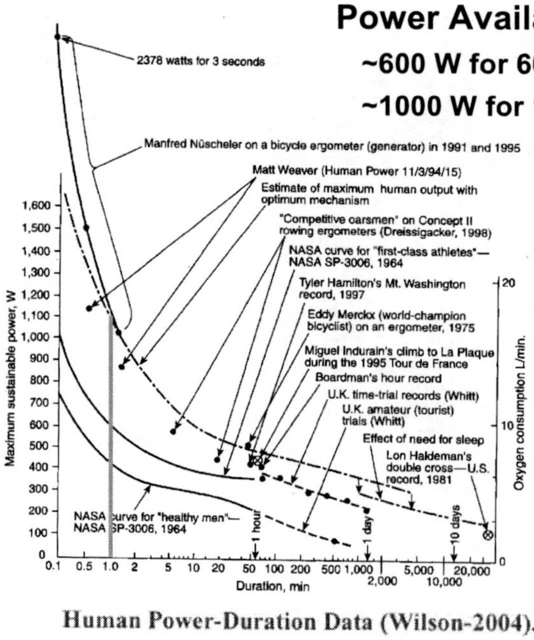

Human Power-Duration Data (Wilson-2004).

6

What original analysis and concept "got wrong"

Weight was assumed

An actual bottoms up design study wasn't performed

Stiffness of superlight structures not considered, or assumed someone would "figure it out"

Controls weren't considered at all

Sensitivity to inflow & "wind" was not considered, nor complications of flying indoors

Design for rapid assembly and disassembly

But does that matter?

7

First Generation Competitors

1963 Attempt by Clifford Davis

"The reality of a human-powered helicopter is that, with the simple formula for static thrust of a rotor, you quickly find you need a disk diameter of 150 feet or so for serious human-powered helicopter flight.

It can be done, but the task is huge, and the dollar prize not worth the time expenditure. There are many more exciting, never-been-done-before challenges that can be accomplished with much less work." - Paul MacCready (quoted on humanpoweredhelicopters.org in a PBS interview with Michael Gibbs

8

Cal Poly 1981 – 1989 DaVinci I - III

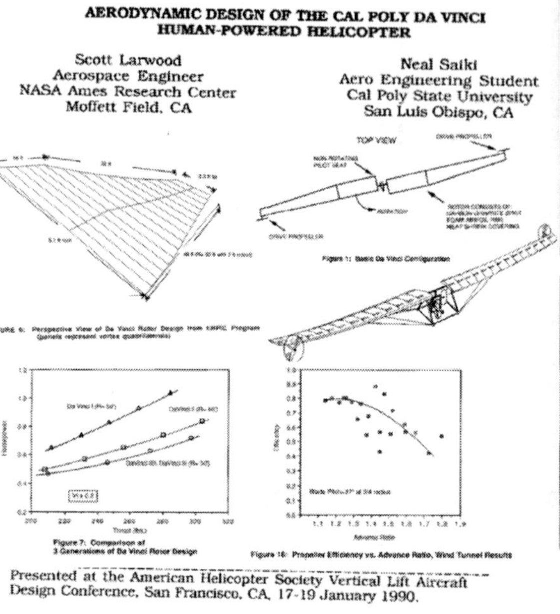

California Polytechnic State-San Luis Obispo

9

Nihon University – Professor Naito 1984 -1986, 1988 - 1992

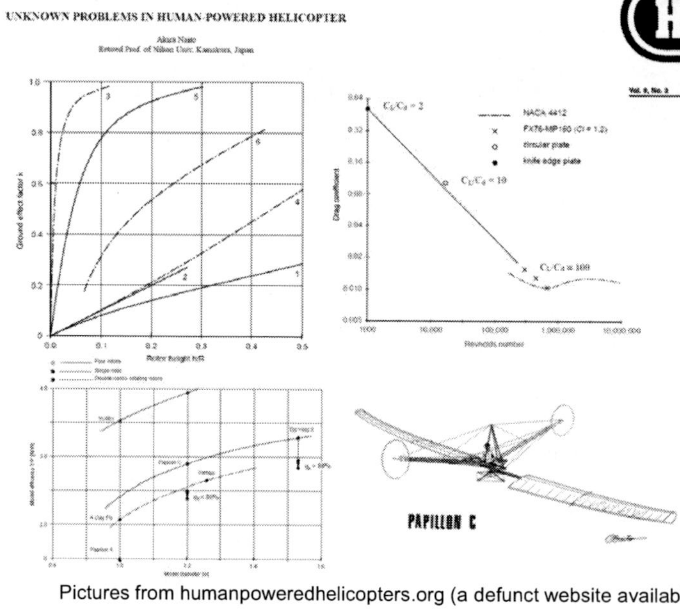

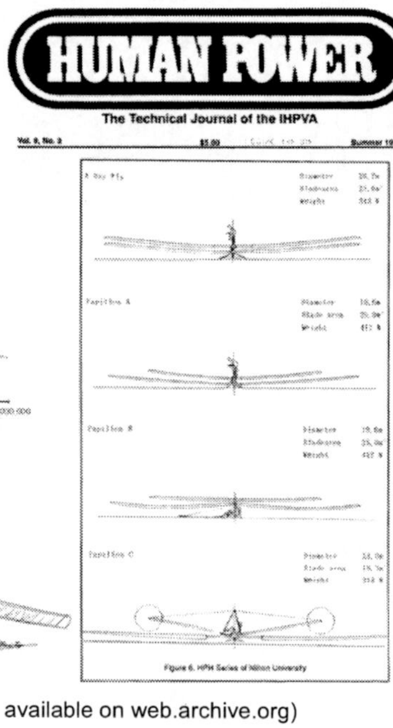

Pictures from humanpoweredhelicopters.org (a defunct website available on web.archive.org)

10

Nihon University – Naito 1986

Pictures from humanpoweredhelicopters.org (a defunct website available on web.archive.org)

11

Naito Team at Nihon University Best flight and world record 1994

Yuri I

	Official		Unofficial
Flight :	03/07/94	Flight :	--
Duration :	19.46 seconds	Duration :	24 seconds
Altitude :	0.2 m	Altitude :	0.7 m

Figure 2. YURI 1 (courtesy Nihon Uni.)

Foundation of the Future Human Powered Helicopter Aircraft

Pictures from humanpoweredhelicopters.org (a defunct website available on web.archive.org)

12

The "dark ages" of HPH

Pictures from humanpoweredhelicopters.org (a defunct website available on web.archive.org)

13

Rough History of HPH

1959	Kremer prize	Henry Kramer
1977 Aug 23	Gossamer Condor	Figure 8 0.5mi 10ft alt, £50K, Mc Cready
1979 Jun 12	Gossamer Albatross	English Channel 22.2mi 2hr49m, £100K, Mc Cready
1980 May	Igor I. Sikorsky Prize	Prize Issued for $10K, increased to $20K later EC Sonny Darling ton, first coordinator
1981 – 1989	CalPoly SL Obispo	Leonardo I&II, III
1984 – 1986	A Day Fly	Nihon Univeristy, Akira Naito
1989 Dec	L. DaVinci III	50ft radius, tip propellers ~8s 8in
1988-1992	Papillion A, B, C	didn't fly, coaxial, coaxial, tip-driven 2-blade
1993	Yuri I	Quad-rotor, First flight, 80 lbs structure, 5m radius, 20m/s
1994	Yuri I	0.2m 19.46s, unofficially 24s 0.7m
2009 May	Sikorsky	Increase prize to $250K USD
2011 May 12	UMD Gamera I	Judy Wexler 4.2s
2011 July 13	UMD Gamera I	Judy Wexler 11.4s
2012 Jun 12	UMD Gamera II	50s
2012 Aug 28	UMD Gamera II	65s
2012 Aug 29	AeroVelo Atlas	First Flight
2013 Feb 27	UMD Gamera II	Baltimore conference flights
2013 Feb 28	AeroVelo Atlas	Second Flight
2013 Apr 18	AeroVelo Atlas	Third Flight, without tip flaps
2013 April 27	UMD Gamera II	Flights at Athletic Center
2013 June 14	AeroVelo Atlas	Wins the Prize
2013 Sep 25	UMD Gamera II	1min 37.5s flight (world record?)

14

Prize Increase to $250K

Options were to make the competition easier, increase the prize money, or do both

Decision made to find a sponsor/s to increase prize money

AHS International Forum in 2009, Sikorsky Aircraft President committed to increase the pledge to $250K

NEWS

Sikorsky
A United Technologies Company

Sikorsky Aircraft Corporation • 6900 Main Street • Stratford, Conn. 06615 • www.sikorsky.com

Prize in Human-Powered Helicopter Contest Rises to $250,000 Thanks to Sikorsky Aircraft Pledge

STRATFORD, Conn., Aug. 27, 2009 – Sikorsky Aircraft Corp. has increased its pledge to provide $250,000 in prize money for the Igor I. Sikorsky Human-Powered Helicopter Competition. Sikorsky previously had pledged $20,000 in prize money for whomever could win the competition by achieving the engineering feat. Sikorsky is a subsidiary of United Technologies Corp. (NYSE:UTX).

The competition, created by the American Helicopter Society International, challenges engineers to develop a human-powered helicopter that can reach an altitude of at least three meters during a hover lasting 60 seconds. No one has accomplished this feat yet. The official world record is held by Professor Akira Naito of Nihon University in Japan, with an altitude of 0.2 meters and flight duration of 19.46 seconds. Sikorsky Aircraft informed AHS International at its annual forum in May that the company would increase its pledged contribution, which AHS would award to the competition winner.

"Igor Sikorsky, founder of Sikorsky Aircraft, believed that individuals provide the spark that moves mankind ahead. This competition continues his legacy by inspiring ingenuity in the next generation of engineers who will design our industry's future," said Mark Miller, Vice President of Research and Engineering at Sikorsky Aircraft. "Our company is built on innovation. We believe strongly in the power of challenge."

More information about the competition is available from the American Helicopter Society at 703-684-6777.

Sikorsky Aircraft Corp., based in Stratford, Connecticut, USA, is a world leader in helicopter design, manufacture and service. United Technologies Corp., based in Hartford, Conn., provides a broad range of high technology products and support services to the aerospace and building systems industries.

# # #

Sikorsky media contacts:

Paul Jackson
Phone: 860-614-3899
Paul.Jackson@sikorsky.com

15

UMD Prework and Analysis 2008 – 2009?

Courtesy of University of Maryland

16

Maryland's Early Attempts

Gamera I	July, 2011
Rotor = 42.6 ft	Time = 11.4 sec
Weight = 106 lbs	Height = 1 ft

Courtesy of University of Maryland

17

Maryland First Flight

Courtesy of University of Maryland

18

Neal Saiki – NTS Works & Cal Poly Tip-prop helicopter

Courtesy of NTS Works

19

Gamera II super high altitude flight

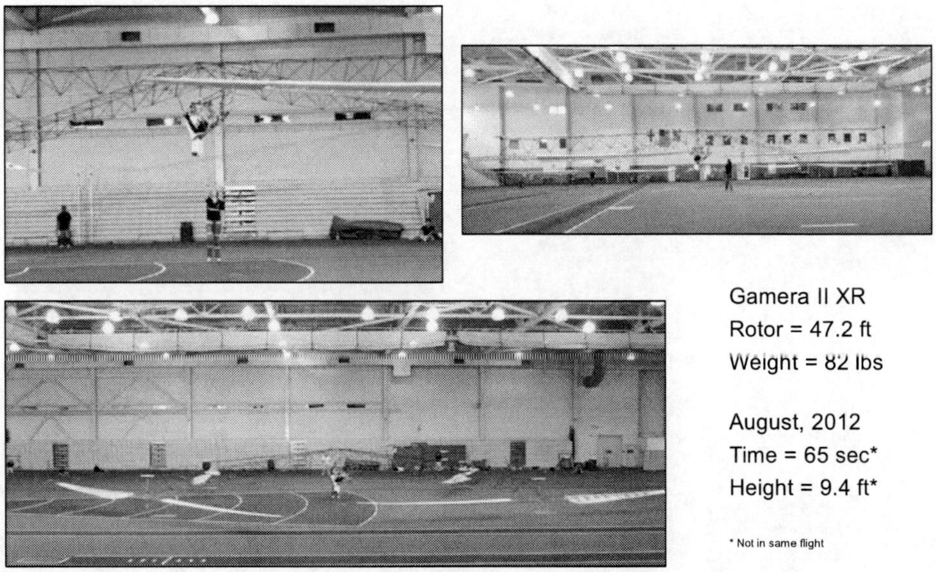

Gamera II XR
Rotor = 47.2 ft
Weight = 82 lbs

August, 2012
Time = 65 sec*
Height = 9.4 ft*

* Not in same flight

Courtesy of University of Maryland

20

Gamera Flying

Courtesy of University of Maryland

21

AeroVelo First Flights & Controls

AeroVelo First Flights/Crash

Size Does Matter

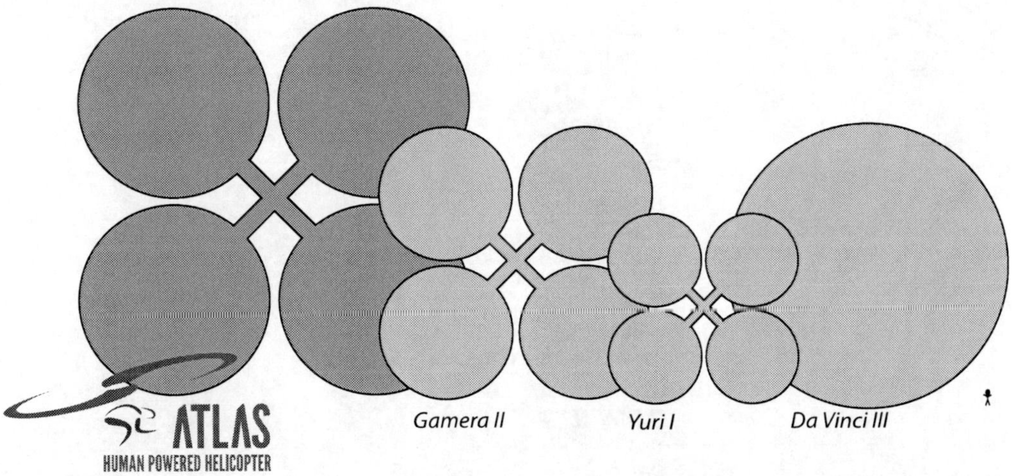

Courtesy of AeroVelo Inc

Just prior to Final Flight

	Da Vinci III / IV	Yuri I	Gamera I	Gamera II	Gamera II XR	Upturn	Atlas
Vehicle	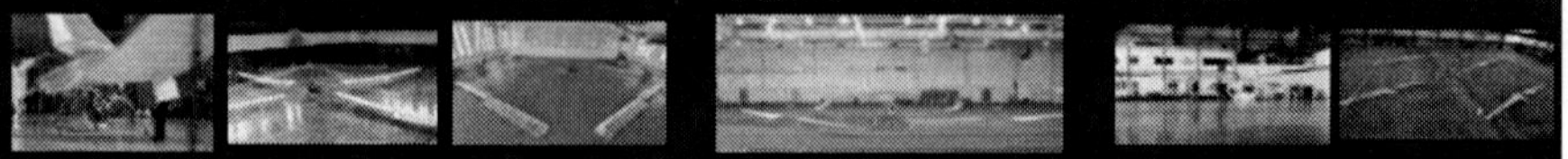						
Team	Cal Poly U.	Nihon U.	U. Maryland	U. Maryland	U. Maryland	NTS Works	AeroVelo
Rotor Type (Nb/Rotor)	Single MR, Propeller (2)	Quad-rotor (2)	Quad-rotor (2)	Quad-rotor (2)	Quad-rotor (2)	Single MR, Propeller (2)	Quad-rotor (2)
First flight	Dec 1989	Dec 1993	May 2011	June 2012	Aug 2012	Jun 2012	Aug 2012
Rotor Diameter	100 ft	33 ft	42.6 ft	42.6 ft	47.2 ft	85 ft	66.9 ft
Max Dimension	100 ft	~ 80 ft	105 ft	105 ft	115 ft	85 ft	190 ft
Empty Weight	97 lbs	83 lbs	106 lbs	76 lbs	82 lbs	95 lbs	120 lbs
Max Endurance	8.6 sec	19.5 sec	11.4 sec	50 sec	65 sec	10 sec	50 sec
Max Altitude	8 in	8 in	1 ft	3.5 ft	9.4 ft	2 ft	~9.8 ft

25

Final Winning Flight

Courtesy of AeroVelo Inc

The Vertical Flight Technical Society

24 HOUR
Hover Challenge

AHS International's Igor I. Sikorsky

24 Hour Hover Challenge

May 10, 2017

Michael Strauss
Sikorsky Innovations
AHS Aircraft Design Technical Committee
AHS Advanced Vertical Flight Technical Committee

June 13, 2013

AeroVelo wins the AHS Igor I. Sikorsky Human Powered Helicopter Challenge 33 Years After Challenge Announcement

2

What Should be the Next AHS Challenge?

- Inspire Next Generation of Engineers

- Increase Public Interest
- Based on the Heart of the AHS Mission and the Benefit of Vertical Lift Technology

3

A Competition at the Heart of AHS: Hover

"If a man is in need of rescue, an airplane can come in and throw flowers on him, and that's just about all.
But a direct lift aircraft could come in and save his life."

Igor I. Sikorsky

4

At a Glance

- $50,000 Prize
- Unoccupied
  (80kg payload to simulate 1 human)
- 24 Hour Controlled Hover
  (OGE, within 20m sphere, winds < 5m/s)
- Real-World Capable
  (Not a Stunt: Transit to 3 Waypoints 1km apart)

24 HOUR
Hover Challenge

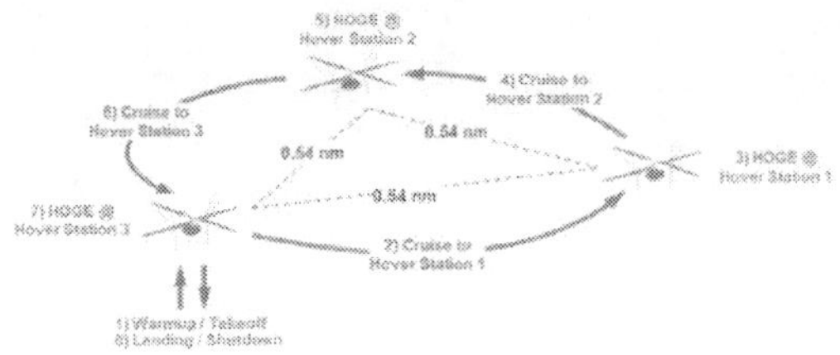

5

Can We Do This With Existing Helicopters?

Mil Mi-26 Halo
The World's Largest Helicopter

Max. Gross Weight: 123,450 lb
Empty Weight: 62,000 lb
Standard Fuel: 21,550 lb

Hover Time w/ 1 Pilot: 4.6 hrs

What If All The Useful Load Were Fuel (61,450 lb) & No Pilot?

Hover Time: 10.4 hrs

Does Scale Matter? Engine Type? Rotor Technology?

6

Can We Do This With Existing Helicopters?

	Mi-26	CH-53K	S-92	S-76C++	AW169	AW139	R66	R44	S-300	R22
Vehicle										
Max. Gross Weight	123,450 lb	84,700 lb	26,500 lb	11,700 lb	10,500 lb	14,110 lb	2,700 lb	2,500 lb	2,050 lb	1,370 lb
Empty Weight	62,000 lb	33,200 lb	15,500 lb	7,000 lb	6,000 lb	8,000 lb	1,400 lb	1,450 lb	1,100 lb	800 lb
Fuel Capacity	21,556 lb	15,620 lb	5,168 lb	1,911 lb	2,033 lb	2,808 lb	500 lb	299 lb	221 lb	115 lb
Max. Fuel Possible	61,450 lb	51,500 lb	11,000 lb	4,700 lb	4,500 lb	6,110 lb	1,300 lb	1,050 lb	950 lb	570 lb
Rotor										
Rotor Radius	52.5 ft	39.5 ft	28.2 ft	22.0 ft	19.9 ft	22.6 ft	16.5 ft	13.4 ft	13.4 ft	12.6 ft
Disk Loading	15.0 psf	18.1 psf	11.2 psf	8.1 psf	8.9 psf	9.2 psf	3.3 psf	4.6 psf	3.8 psf	2.9 psf
Engine										
Type	Lotarev D-136	GE 38-1B	CT7-8A	2S2	PW210A	PT6C-67C	RR300	IO-540-AE1A5	IO-360	O-320-A2B
Total Installed Power	22,800 hp	22,500 hp	5,000 hp	1,844 hp	2,020 hp	3,060 hp	270 hp	245 hp	190 hp	124 hp
Max. Thermal Efficiency	33.9%	36.4%	30.1%	27.7%	28.0%	29.4%	23.9%	30.0%	30.0%	30.0%
SFC @ Max. Power	0.408 lb/hp/hr	0.380 lb/hp/hr	0.460 lb/hp/hr	0.500 lb/hp/hr	0.494 lb/hp/hr	0.471 lb/hp/hr	0.578 lb/hp/hr	0.463 lb/hp/hr	0.463 lb/hp/hr	0.463 lb/hp/hr

7

Can We Do This With Existing Helicopters?

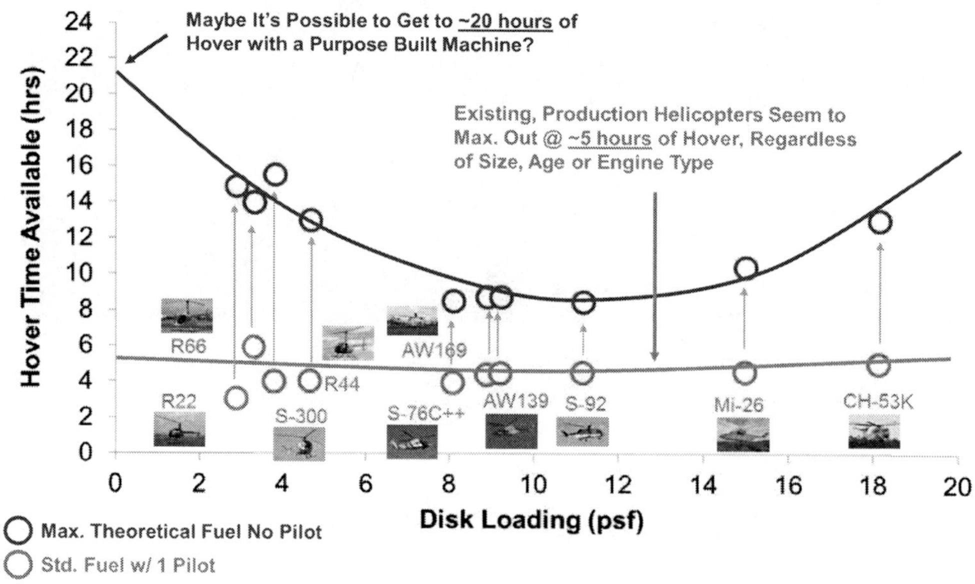

8

Idealized Hover

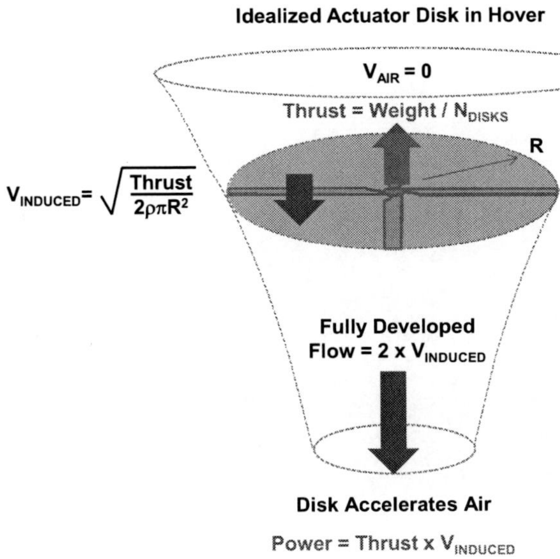

Disk Loading:

Thrust / Area

As Disk Loading Reduces:

Velocity Reduces
Power Reduces

9

Disk Loading & Power Loading

Disk Loading & Power Loading

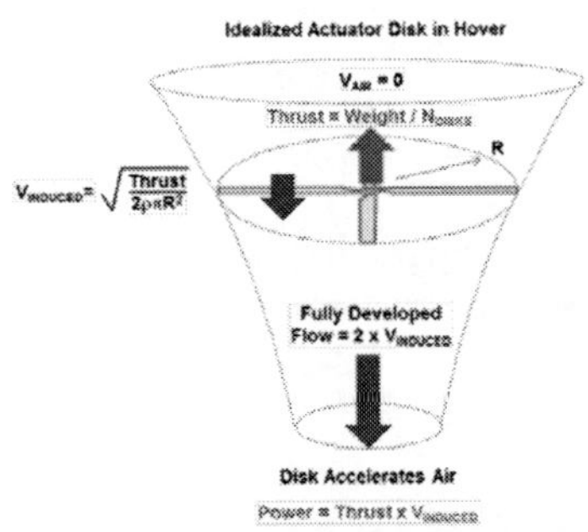

Disk Loading = GW / N_{DISKS} / Area

Power Loading = GW / $Power_{TOTAL}$

Large Amount Of Air at Low Velocity

Small Amount Of Air at Very High Velocity

Disk Loading (psf)	Power Loading (lb/hp)
10	5
22	3
1,000+	<< 1

10

Drivers of Hover Efficiency

Ideal Power / Actual Power

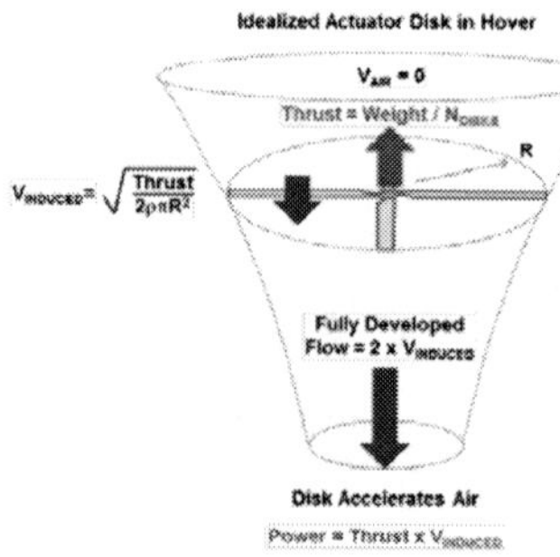

$$P_{IDEAL} = N_{DISKS} \cdot \frac{(T_{ROTOR})^{3/2}}{\sqrt{2\rho}\,\pi R^2}$$

(1) Increase Disk Area

(3) Reduce Airframe Blockage

$$P_{ROTOR} = P_{IDEAL}\,\frac{1}{FM_{ROTOR}}\left[1+\frac{D_V}{T_{ROTOR}}\right]^{3/2}$$

(2) Increase Rotor Efficiency

$$\eta_{HOVER} = \frac{P_{IDEAL}}{P_{SYSTEM}} = \frac{P_{IDEAL}}{P_{ROTOR} + P_{ANTI-TORQUE} + P_{XMSN\_LOSS} + P_{ACC.}}$$

(4) Eliminate Additional Power "Losses"

P_{IDEAL} is the power of an actuator disk of equivalent area given by momentum theory (hp). P_{ROTOR} is the power consumed by the rotor (hp). Actual rotor power also uses a modified rotor thrust that includes the additional thrust required due to the vertical drag, D_V, developed from vehicle components enveloped in the rotor downwash (which are not present in the idealized actuator disk). The rotor hovering efficiency, FM_{ROTOR}, represents the ratio of actual rotor power required to the idealized actuator disk of the same area. The remainder of the aircraft power, P_{SYSTEM}, is comprised of any other power consumption required to keep the aircraft aloft and include anti-torque power required, $P_{ANTI-TORQUE}$, transmission losses in transmitting engine power through the drivetrain to the rotor system, P_{XMSN_LOSS}, and accessory power, P_{ACC}, consumed by electrical generators, hydraulic pumps and any other ancillary systems like thermal management blowers.

11

Getting Power to the Rotors

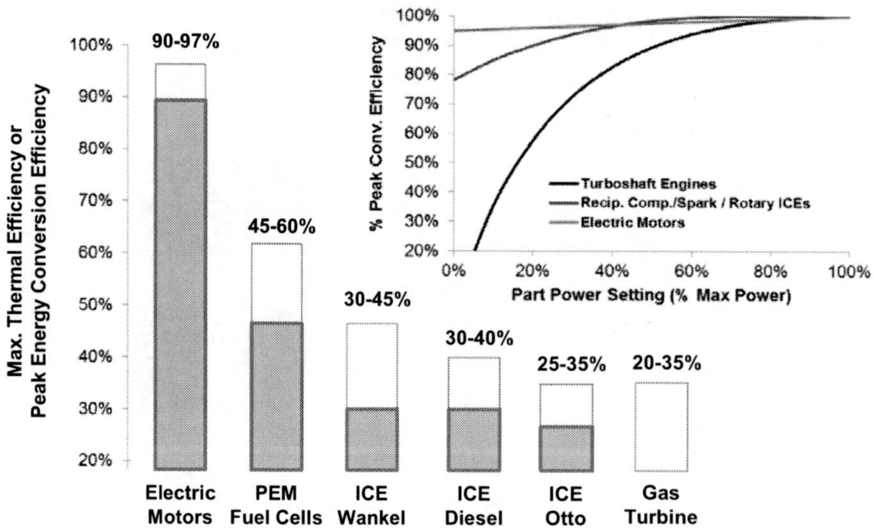

12

Engine Power to Weight

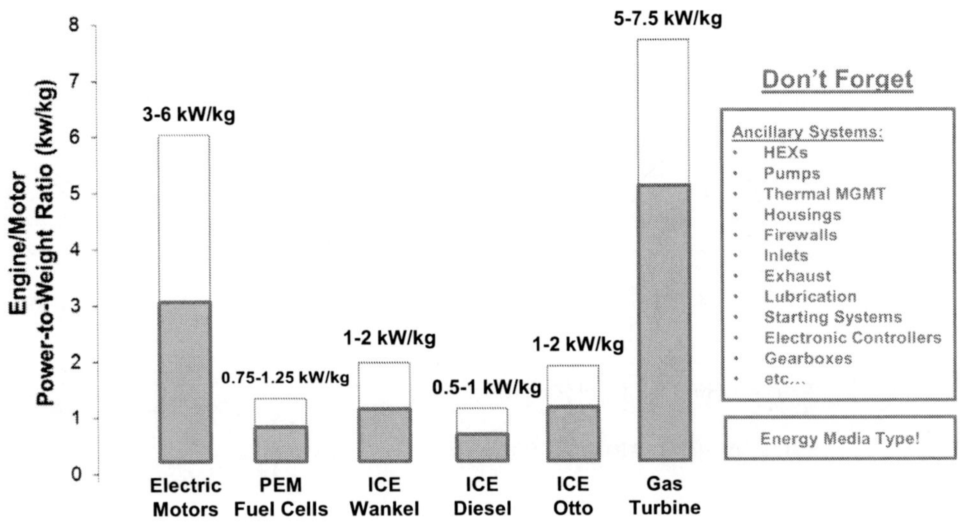

13

Energy Sources

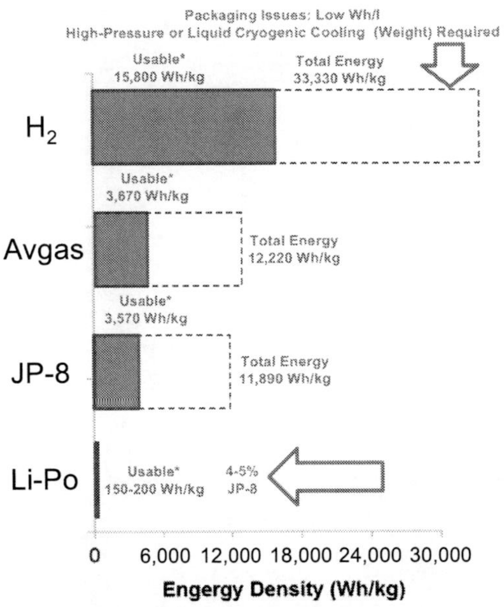

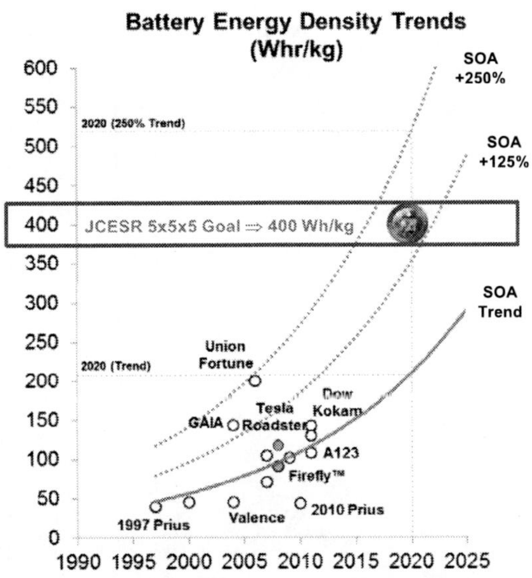

* PEM Fuel Cell Conversion Efficiency = 50%
Electric Motor Conversion Efficiency = 95%
Thermal Engine Conversion Efficiency = 30%

14

Is This Really Feasible?

Explore the Design Space:

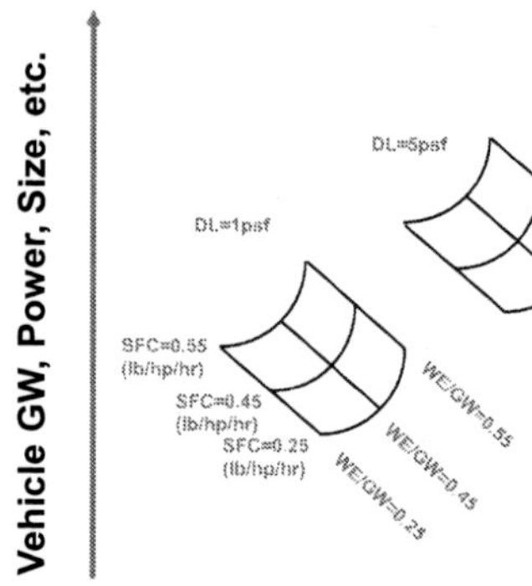

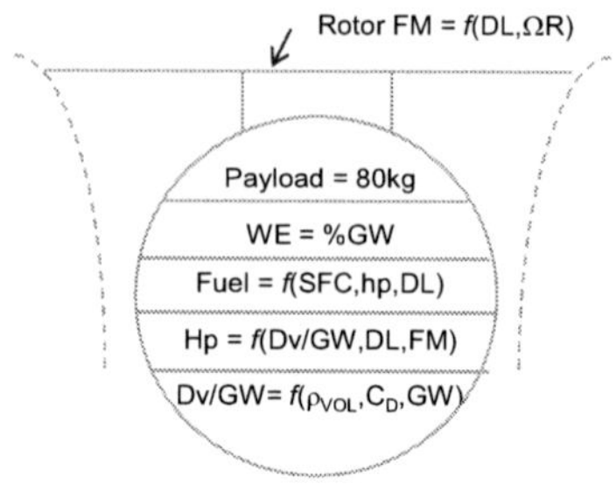

1. **Engine SFC** => Fuel Required
2. **WE Empty Fraction** => Gross Weight
3. **Disk Loading** => Rotor Power & Efficiency

<u>State of the Art Values:</u>
SFC = 0.35 (Diesel)
SFC = 0.45-0.55 (turboshaft)
DL = 5-7 psf
WE/GW = 0.55-0.70

15

First Pass Sizing

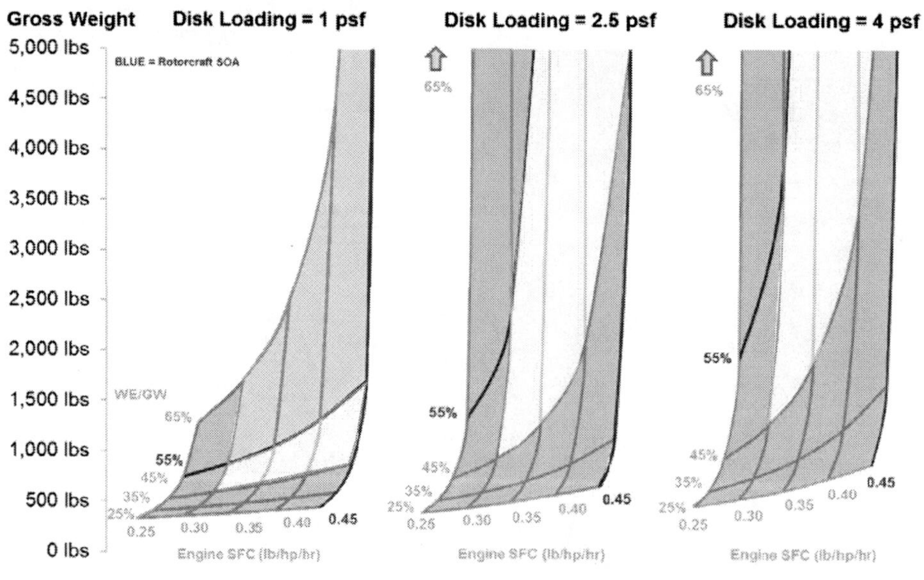

Unlikely Feasible | Likely Feasible | Feasible

16

First Pass Sizing

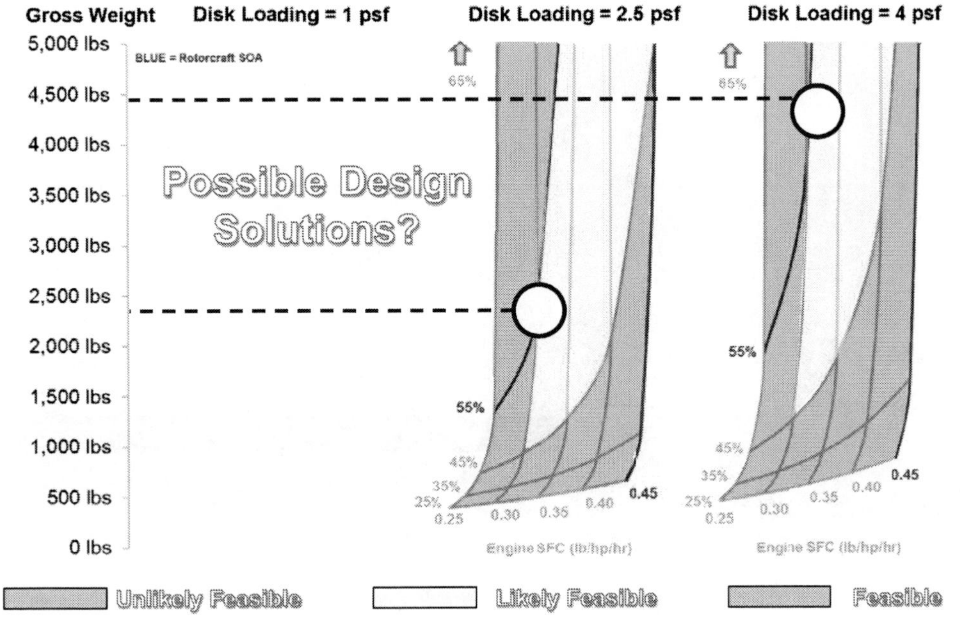

17

Observations

Disk Loading

- Conventional rotorcraft do not have disk loadings below 2 psf. DL scales w/ GW.
- Lower disk loading requires lower rotor rpm, which makes rotor controllability very difficult and causes autorotative issues.

Fuel Consumption (SFC)

- Diesel engines with SFCs of 0.35 have power to weight ratios of 0.5 hp/lb, while gas turbine engines with SFCs of 0.45-0.55 have power to weight ratios of 4-6 hp/lb.
- A propulsion system invention is necessary to get the SFC and P/W to match the WE/GW required
- The AeroVelo Altas HPH propulsion system (T. Reichert) included the energy and motive system

Weight Empty Fraction

- The lowest WE fraction on a real helicopter is 50% (Mosquito Aviation XE)
- The AeroVelo Altas HPH, with peak power of 1-2 hp, and a WE fraction of ~50% and broke apart on several occasions, making flights of <60 seconds.
- An invention is necessary to get the high-strength, lightweight structure and that can withstand a helicopter vibratory environment for 24hrs and meet the WE/GW required

18

Summary

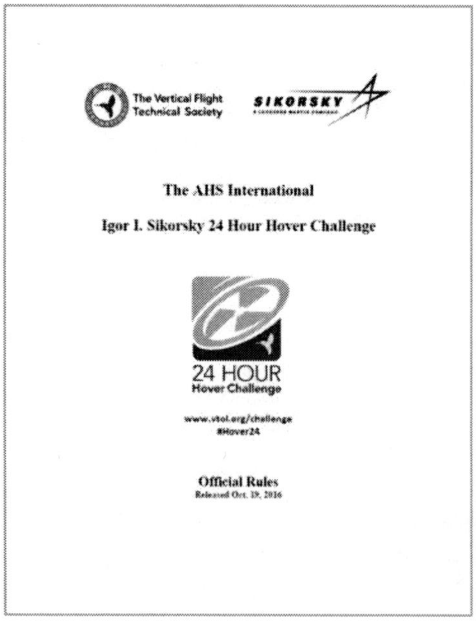
The Vertical Flight Technical Society

SIKORSKY

A LOCKHEED MARTIN COMPANY

The AHS International

Igor I. Sikorsky 24 Hour Hover Challenge

24 HOUR

Hover Challenge

www.vtol.org/challenge

#Hover24

Official Rules

Released Oct. 19, 2016

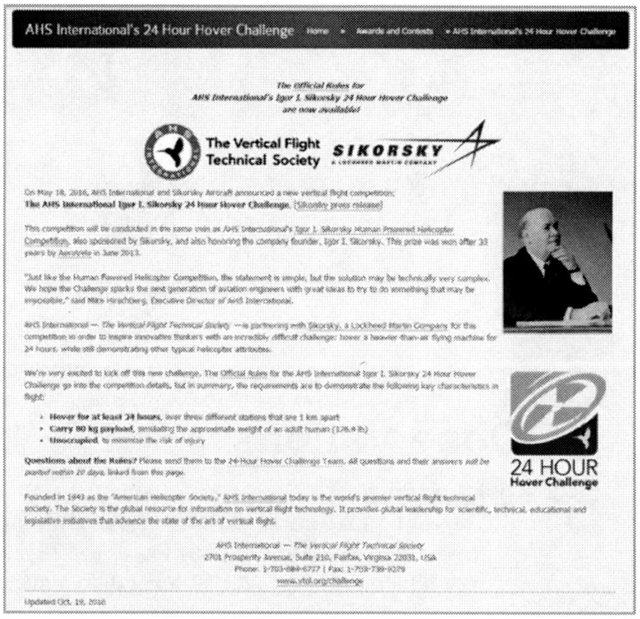
AHS International's 24 Hour Hover Challenge   Home » Awards and Contests » AHS International's 24 Hour Hover Challenge

The Official Rules for

AHS International's Igor I. Sikorsky 24 Hour Hover Challenge

are now available!

The Vertical Flight Technical Society   SIKORSKY A LOCKHEED MARTIN COMPANY

On May 18, 2016, AHS International and Sikorsky Aircraft announced a new vertical flight competition:
The AHS International Igor I. Sikorsky 24 Hour Hover Challenge. [Sikorsky press release]

This competition will be conducted in the same vein as AHS International's Igor I. Sikorsky Human Powered Helicopter Competition, also sponsored by Sikorsky, and also honoring the company founder, Igor I. Sikorsky. This prize was won after 33 years by AeroVelo in June 2013.

"Just like the Human Powered Helicopter Competition, the statement is simple, but the solution may be technically very complex. We hope the Challenge sparks the next generation of aviation engineers with great ideas to try to do something that may be impossible," said Mike Hirschberg, Executive Director of AHS International.

AHS International — *The Vertical Flight Technical Society* —is partnering with Sikorsky, a Lockheed Martin Company for this competition in order to inspire innovative thinkers with an incredibly *difficult* challenge: hover a heavier-than-air flying machine for 24 hours, while still demonstrating other typical helicopter attributes.

We're very excited to kick off this new challenge. The Official Rules for the AHS International Igor I. Sikorsky 24 Hour Hover Challenge go into the competition details, but in summary, the requirements are to demonstrate the following key characteristics in flight:

- **Hover for at least 24 hours**, over three different stations that are 1 km apart
- **Carry 80 kg payload**, simulating the approximate weight of an adult human (176.4 lb)
- **Unoccupied**, to minimize the risk of injury

Questions about the Rules? Please send them to the 24-Hour Hover Challenge Team. All questions and their answers *will be posted within 20 days*, linked from this page.

24 HOUR

Hover Challenge

Founded in 1943 as the "American Helicopter Society," AHS International today is the world's premier vertical flight technical society. The Society is the global resource for information on vertical flight technology. It provides global leadership for scientific, technical, educational and legislative initiatives that advance the state of the art of vertical flight.

AHS International — *The Vertical Flight Technical Society*

2701 Prosperity Avenue, Suite 210, Fairfax, Virginia 22031, USA

Phone: 1-703-684-6777 | Fax: 1-703-739-9279

www.vtol.org/challenge

Updated Oct. 19, 2016

https://vtol.org/awards-and-contests/24-hour-hover-challenge

19

1

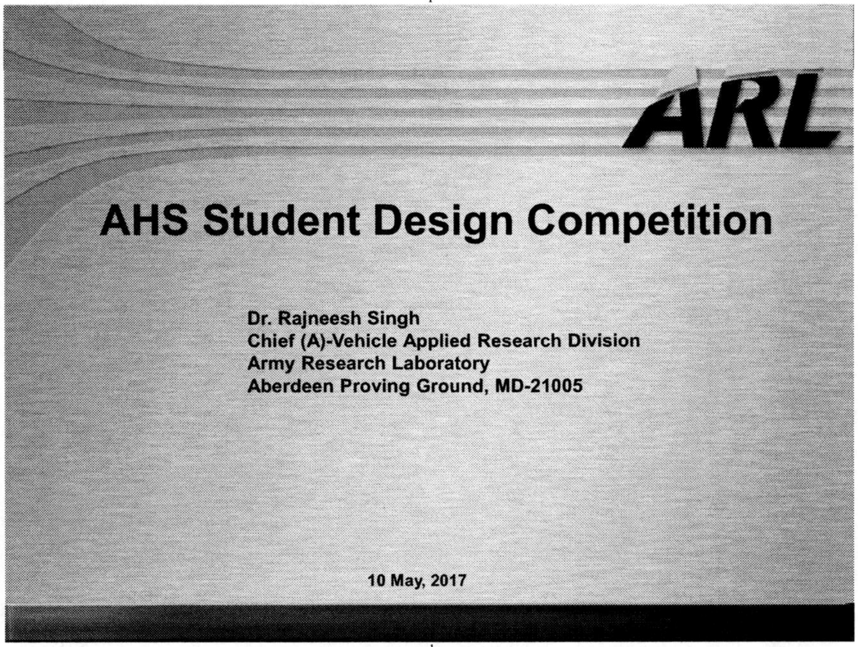
ARL
AHS Student Design Competition
Dr. Rajneesh Singh
Chief (A)-Vehicle Applied Research Division
Army Research Laboratory
Aberdeen Proving Ground, MD-21005
10 May, 2017

1

2

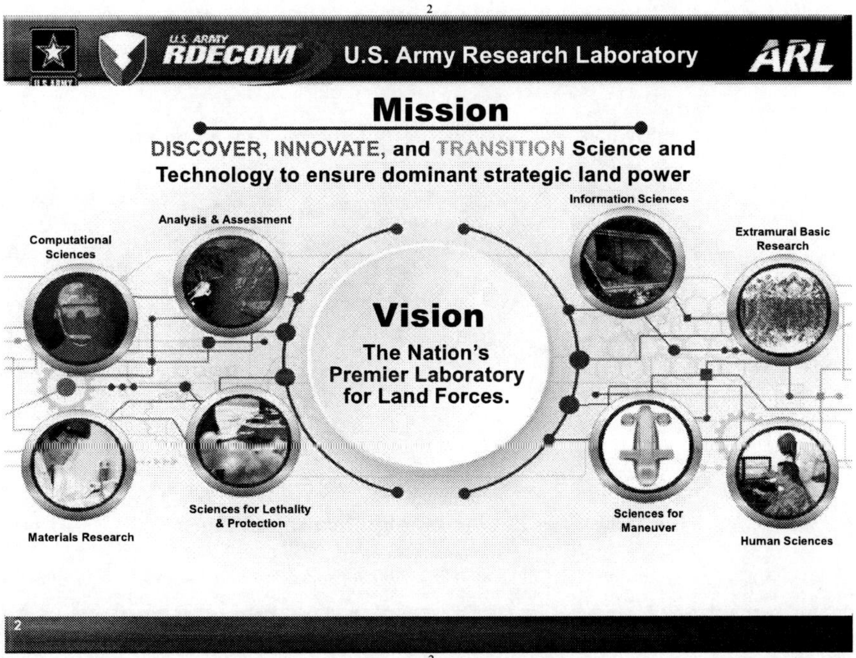
U.S. ARMY
RDECOM
U.S. Army Research Laboratory
ARL
Mission
DISCOVER, INNOVATE, and TRANSITION Science and Technology to ensure dominant strategic land power
Vision
The Nation's Premier Laboratory for Land Forces.
Computational Sciences
Analysis & Assessment
Information Sciences
Extramural Basic Research
Materials Research
Sciences for Lethality & Protection
Sciences for Maneuver
Human Sciences
2

2

3

U.S. ARMY RDECOM **Presentation Scope** ARL

AHS Student Design Competition

3

3

4

U.S. ARMY RDECOM **Jugaad**
(Colloquial Hindi word) ARL

Innovative fix or a work-around using limited resources

Jugaad works but incurs significant penalties

4

4

5

U.S. ARMY RDECOM

Formula-1

ARL

Innovation complements resources

Extremely high-level performance

5

5

6

U.S. ARMY RDECOM

Key challenges

ARL

1. **Performance (Payload, Range/Endurance, Speed)**
2. **Flight Stability & Control**
3. **Handling Qualities, Maneuverability**
4. **Noise**
5. **Safety, reliability**

6

6

7

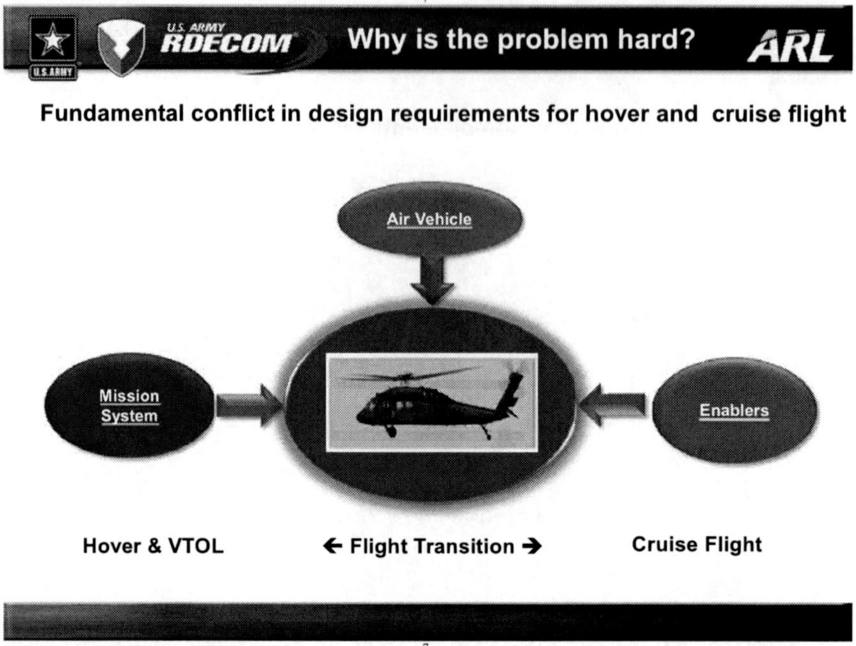
U.S. ARMY
RDECOM
Why is the problem hard?
ARL
U.S. ARMY
Fundamental conflict in design requirements for hover and cruise flight
Air Vehicle
Mission System
Enablers
Hover & VTOL
← Flight Transition →
Cruise Flight

7

8

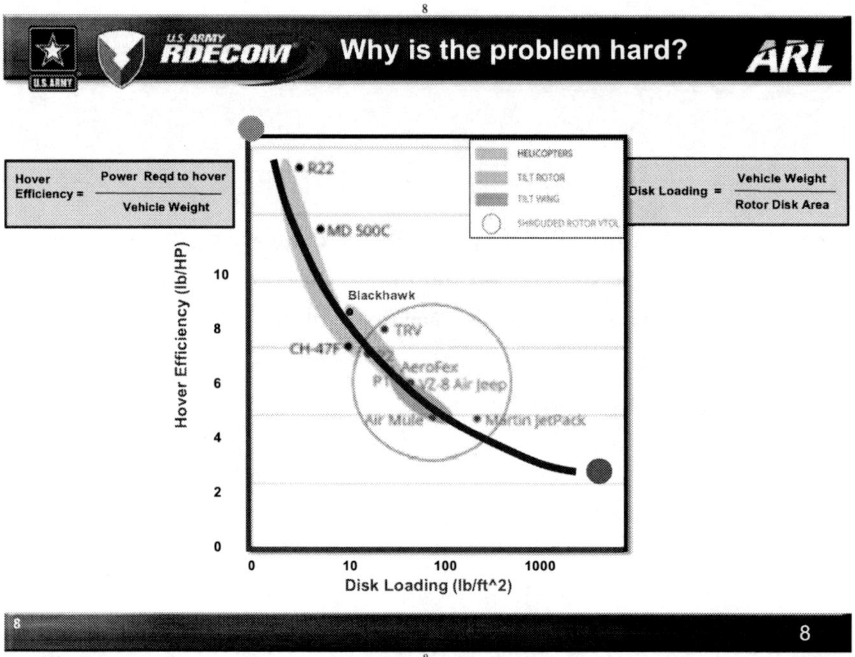
U.S. ARMY
RDECOM
Why is the problem hard?
ARL
U.S. ARMY
Hover Efficiency = Power Reqd to hover / Vehicle Weight
Disk Loading = Vehicle Weight / Rotor Disk Area
HELICOPTERS
TILT ROTOR
TILT WING
SHROUDED ROTOR VTOL
R22
MD 500C
Blackhawk
TRV
CH-47F
AeroFex
VZ-8 Air Jeep
Air Mule
Martin JetPack
Hover Efficiency (lb/HP)
0
2
4
6
8
10
0
10
100
1000
Disk Loading (lb/ft^2)
8
8

8

9

U.S. ARMY RDECOM — Student Design Competition — ARL

- **Student teams design a VTOL vehicle to meet the goals and restriction set in RFP.**
- **50-100 page proposal submitted by each team to AHS by ~end of school year.**
  - Technical Content 40%
    - Design meets RFP goals, Understanding of tools and underlying theory
    - Major technical issues considered, Substantiation of design
  - Application & Feasibility 25%
    - Affordability, Maintainability, Manufacturing Methods, Vehicle Integration
  - Originality 20%
    - Innovation, Design aesthetics, Imagination
  - Organization & Presentation 15%
    - Executive Summary which presents a compelling case
    - Professional quality graphics, drawings, writing
- **Beneficial learning experience for students**
  - Application of studies to a real world problem
  - Learn many disciplines of VTOL design
  - Participate in professional design reviews and develop business case

AgustaWestland
A Finmeccanica Company
BOEING
Bell Helicopter
A Textron Company
SIKORSKY
A LOCKHEED MARTIN COMPANY
ARL

9

10

U.S. ARMY RDECOM — Student Design Competition
Past challenges — ARL

- **Request for proposal (RFP) determined by sponsoring organization and announce in late summer**

2014: DARPA X-plane (Ga Tech)

2015: Distributed Logistics in an Urban Setting

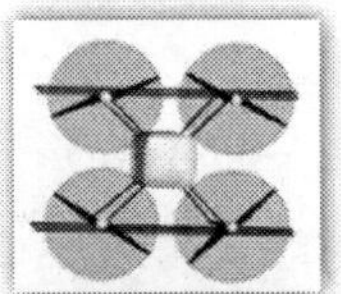

2016: Air Launched Disaster Relief Delivery Vehicle

2017 (Current): 24hr Hovering Vehicle

10

11

Student Design Competition

Global competition open to any graduate and undergraduate students at any university or college.

Georgia Institute of Technology, Pennsylvania State University, Rensselaer Polytechnic Institute, University of Maryland

Israel: Technion – Israel Institute of Technology
Turkey: Middle East Technical University
UK: University of Liverpool

Australia: Royal Melbourne Institute of Technology
China: Nanjing Univ. of Aeronautics and Astronautics
India: Indian Institute of Technology Kanpur

11

12

Student Design Competition
Sample Winning Submission

2016 1st Place Winner: U. Maryland

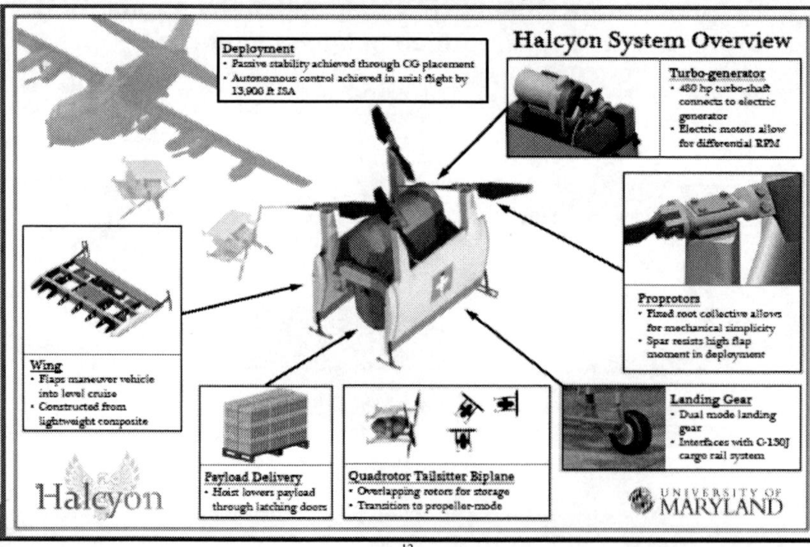

12

13

Student Design Competition
Judging & Awards

U.S. ARMY RDECOM — ARL

- **Entries are judged by Student Design Competition Committee and sponsoring organization.**

Undergraduate	Graduate	Best first time entrant	Hardware Validation Bonus
1st place: $1,850	1st place: $2,500	$750 (Graduate)	$500 (Graduate)
2nd place: $1,200	2nd place: $1,750	$500 (Undergraduate)	$500 (Undergraduate)
3rd place: $500	3rd place: $950		

- **Team is invited to give presentation at the AHS Annual Forum and Tech Display**
- **They are also provided up to $1,000 to defray travel costs.**

13

14

Student Design Competition
2018 Challenge

U.S. ARMY RDECOM — ARL

RFP to be announced in late summer

Additional Incentives:

- Winning team to be invited to give presentation and visit Army Research Lab at Aberdeen Proving Ground
- Follow-up research collaboration to further develop vehicle concept or sub-system technologies

More Information at:

https://www.vtol.org/education/student-design-competition

14

15

U.S. ARMY RDECOM

Final Thoughts

ARL

Wright Flyer, 1903

- Lift: wings, airfoils
- Thrust: propeller
- Propulsion: liquid fuel with piston engine
- Flight Control: flaps, linkages, human, knowledge

Sikorsky VS-300, 1939

- Lift: rotor pitch collective
- Thrust: rotor cyclic pitch
- Propulsion: liquid fuel with piston engine
- Flight Control: main and tail rotor, human, knowledge

VTOL Vehicles, 2010s

- Lift: multi-rotors
- Thrust: vehicle pitch and roll
- Propulsion: electric or electric hybrid
- Flight Control: sensors, variable speed motors, flight control logic

SDC has a key role as we move from Jugaad to F1 performance at this dawn of new era of aviation

15

16

U.S. ARMY RDECOM

ARL

Questions ?

16

16

1

Join us in making the impossible, possible.

GO FLY
PRIZE

1

2

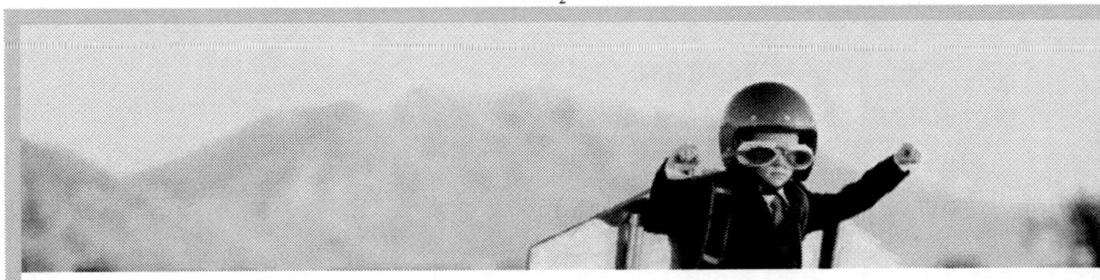

Remember when you were a child and
***wanted to fly*?**

GO FLY
PRIZE

2

3

We are going to make that dream a reality.

Together, we are going to make humans fly.

3

4

A Special Thank You

Eric Allison
Richard Pat Anderson
Ashish Bagai
Eric Bartsch
Gregg Beeber
Paul Bevilaqua
Joe Ben Bevirt
Rob Bulaga
Carlos Cesnik
Mark Costello
John Del Frate
Mark Drela
Mike Duffy
Sean Elliot
Steve Ericson
Mark Forger
Brian German
Tony Ginn
Martin Gomez
Mike Hirschberg
Douglas Jarmusz
David Josephson
Jeff King
Ilan Kroo
Earl Lawrence
Dimitri Mavris
Mag Dean Mirmirani
Jerry Montague
Mark Moore
Dan Newman
Rich Ouellette
Mark Page
Bob Pearce
Darryll Pines
Blaine Rawdon
Stephen Rizzi
Sky Sartorius
Scott Schultz
Tony Tether
Ben Tigner
John Tracy
Mike Ward
James Wong
Chip Yates
Hal Youngren

4

5

Objective

To build a **safe, ultra-compact, quiet, socially acceptable, urban-compatible, near-VTOL** individual flying device **capable of carrying one 200 lb. person a distance of 20 miles without refueling or recharging.**

GO FLY PRIZE

5

6

GO FLY PRIZE

Vision: Anyone, Anywhere

CHALLENGE LANGUAGE:

"What the device looks like or how it works is up to you. We welcome revolutionary design. The device should function safely in both crowded cities and rural areas; it should be lightweight enough and maneuverable enough so that anyone can move it around, and it should be quiet not only for the user, but also for the general public. We are propulsion agnostic, but like all great inventions the device should be user-friendly, almost an extension of the user's body, and provide the thrill of flight."

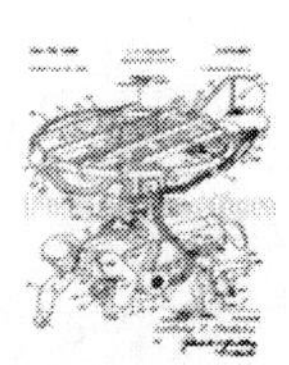

6

7

The GoFly Ecosystem

	HUB TO HUB	HUB TO HUB	DOOR TO DOOR
SKY	BOEING 747 	UBER AIR SHUTTLE 	GOFLY GO FLY PRIZE
LAND	BUS 	CARPOOL VANS/SHUTTLES 	MOPED, MOTORCYCLE, BIKE SCOOTER, SEGWAY, SKATEBOARD 

7

8

Demographic Basis for Vision

- Urbanization
- Failing Infrastructure
- Commuting Statistics

8

Urban Migration 1950-2050

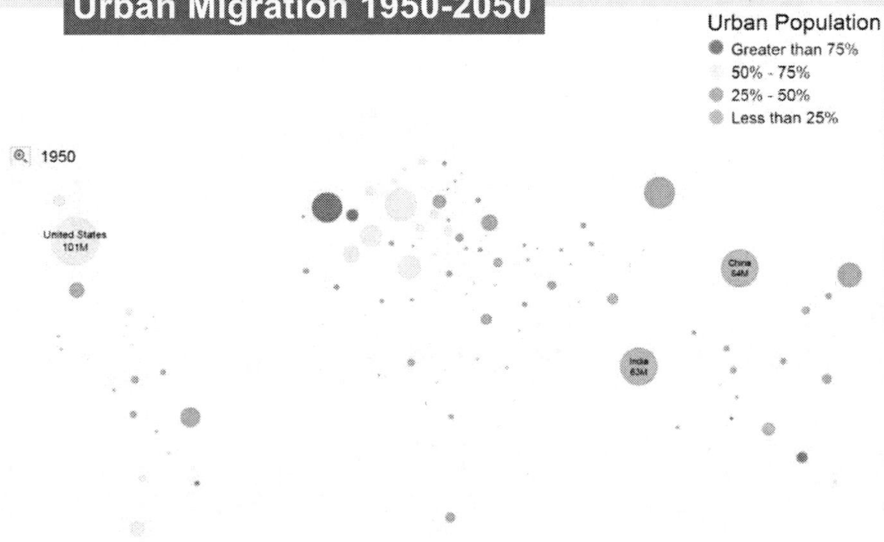

http://www.unicef.org/sowc2012/urbanmap/

9

10

Failing Infrastructure

- American Society of Civil Engineers estimates the US needs to spend $2 Trillion over the next five years to make necessary infrastructure repairs--- and that is just repairs, not modernization.
- 25% of US bridges are structurally deficient or functionally obsolete.
- By 2023, 1 in 4 bridges, which are typically designed to serve for 50 years, will be more than 65 years old.

10

11

Commuting 85% of world travels *20 miles* or less to get to work

Worldwide we spend a year of our lives commuting.

Region	Vehicles on Road in millions	Number of Commuters in millions	Average Commute Time in minutes	Commute Distance in miles	% of Car Commuters
USA	253	128.3	25.4	25	75.7
UK	35	11.3	54	9.3	60

Region	Vehicles on Road (mil)	Average Commute Time	% of Car Commuters
Germany	53.6	21	58
China	240	39	15
Japan	75	45	24
India	72.7	29	67
Brazil	80	33	20*
Russia	48	52	40
Worldwide	**1000**	**37**	**64**

11

12

Now is the Time

Convergence of 4 Breakthrough Technologies:

1. **Stability and Control Systems**:
   - Commercial and consumer driven access to autopilots and control systems make otherwise unflyable concepts easy to operate.
2. **Electrics:**
   - Advances in electric components (batteries, capacitors, motors, controllers) open the door to new hybrid and all-electric concepts.
3. **Materials and structures innovations**
4. **Advanced manufacturing**
   - Offers quick and low-cost prototyping.

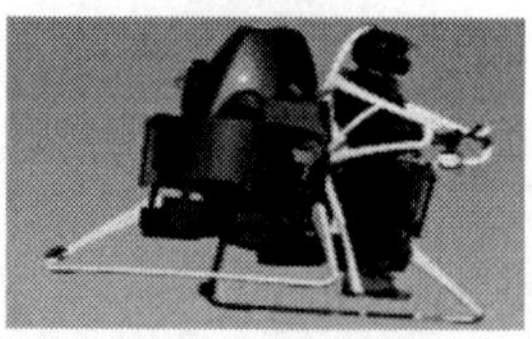

12

13

Why People Participate?

1. The Glory
2. The Prize Money
3. The Post-Prize Opportunities

Opens up a whole new world to winners and creates a new future.

13

14

Prizes Create the Future

Orteig Prize ushers in the age of Airmail and the Commercial Airline Industry

BLUE ORIGIN

GALACTIC

SPACEX

Ansari XPRIZE (space) creates the Commercial Space Industry. (Virgin bought the winner)

DARPA GRAND CHALLENGE

DARPA Grand Challenge creates Autonomous Cars. (Google bought the winner)

14

15

Strategy:

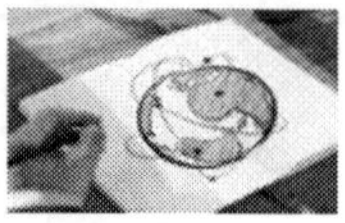

GLOBAL INCENTIVE COMPETITION

The GoFly Prize is a $2M, two year incentive challenge to create personal flying devices, providing global exposure and large financial rewards.

INNOVATION INCUBATOR MODEL

GoFly empowers innovator teams, giving them access to experienced mentors in design, engineering, finance, law, branding, and marketing.

SCHOOL STEM OUTREACH

GoFly offers high school and middle school students the opportunity to connect STEM to the real world by participating in our programs.

15

16

The Incubator - Mentor Program

Technical Mentors

- Renowned experts from leading aerospace corporations and universities provide guidance in areas of VTOL, propulsion, noise mitigation, electrics, and others.

Funding Mentors

- GoFly Network of financial professionals work with teams to introduce them to funding sources.
- Services may include business plan development, strategic consulting, presentation skills.

Legal Mentors

- Mentors may assist with IP filing, corporate formation documents, insurance procurement.

Branding & Marketing Mentors

- Experts help with messaging and pitch books.

In-Kind Product Acquisition

- Procurement of corporate in-kind product for all teams.

16

17

Challenge Details

Time Frame:	24 Month Competition: Completion Summer 2019
3 Stages:	Stage 1: Technical Specs with ten $20k prizes Stage 2: Prototype with four $50k prizes Stage 3: Fly-Off with one $1.5M prize Innovation Prize: $100,000
Innovators:	4 Types: Universities, Backyard Engineers, Small and Medium Sized Companies

17

18

Attributes of a Good Rules Package

- State everything once and only once.
- No requirement can be ignored and the entry remain viable.
- No strong incentives for unsafe design or operation.
- Nothing that unfairly favors teams with abnormally skilled or brave pilots.
- Nothing existing can achieve a prize-eligible score.
- Design space not overly restricted by requirements that have traditionally burdened aerospace.

18

19

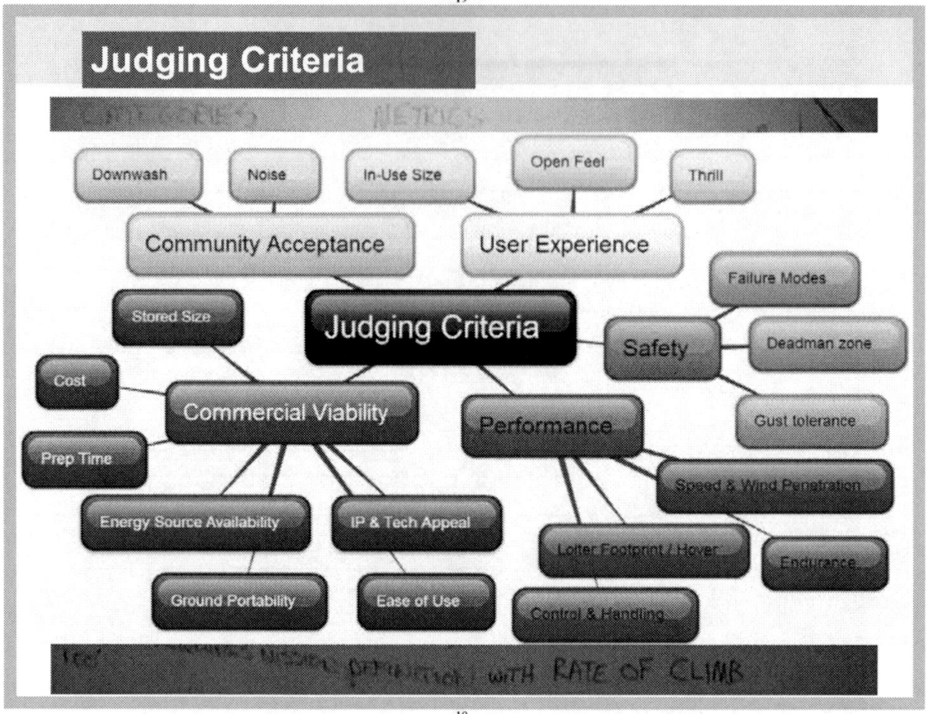

19

20

The Journey to the Rules

The Process:
Close to 100 SME interviews.
15+ Drafts of Rules and Guidelines.
Over 50 private one-on-one technical expert commentaries.
Multiple full day wire-brushing sessions. *Listening.*
Revisions, comments, revisions, comments, revisions, comments.
Thank you, thank you, thank you.

Obstacles we have overcome (and some we are still dealing with):

- Fighting against other visions and the efficiency debate.
- Understanding and capturing noise.
- Incentivizing safety without "owning" it.
- Allowing unmanned; equating manned and unmanned for scoring.
- Regulatory hurdles.

20

21

And the Winner Is…

FINAL SCORE = $\theta_{Size} + \theta_{Noise} + \theta_{Speed}$

21

22

Fly-Off Tasks & Scoring Measurements

Size Measurement → Scoring Parameter θ_{Size}

Threshold: 8.5 ft diameter *(to be finalized)*

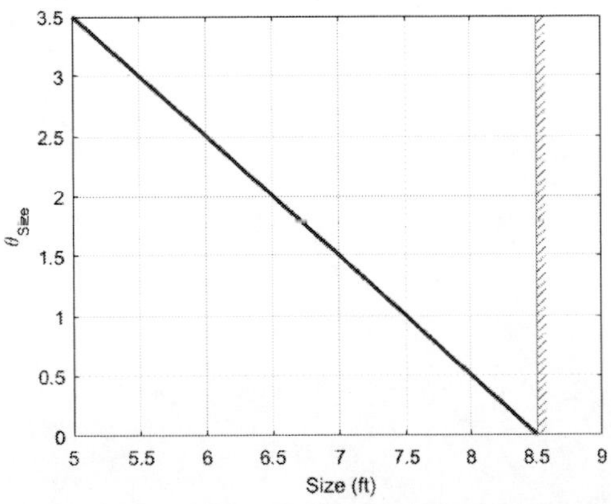

22

23

Fly-Off Tasks & Scoring Measurements

Noise Measurement → Scoring Parameter θ_{Noise}
Measured 50’ from envelope center for takeoff and landing
Threshold: 87 dBA

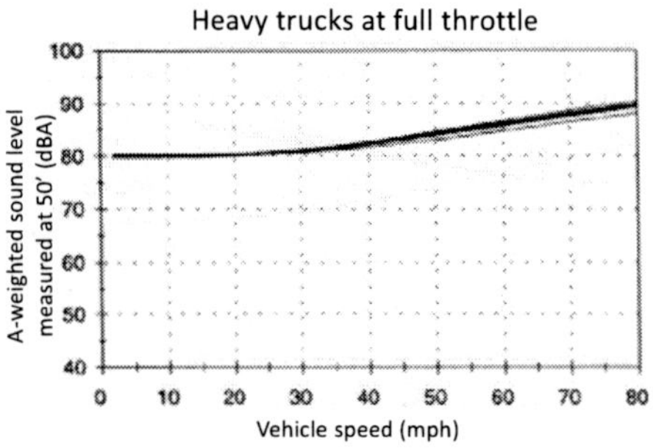

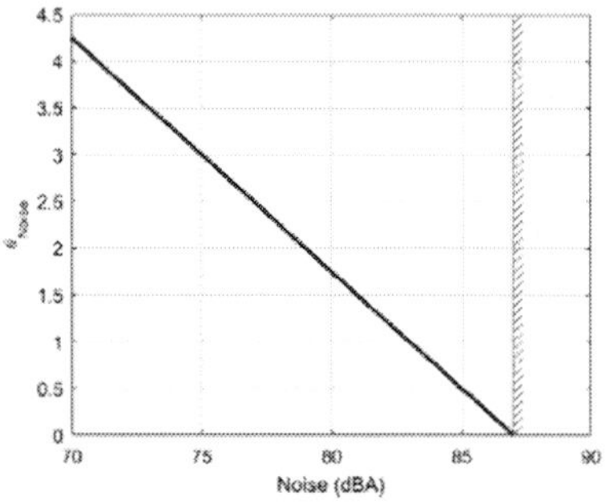

23

24

Fly-Off Tasks & Scoring Measurements

Speed Measurement → Scoring Parameter θ_{Speed}
Threshold: 30 kts average

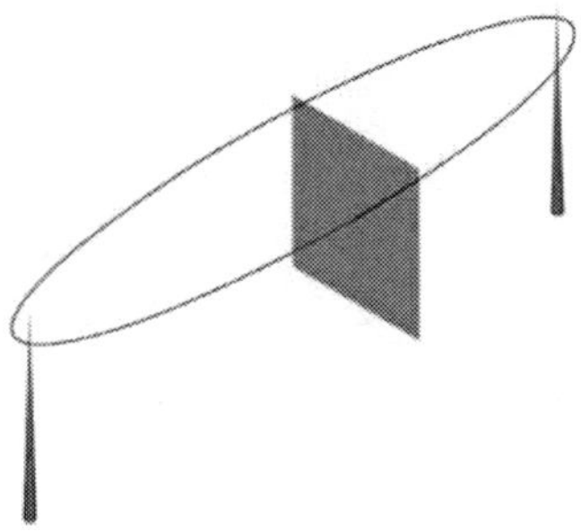

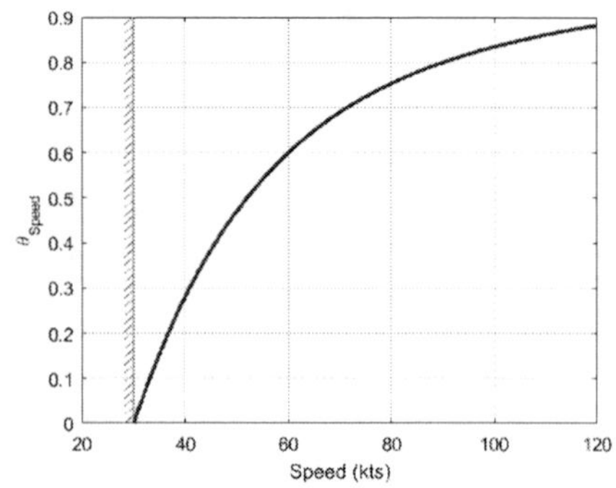

24

Fly-Off Tasks & Scoring Measurements

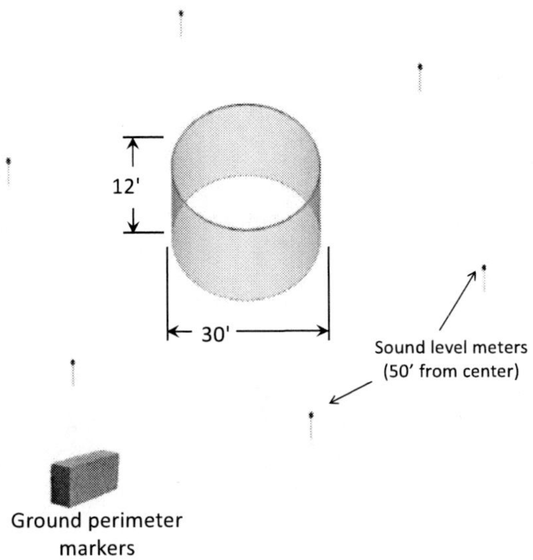

Flight Demonstration

1. Takeoff and climb without violating the takeoff/landing envelope.
2. Conduct a speed run of six laps around a 1 nmi course.
3. Demonstrate the capability to abort a landing by performing a touch & go without violating the takeoff/landing envelope.
4. After loitering such that total flight demonstration endurance is greater than **20 minutes**, descend and land without violating the takeoff/landing envelope.

Fly-Off Tasks & Scoring Measurements

Sightlines – enforce 90° field of view
- Incentive for open, thrilling flying experience
- Disincentive for overhead open rotors

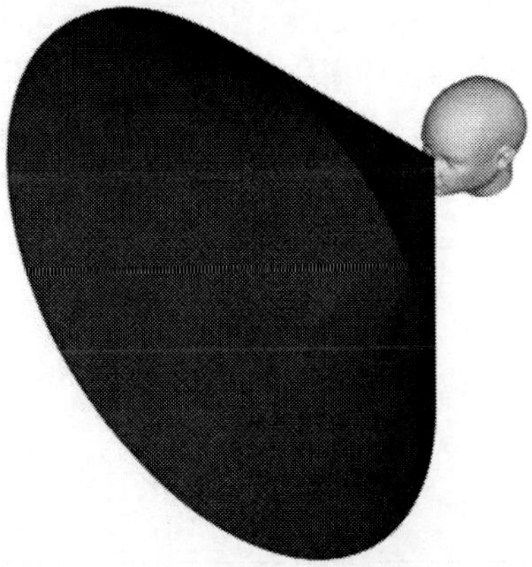

27

Safety

Safety is ultimately the responsibility of the teams.

However, the intent is to develop technologies that can be rated for people to safely fly in the future. Consequently, the competition requires that all device architectures are, at the fundamental concept level, "human-ratable," i.e.,

A device that has no systems or components, except for primary structure, in which a single point failure could result in loss of an operator's life or limb.

Reporting and documentation of safety aspects *are* required.

27

28

Unmanned Option

Autonomous or remotely piloted with "dummy" operator.

Creates extra regulatory and operational challenges for both team practice and the fly-off.

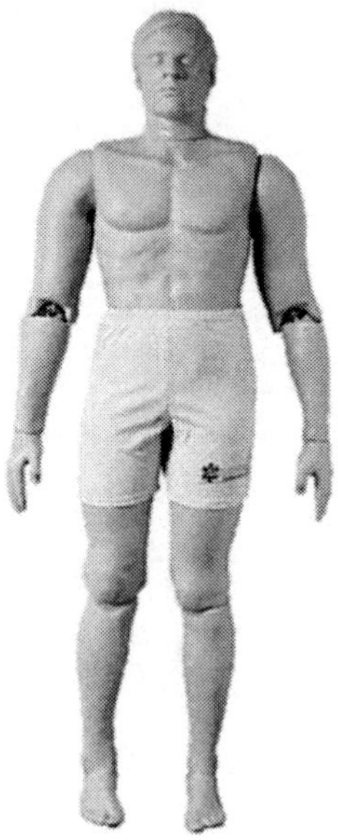

28

29

Where We Are Now

Almost Ready to Launch…

We would welcome your expertise and guidance.
Any and all comments and criticisms are welcome.
… and of course as you are the brightest minds in VTOL, we would love for you to join us.

1. Come help us as an Advisor to the GoFly Prize.
2. Be one of our Incubator Technical Mentors helping the teams create.
3. Form a Team and win $2,000,000 in prizes.

lighter@goflyup.com

GO FLY PRIZE

29

30

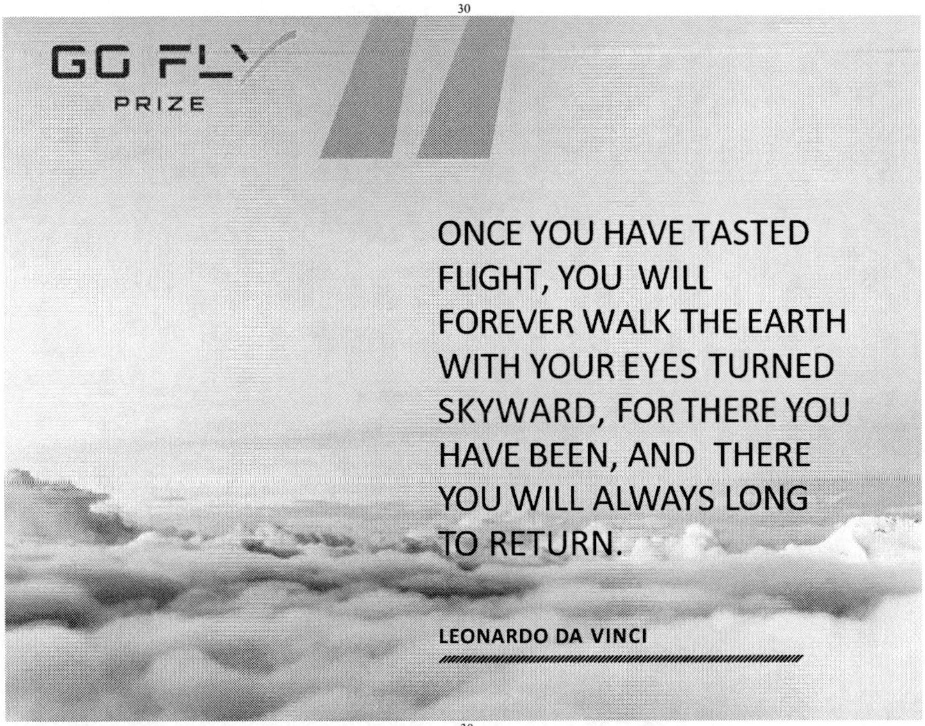

30

31

Join us in making the impossible, possible.
GO FLY
PRIZE

31

32

GO FLY
PRIZE

32

Scoring

Add vs. multiply

Product-based schemes carry the danger that two teams with very close scoring parameters in one area can get wildly different scores, overpowering much more significant differences in another scoring parameter

Which team should win, red or blue?

Case A

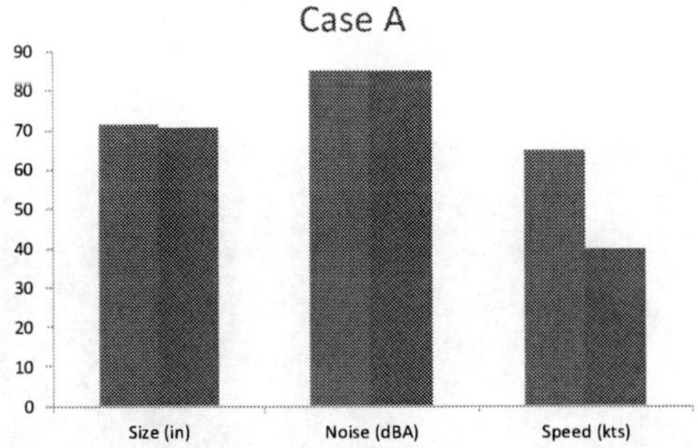

Case B

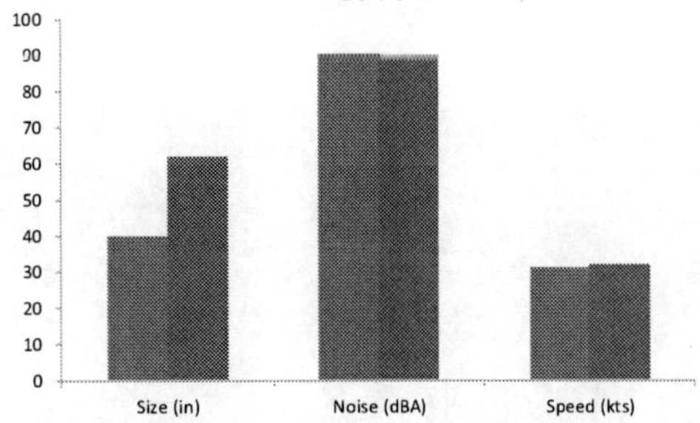

33

Pixelmonger
Studios
Scott@PixelmongerStudios.com

DRONE
WAR 1

READY TO LAUNCH
DRONE WAR 1, THE INAUGURAL DRONE WAR CHAMPIONSHIP EVENT, IS A TURNKEY OPERATION READY TO LAUNCH IN 6 MONTHS. WE HAVE BUILT THE IDEAL STRATEGIC PARTNERSHIPS TO LAUNCH THE WORLD'S PREMIERE DRONE BATTLE LEAGUE.
We have signed elite engineers from universities and the private sector to build custom, blue-sky prototypes. For institutions dedicated to drone innovation, DRONE WAR 1 is equal parts a recruiting bonanza and an R&D playground, all the while crafting an inclusive, exciting message about drone technology for the public at large.

KILBURN MEDIA
A LIVE ENTERTAINMENT COMPANY WITH A PROVEN RECORD OF SUCCESS
We produce world-class live shows, from concerts to magic to circus to competitions. Our productions light up Broadway, Las Vegas, London and touring locations around the world.
JURASSIC WORLD
THE EXHIBITION
THE
ILLUSIONISTS
WITNESS THE IMPOSSIBLE
BACKSTREET BOYS
LARGER THAN LIFE
LAS VEGAS
THE DARK SIDE OF CIRQUE
LE NOIR
PRISMATIC
WORLD TOUR
THE GOLDEN AGE OF CIRCUS
CIRCUS 1903

DRONE WAR
Scott Billups 6/14/2017

THE UNIVERSITIES
WE PROVIDE A PROVING GROUND FOR THE FUTURE OF DRONE ENGINEERING
Faculty advisers and students alike are thrilled at the opportunity to compete in the DWC. UAV clubs are a fast-growing phenomenon on campuses nationwide, and actual drone engineering programs are booming. University club participation in Drone War 1 will build momentum toward a national collegiate competition, tied to official curricula, which can subsequently feed into the DWC pro league. This is an opportunity for young geniuses to apply their designs in a real-world scenario.
INTERESTED SCHOOLS
BOSTON UNIVERSITY
Berkeley UNIVERSITY OF CALIFORNIA
UND NORTH DAKOTA
MIDDLE TENNESSEE STATE UNIVERSITY
West Virginia University

THE PROS
WE CONTRACT ELITE DRONE ENGINEERS / PILOTS TO INNOVATE AND COMPETE
We have made seed investments in their cutting-edge drones, in exchange for the exclusive right to showcase those drones in the Drone War Championship league. These pros represent a talent pool on the leading edge of commercial drone construction. We remain open to revising the final list of pros as we meet new experts.
PROS INCLUDE
GARY + KYLE ETTINGER Father and son recognized as the top drone combatants in the world.
ANDY SHEN The world's top designer of racing drones. The creator of the world's only mass-market shooting drone.
STEVE COHEN + TEAM STEM educator and president of the Drone User Group Network, the largest such advocacy group in the U.S.
NEW JERSEY DRONE ACADEMY Institute dedicated to teaching the public how to build, program and fly drones.

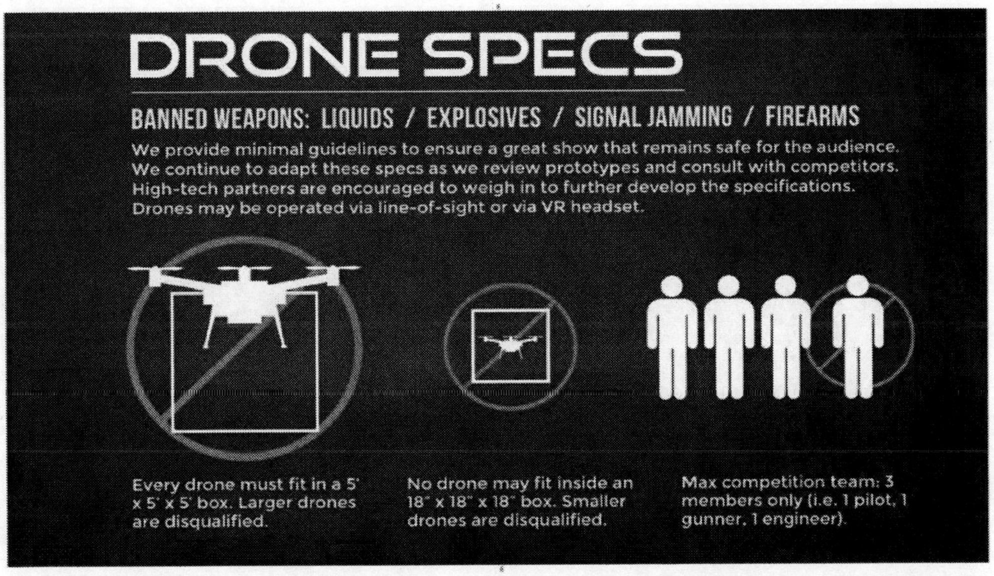
DRONE SPECS
BANNED WEAPONS: LIQUIDS / EXPLOSIVES / SIGNAL JAMMING / FIREARMS
We provide minimal guidelines to ensure a great show that remains safe for the audience. We continue to adapt these specs as we review prototypes and consult with competitors. High-tech partners are encouraged to weigh in to further develop the specifications. Drones may be operated via line-of-sight or via VR headset.
Every drone must fit in a 5' x 5' x 5' box. Larger drones are disqualified.
No drone may fit inside an 18" x 18" x 18" box. Smaller drones are disqualified.
Max competition team: 3 members only (i.e. 1 pilot, 1 gunner, 1 engineer).

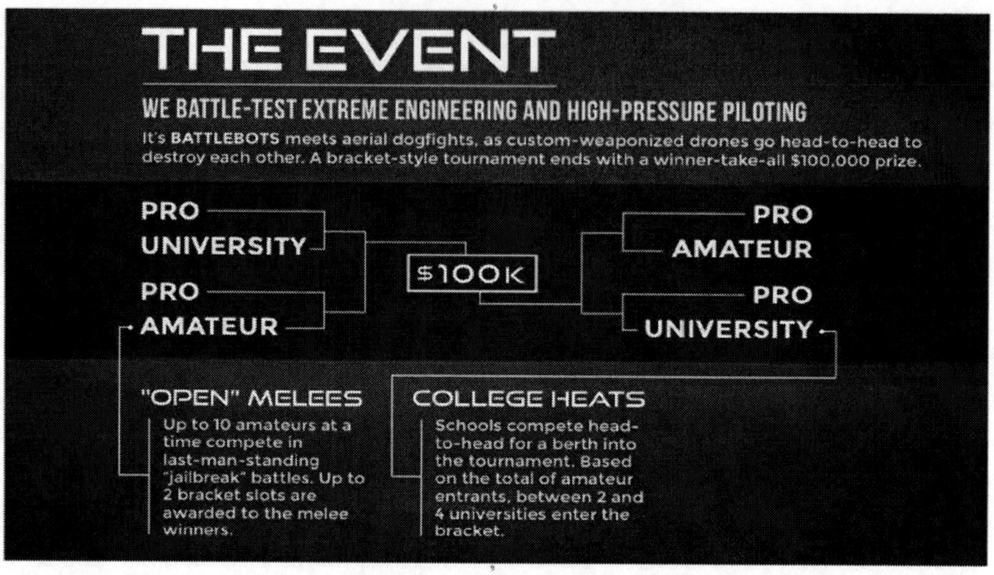
THE EVENT
WE BATTLE-TEST EXTREME ENGINEERING AND HIGH-PRESSURE PILOTING
It's BATTLEBOTS meets aerial dogfights, as custom-weaponized drones go head-to-head to destroy each other. A bracket-style tournament ends with a winner-take-all $100,000 prize.
PRO
UNIVERSITY
PRO
AMATEUR
$100K
PRO
AMATEUR
PRO
UNIVERSITY
"OPEN" MELEES
Up to 10 amateurs at a time compete in last-man-standing "jailbreak" battles. Up to 2 bracket slots are awarded to the melee winners.
COLLEGE HEATS
Schools compete head-to-head for a berth into the tournament. Based on the total of amateur entrants, between 2 and 4 universities enter the bracket.

BOEING
LOCKHEED MARTIN
GENERAL DYNAMICS
NORTHROP GRUMMAN
$100K

Pixelmonger
Studios
Scott@PixelmongerStudios.com

1

XV-24A LightningStrike

*"a new species in the taxonomy of vertical flight"**

*Dr. Ashish Bagai, DARPA PM

Presentation to the

AHS International 73rd Annual Forum & Technology Display

May 9 – 11, 2017, Fort Worth, Texas

Carl G. Schaefer, Jr.
Program Manager, XV-24A LightningStrike Program
Aurora Flight Sciences Corporation
Manassas, VA

This research was developed with funding from the Defense Advanced Research Projects Agency (DARPA).
The views, opinions and/or findings expressed are those of the author and should not be interpreted as representing the official views or policies of the Department of Defense or the U.S. Government.

1

2

What is the XV-24A LightningStrike?

- Aurora's concept for the DARPA VTOL X-Plane program.
- Powered by a unique AC synchronous hybrid electric distributed propulsion system
- 3 MW and over 12,000 amps flow through this aircraft: enough to power a home subdivision
- Three phase program:
  - Phase 1: 22 month preliminary design phase; PDR in September 2015
  - Phase 2: 24 month detail design, fabrication, and integrated test.
  - Completed CDR in March 2017.
  - Phase 3: 6 month flight test phase

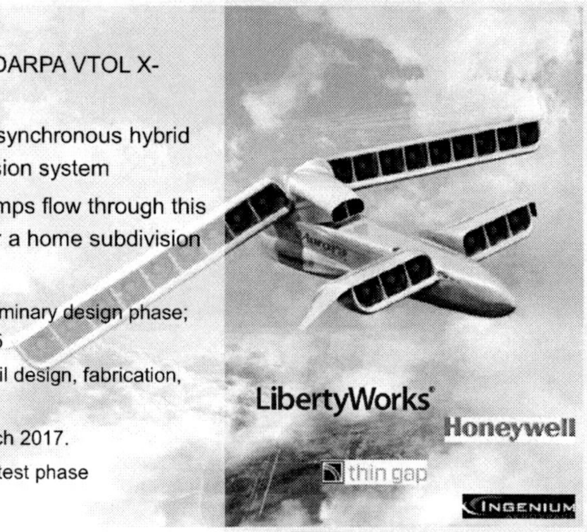

2

DARPA VTOL X-Plane

DARPA VTOL X-Plane program was formed to develop innovative VTOL platforms that leverage modern technologies and engineering tools.

The program established four aggressive performance objectives (below).

Enables new design trade spaces unconstrained by conventional paradigms

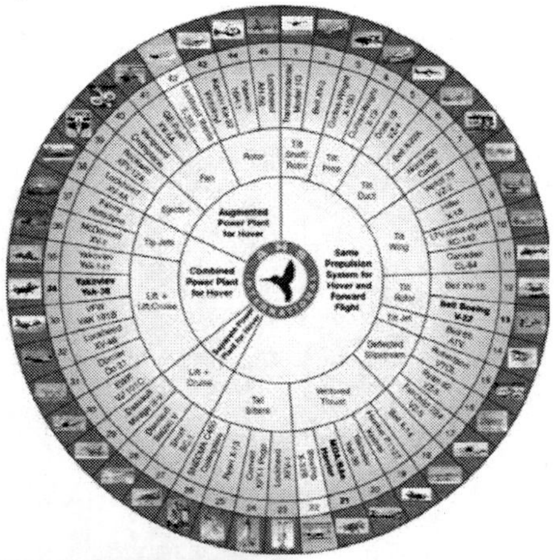

Performance Objectives	State-of-the-Art (Helicopter)	VTOL X-Plane
Sustained max. speed	150-170 kt	≥ 300 kt
Hover efficiency (Aircraft FM)	60%	≥ 75%
Cruise efficiency (Aircraft L/De)	4-5	≥ 10
Useful Load	35-40% GW	≥ 40% GW

XV-24A Aircraft 3-View

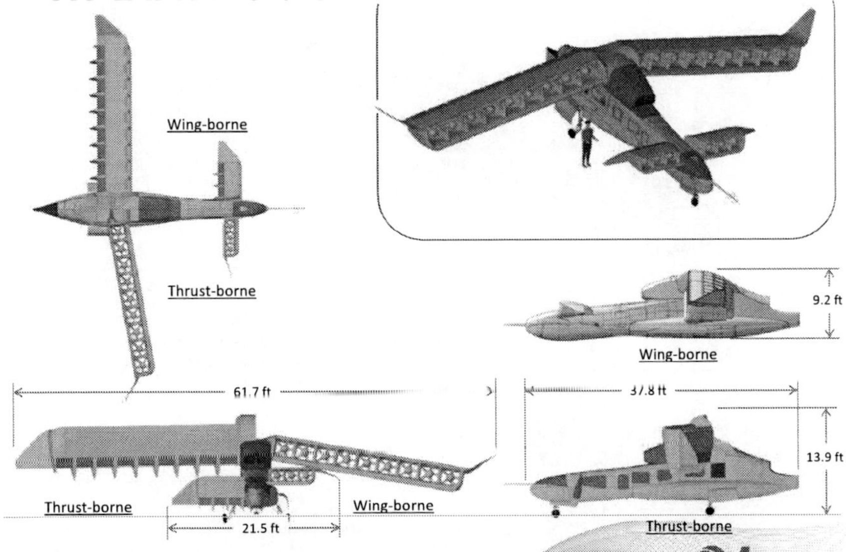

5

Fuselage General Arrangement

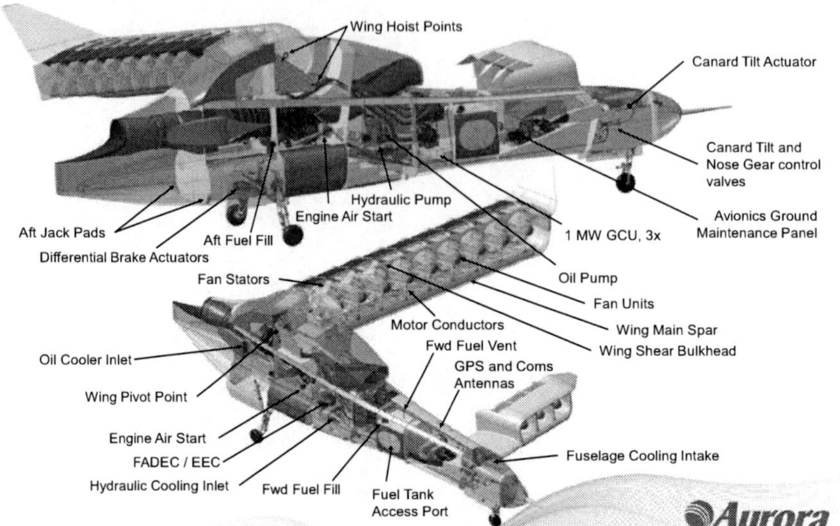

DISTRIBUTION A. APPROVED FOR PUBLIC RELEASE; DISTRIBUTION UNLIMITED

5

6

Fuselage Structure

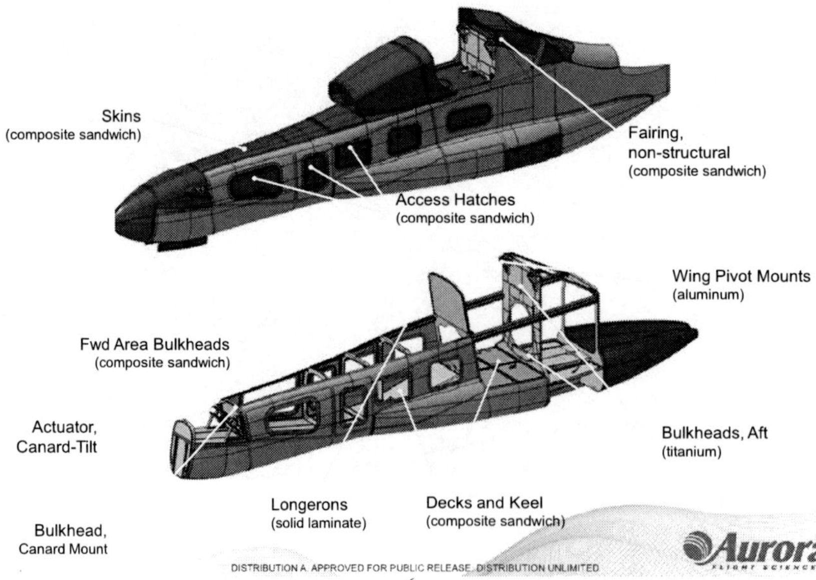

DISTRIBUTION A. APPROVED FOR PUBLIC RELEASE; DISTRIBUTION UNLIMITED

6

7

Power-Generation System (PGS)

- Off-the-shelf equipment
  - AE1107C turboshaft engine, V-22 and C-130 heritage
  - "Hotel" generators (45 kVA, each) – non-propulsive power
  - Hydraulic and oil cooling pumps
- Modified / new equipment
  - Driveshaft
  - Gearbox
  - 1 MW generators
  - Engine inlet

DISTRIBUTION A. APPROVED FOR PUBLIC RELEASE. DISTRIBUTION UNLIMITED

Aurora FLIGHT SCIENCES

7

8

Side-View of the PGS

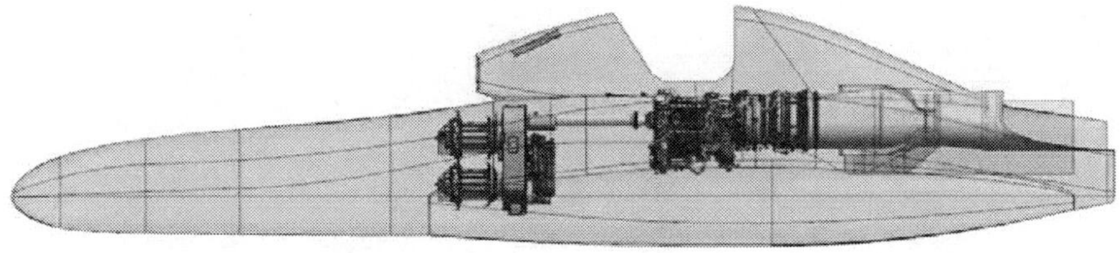

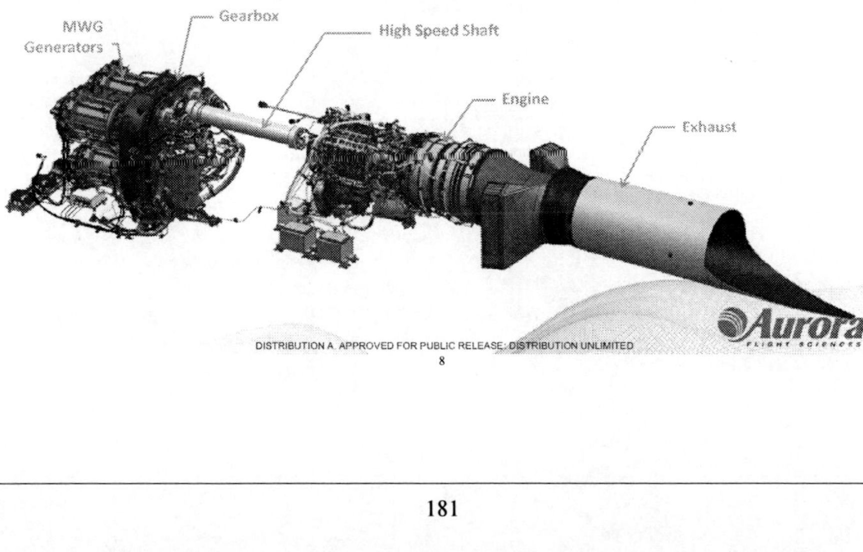

DISTRIBUTION A. APPROVED FOR PUBLIC RELEASE: DISTRIBUTION UNLIMITED

Aurora FLIGHT SCIENCES

8

9

Electric Distributed Propulsion (EDP)

- EDP is the "nervous system" of the XV-24A LightningStrike: power distribution (conductors), electric motors, and variable fan pitch control
- Highly distributed—and integrated—propulsion system
- 18 wing and 6 canard brushless fan motors
- 5-blade, variable-pitch ducted-fan rotors optimized for hover and cruise

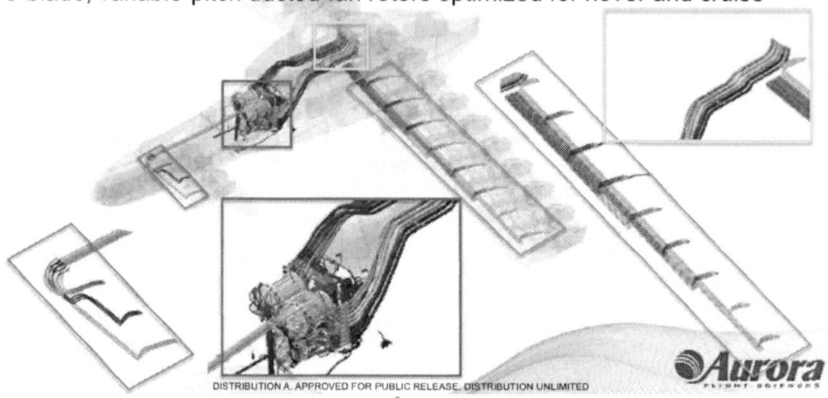

Aurora FLIGHT SCIENCES

DISTRIBUTION A. APPROVED FOR PUBLIC RELEASE; DISTRIBUTION UNLIMITED

9

10

EDP Analogous to the National Electric Grid

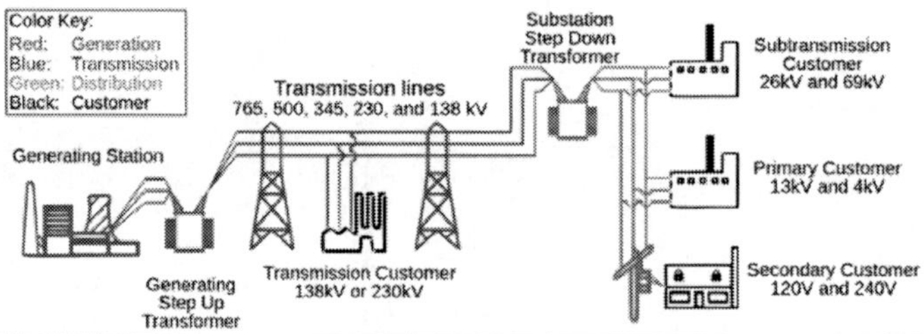

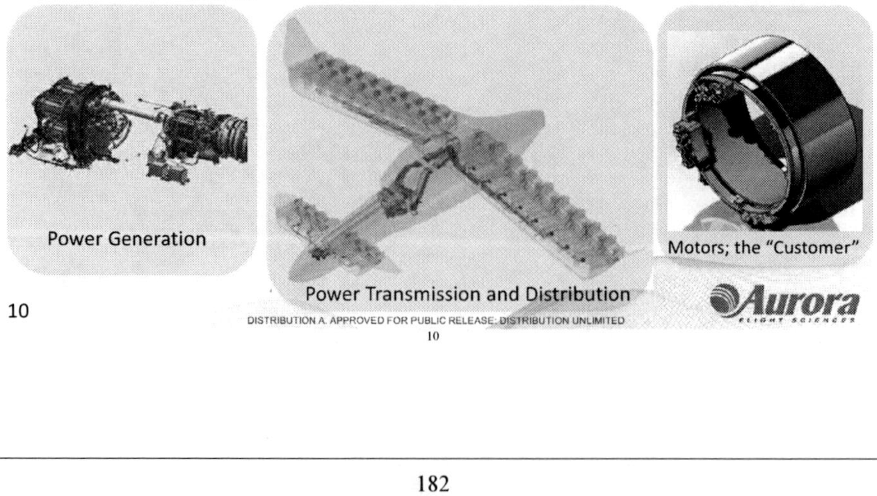

10

Aurora FLIGHT SCIENCES

DISTRIBUTION A. APPROVED FOR PUBLIC RELEASE; DISTRIBUTION UNLIMITED

10

Flight Control System

Over-actuated system – 74 primary effectors available for flight control!

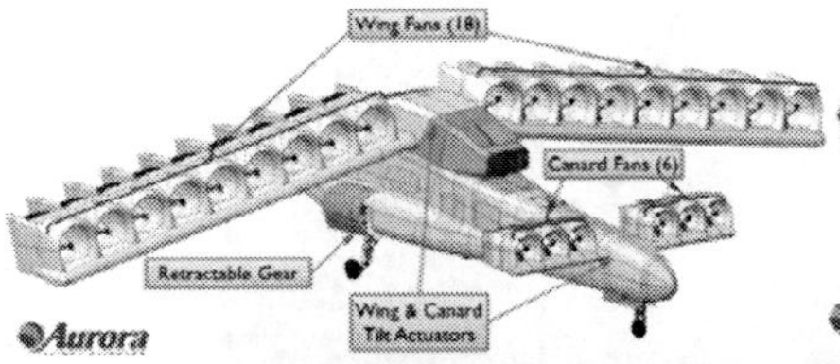

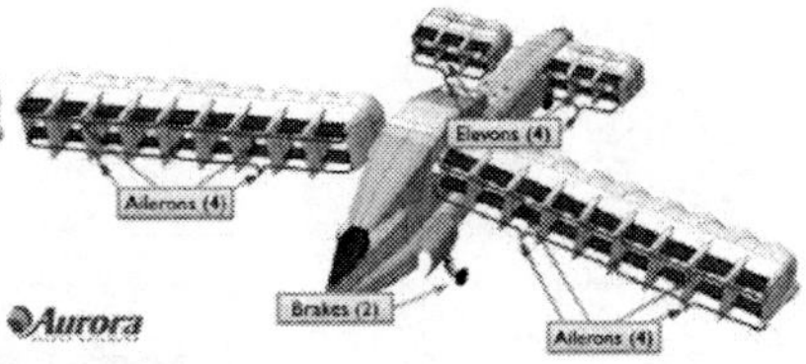

Aurora	Thrust-borne Flight	Transition Flight	Wing-borne Flight
Tilt Orientation			
Pitch Control	Symmetric canard fan thrust via variable fan pitch	Symmetric canard fan thrust via variable fan pitch & symmetric elevon control surfaces	Symmetric elevon control surfaces
Roll Control	Asymmetric wing fan thrust via variable fan pitch	Asymmetric wing fan thrust via variable fan pitch & Asymmetric aileron control surfaces	Asymmetric aileron control surfaces
Yaw Control	Asymmetric aileron control surfaces	Asymmetric aileron control surfaces & asymmetric wing fan thrust via variable fan pitch	Asymmetric wing fan thrust via variable fan pitch
Lift Control	Wing and canard fan thrust via variable fan pitch	Wing and canard fan thrust via variable fan pitch & ganged wing and canard tilt angle	Wing aerodynamics

Aurora FLIGHT SCIENCES

DISTRIBUTION A. APPROVED FOR PUBLIC RELEASE; DISTRIBUTION UNLIMITED

11

XV-24A Design and Fabrication Philosophy

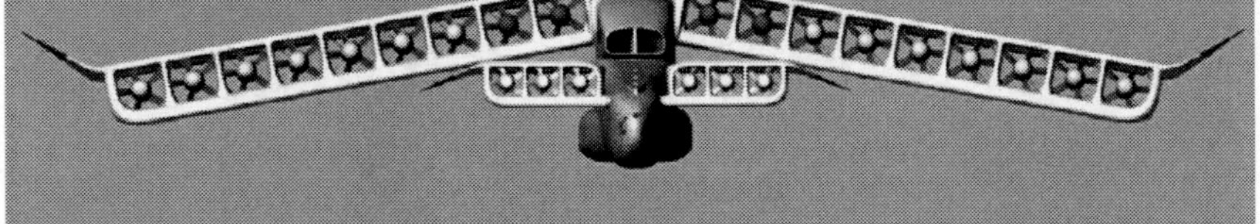

- Concurrent engineering/manufacturing
- Rapid prototyping methods, including additive manufacturing
- COTS, where possible
- Vertical integration (Design and FAIT)
- Scale prototype vehicle and subsystem test articles
- SCRUM/Agile management approach

Aurora FLIGHT SCIENCES

DISTRIBUTION A. APPROVED FOR PUBLIC RELEASE; DISTRIBUTION UNLIMITED

12

13

"Reuse what you can, design what you must"

Centaur
Ground Control Station

F-22 Hydraulics Common
Components

CH-53K Fuel System Common
Components

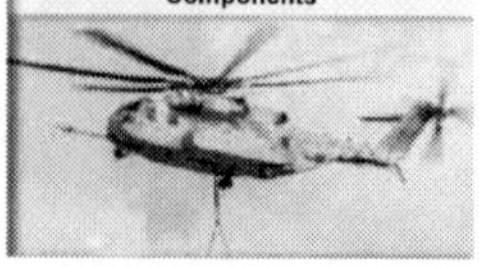

Orion Triplex
Avionics System

MV-22 AE-1107
Turboshaft Engine

S-76
Main and Nose Landing Gear

USN SSC
Engine FADEC

F-35 Oil Cooling Common
Components

DISTRIBUTION A. APPROVED FOR PUBLIC RELEASE; DISTRIBUTION UNLIMITED

13

14

Concurrent Engineering & Manufacturing

DISTRIBUTION A. APPROVED FOR PUBLIC RELEASE; DISTRIBUTION UNLIMITED

14

15

Additive Manufacturing

Direct Metal Laser Sintering Example

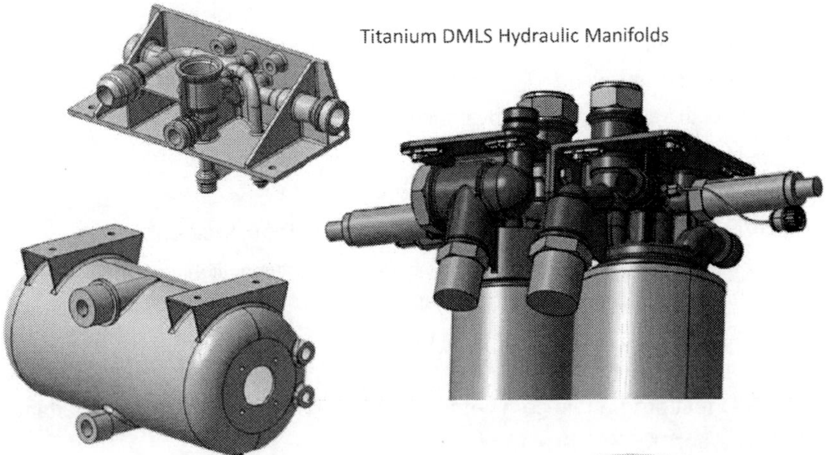

Aluminum DMLS Header Tanks

15

16

Subscale Vehicle Demonstrator (SVD)

Gross Takeoff Weight:	325 lbs
Max Speed:	100 knots
Endurance:	5 - 8 minutes
Max Altitude:	500 ft AGL
Wing Span:	10.7 ft
Fuselage Length:	7.8 ft
Power:	Lithium batteries

16

17

Successful SVD Flight Test Program

- The subscale aircraft validated a number of key technologies for the full-scale XV-24A:
  - The unique tilt wing/tilt canard configuration
  - Aerodynamics and hyperdimensional aerodatabase
  - Control allocation for an over-actuated aircraft
  - Avionics architecture
  - Flight control algorithms and software
- SVD flew a total of ten autonomous flights without incident:
  - Phase 1: auto-takeoff, sustained hover, translational flight (lateral and rearward flight), waypoint navigation, lost-link, and auto-landing
  - Phase 2: "regression flights" to validate new software load, lost-link, and transition to wing-borne flight
- SVD test program is complete but aircraft will remain in flight test status to support XV-24A program

17

DISTRIBUTION A. APPROVED FOR PUBLIC RELEASE. DISTRIBUTION UNLIMITED

17

18

XV-24A Program Summary

- Unique configuration defined by the program objectives outlined by DARPA.
- XV-24A unconstrained by conventional design concepts (the VTOL "Wheel of Misfortune")
- Removes many design constraints for future CTOL and VTOL air vehicles
- Design unlike other aircraft;
  - Hybrid, electric distributed propulsion system
  - Highly redundant, over-actuated control system
  - Enables transmission agnostic configurations
- High confidence in the XV-24A design;
  - Very successful flight test campaign of the subscale demonstrator
  - Key risk reduction tests of major subsystems retires major risks before flight test program begins. MWG test; Copper Bird test; Iron Bird test
  - COTS products where possible; rapid prototyping
- XV-24A is being fabricated and assembled now. First flight is scheduled for late 2018.

18

DISTRIBUTION A. APPROVED FOR PUBLIC RELEASE: DISTRIBUTION UNLIMITED

18

19

Questions?

19

DISTRIBUTION A. APPROVED FOR PUBLIC RELEASE: DISTRIBUTION UNLIMITED

19

1

NATO Industry Advisory Group (NIAG)

Next Generation Rotorcraft Capability (NGRC) Study Group

NATO
OTAN

NIAG NGRC Study Group

AHS Forum 9-11 May, 2017

Daniel Newman, SG-219 Chair

1

2

Next Generation Rotorcraft Capability (NGRC) Background

- After focus past decade Out-of-Area Operations, events in Ukraine were a "Wakeup Call" for Alliance
- Ability to counter peer nation diminished, becoming obsolete
- Alliance earmarked rotorcraft aging, need replacement ~2040
- New complex system acquisitions (design, develop, test, produce) are lengthy and costly, so capability/requirements process must start now
- JCG-VL Vision *(Joint Capability Group – Vertical Lift)*
  - Define future rotorcraft operational capability
  - Shorten acquisition cycle
  - Control cost and manpower
  - Leverage rapidly evolving technology
  - Identify enhanced operational capability next generation rotorcraft offer

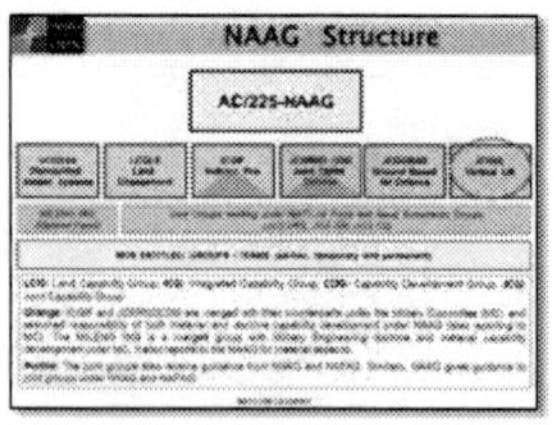

2

3

NATO OTAN NGRC Team of Experts (TOE)

- JCG- VL established a Team of Experts (TOE) to define the "Next Generation Rotorcraft Capability (NGRC)"
  - NATO Organizations: ACT, JAPCC, STO, NIAG
  - NATO Nations
  - Two-year effort ending in Jul 18
- NGRC TOE Mission:
  - To identify and assess current and evolving applicable rotorcraft technologies, force structure implications, force capabilities, as well as operational concepts leading to the development of an integrated next generation capability
- NGRC TOE Responsibility:
  - Act as the single integrating organization with the CNAD for future rotorcraft S&T, force operational concepts to inform defense investment planning
  - TOE Report an input to 2018 NDPP, assist defining Long Term Capability Reqts
  - JCG-VL will use data to develop NATO Staff Requirement to create national interest in a new rotorcraft development program

3

4

NATO OTAN NIAG NGRC Advisory Study Charter

Nr. 3 NAAG JCGVL

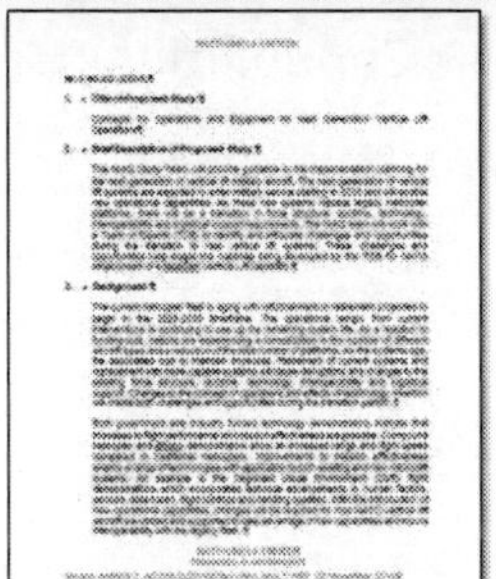

1. **Title of Proposed Study:**
   Concepts for Operations and Equipment for Next Generation Vertical Lift Operations
2. **Brief Description of Proposed Study:**
   The NIAG Study Team will provide guidance to the implementation planning for the next generation of vertical lift military aircraft. The next generation of vertical lift systems are expected to enter military service starting in 2035 and will enable new operational capabilities. As these new systems replace legacy helicopter platforms, there will be a transition in force structure, doctrine, technology, interoperability and logistical support requirements. The NIAG team will work with a Team of Experts (TOE) to identify and articulate challenges and opportunities during the transition to new vertical lift systems. These challenges and opportunities help shape the roadmap being developed by the TOE for NATO employment of a NextGen Vertical Lift capability.

4

5

NATO OTAN NATO Industry Advisory Group (NIAG) NGRC Support

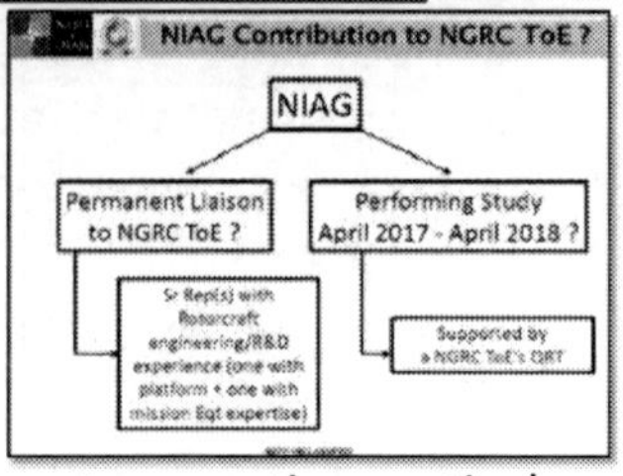

- NGRC TOE Liaison Group
  - Chair: Mr. Daniel Newman, The Boeing Co (USA)
  - Deputy/Mission Systems:
    Dr. Thomas Münsterer, Hensoldt (DEU)
  - Deputy Platform Domain:
    Mr. Luigi Ricci Moretti, Leonardo Helicopter (ITA)
- NIAG Advisory Study:
  Concepts for Operations and Equipment for next generation vertical lift operations (or "Next Generation Rotorcraft Capability")
  - Sponsor: NAAG/JCGVL
  - Reference: AC/259-D(2016)0043 (INV) & NIAG-D(2016)0024 (INV)
  - Approved by Conference of National Armaments Directors (CNAD) on 6 December, 2016 (Silence Procedure) ... Open to Partners
  - One-year effort to provide industry input began 9 Mar 17
  - Leadership team (Chair, Deputy, Rapporteur) elected 9 Mar 17

5

6

NATO OTAN NIAG NGRC Study SG 219 Kickoff Study Approach – Capabilities & Data (1/2)

- Capabilities

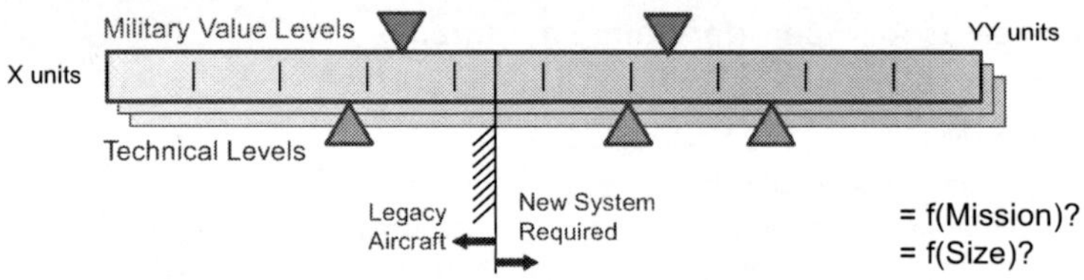

- Surveying Industry
  - No flag, no badge!
  - Utilize Industry Trade Groups (e.g. NDIA in USA)
  - Utilize Industry Events (e.g. AHS Forum, ERF)

6

7

NIAG NGRC Study SG 219 Kickoff
Teams/Leads &* Meeting Plan

Study Sub-Teams

A. Performance, Noise & Vibration — Lead: Luigi Ricci-Moretti
B. Mission Systems, Avionics and Human Factors — Lead: Thomas Muensterer
C. Survivability & Mission Operations/Teaming — Lead: Mario de Lucia
D. Manufacturing, Operations & Support Technologies — Lead: Jose Dias
E. Interdependencies, System Integration & Force Structure — Lead: Julian Lee

Meetings

#	Scope	Location	Date(s)
	Launch	Brussels	9 March '17
1	Kickoff	Ft Worth	11-12 May '17
2	Core	Milan, Italy	11-12 July '17
3	Core	*Amsterdam*	Oct '17
4	Core	TBD	Jan '17
6	Edit	Brussels	Mar '18

To join SG-219, information available at AHS Forum 73 Novelty Desk

7

8

8

1

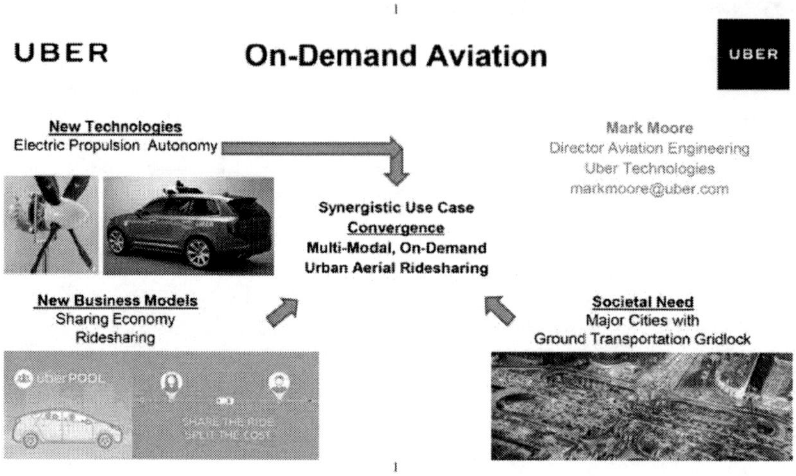
UBER
On-Demand Aviation
UBER
New Technologies
Electric Propulsion Autonomy
Mark Moore
Director Aviation Engineering
Uber Technologies
markmoore@uber.com
Synergistic Use Case
Convergence
Multi-Modal, On-Demand
Urban Aerial Ridesharing
New Business Models
Sharing Economy
Ridesharing
Societal Need
Major Cities with
Ground Transportation Gridlock
uberPOOL
SHARE THE RIDE
SPLIT THE COST

1

2

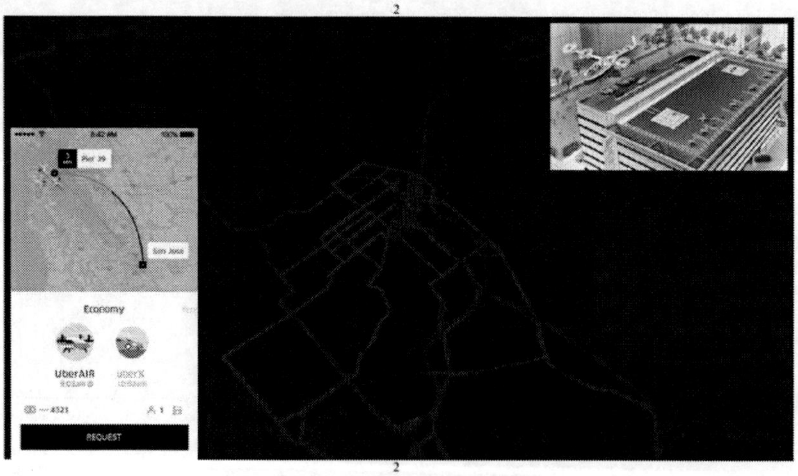
Economy
UberAIR
uberX
REQUEST

2

3

UBER

Helicopters

UBER

- Too Noisy
- Not Safe Enough
  - Single Part Failure Modes
- Inefficient
  - Poor Aerodynamic Efficiency
  - Poor Propulsion Efficiency
- Too Expensive
  - High Energy Use
  - High Maintenance Cost
  - High Acquisition Price
  - ~300 Hours Annual Utilization

3

4

UBER

eVTOL

UBER

Electric Vertical Takeoff and Landing Aircraft

- Noise: 15 to 20 dB Lower
- Safety: 2x Safer than Autos
  - No Single Part Failure Modes
  - Ballistic Recovery Parachutes
- Efficient: ~10x Lower Energy Use
  - 3x Higher Aerodynamic Efficiency
  - 3.3x Higher Propulsion Efficiency
- Low Operating Cost
  - Low Energy Use
  - Low Maintenance Cost
  - Higher Volume Production
  - ~2000 Hours Annual Utilization

EMBRAER

Bell Helicopter

PIPISTREL

4

5

UBER

Multicopters

Too Inefficient – Too Slow

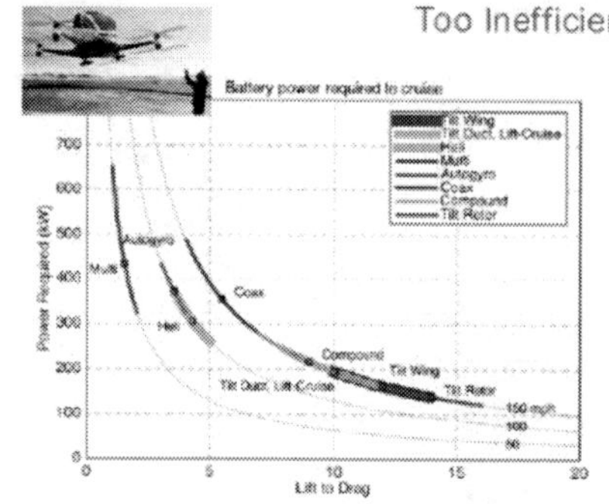

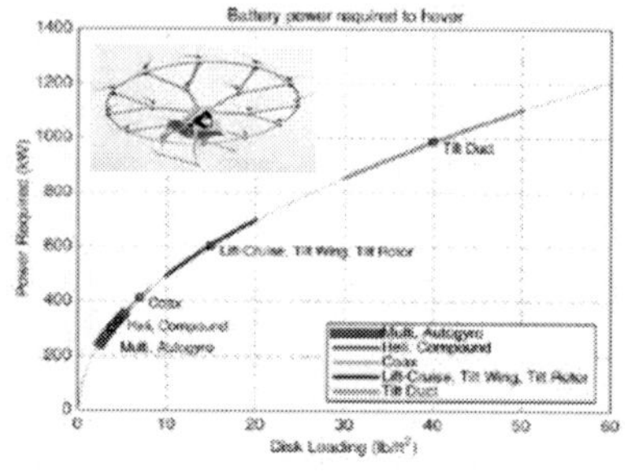

Gross Weight = 5000 lb

5

6

UBER

Fast Forwarding eVTOL

Why Uber?

Focus on city productivity
On-demand travel leader
$68B company
>60M active riders
>500 cities

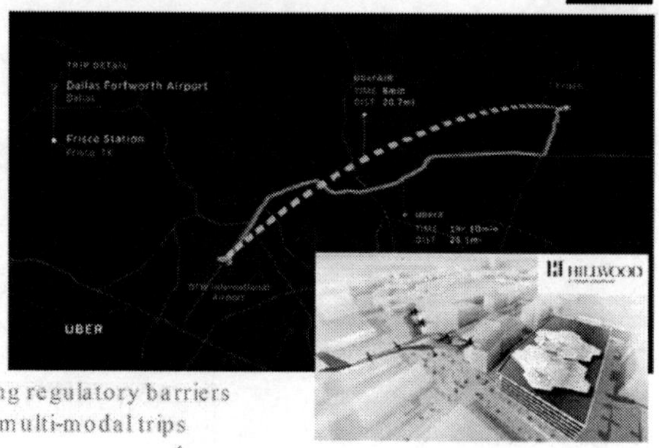

Uber Capabilities

Integrated software
Network operations
Efficient trip pooling
Track record of overcoming regulatory barriers
Able to provide seamless multi-modal trips

6

7

UBER

Fast Forwarding eVTOL

UBER

7

8

UBER

Fast Forwarding eVTOL

UBER

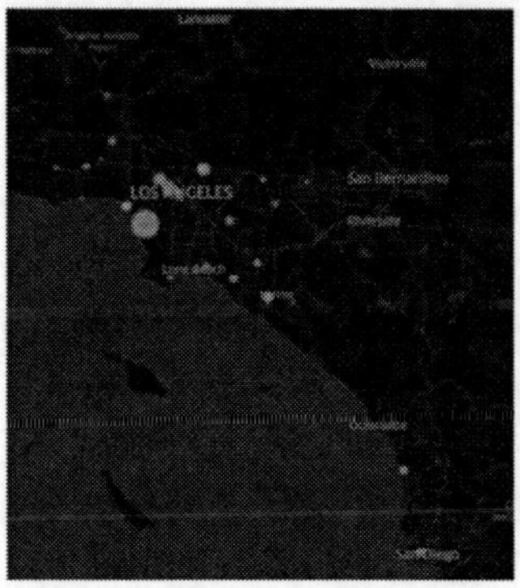

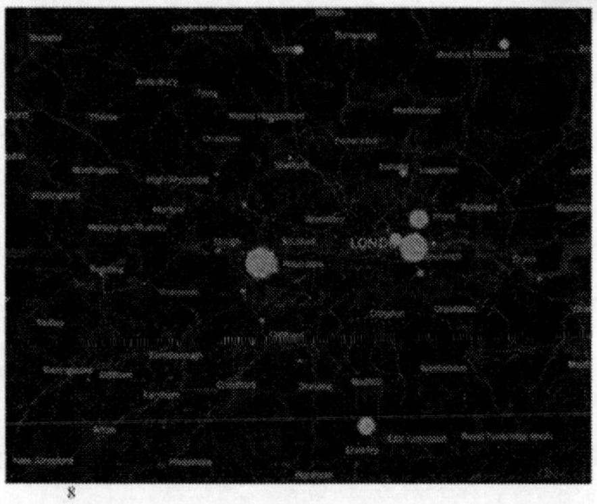

8

9

UBER Fast Forwarding eVTOL UBER

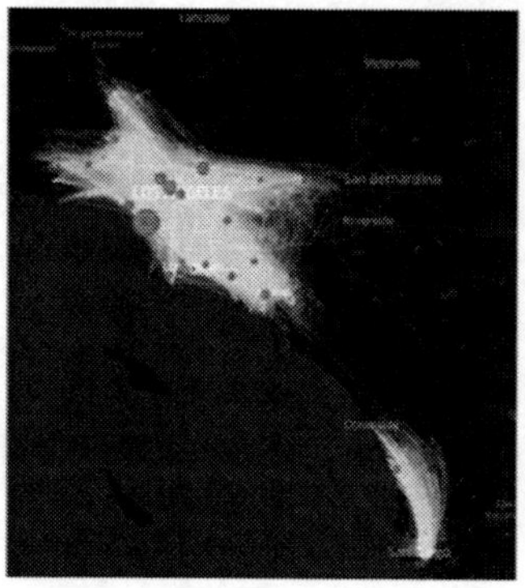

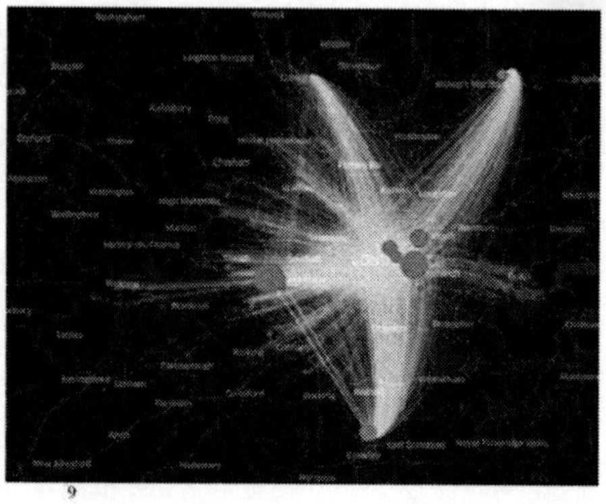

9

10

UBER Fast Forwarding eVTOL UBER

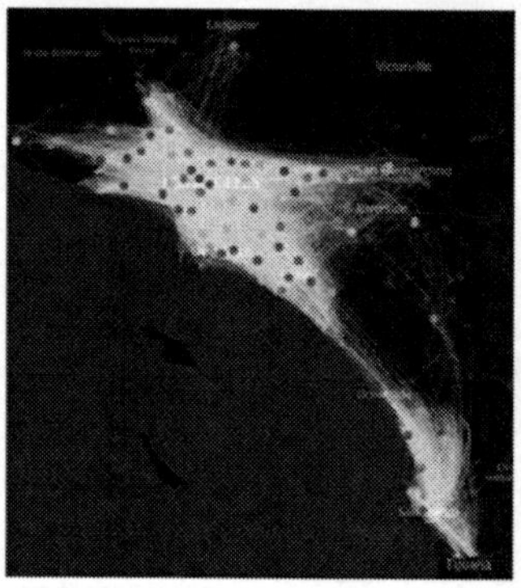

10

11

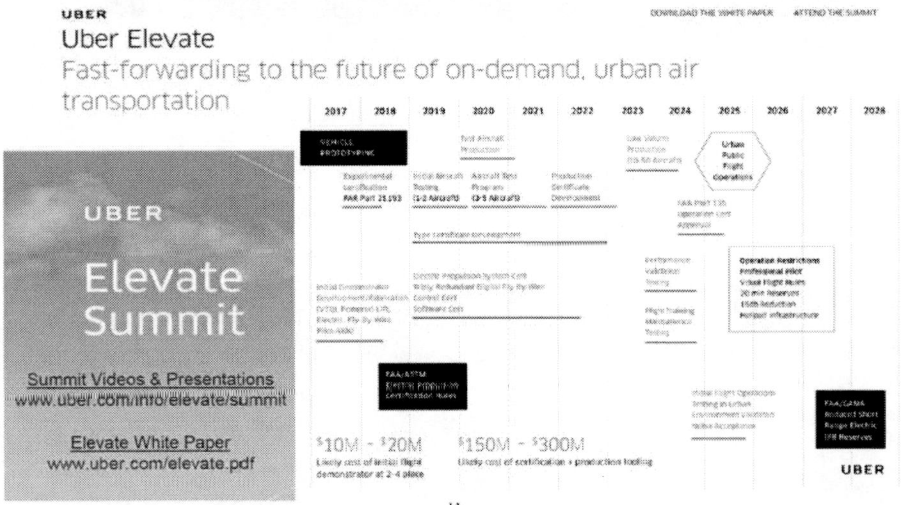
UBER
DOWNLOAD THE WHITE PAPER
ATTEND THE SUMMIT
Uber Elevate
Fast-forwarding to the future of on-demand, urban air transportation
UBER
Elevate Summit
Summit Videos & Presentations
www.uber.com/info/elevate/summit
Elevate White Paper
www.uber.com/elevate.pdf
2017
2018
2019
2020
2021
2022
2023
2024
2025
2026
2027
2028
Experimental Certification
Urban Public Flight Operations
Operation Restrictions
Professional Pilot
Visual Flight Rules
20 min Reserves
Hotspot Infrastructure
$10M - $20M
Likely cost of initial flight demonstrator at 2-4 place
$150M - $300M
Likely cost of certification + production tooling
UBER

11

1

1

2

Why is A³ Here

Disrupt Airbus and the rest of the industry before someone else does.

Recognize significant technological inflections and the future they enable, and identify major obstacles to making that future possible.

Retire technological risks via fast-paced, aggressive projects that demonstrate the future of flight today.

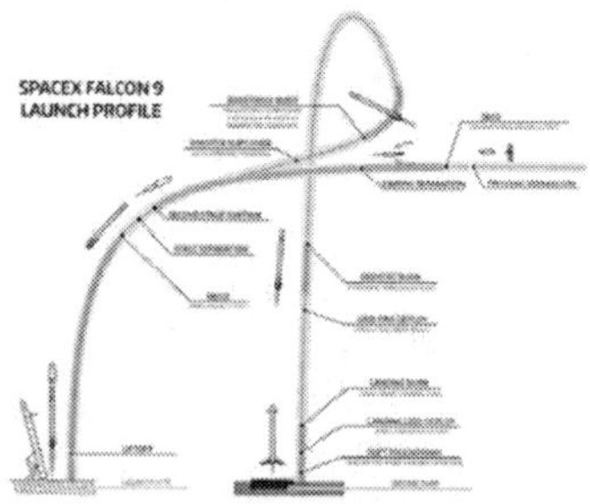

8 April 2016

2

3

What is an A³ Project

Disruptive to our core business
Productizable demonstrator in 2-4 years
Airbus is not willing or able

Rigorous
Ambitious
Risky

3

4

Problem

Typical Rush Hour Traffic

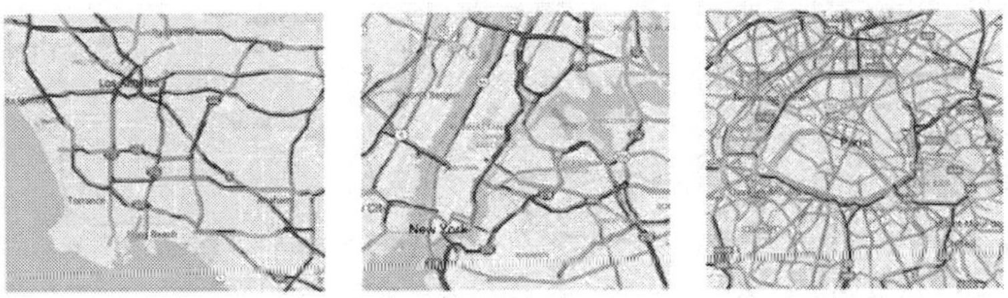

Los Angeles | New York City | Paris

4

5

Why now?

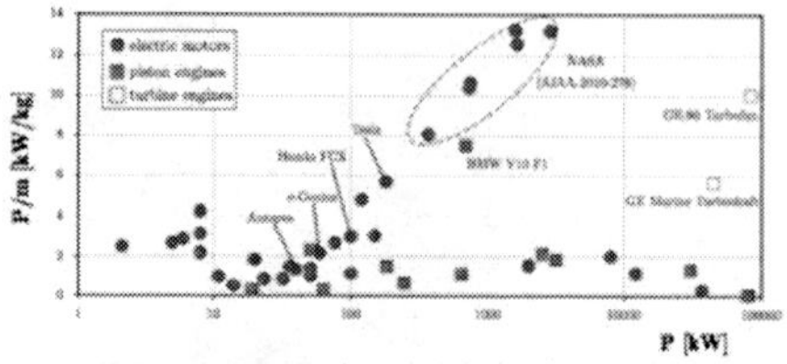

Source: "Electric Flight – Potential and Limitations" by Martin Hepperle

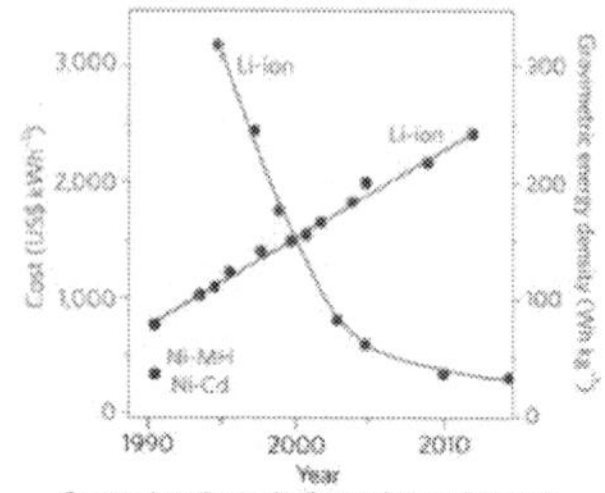

Source: Joint Center for Energy Storage Research

- High specific energy batteries
- High specific power motors and controllers
- Low cost sensors driven by consumer electronics
- Composite automotive manufacturing techniques

5

6

Solution Space

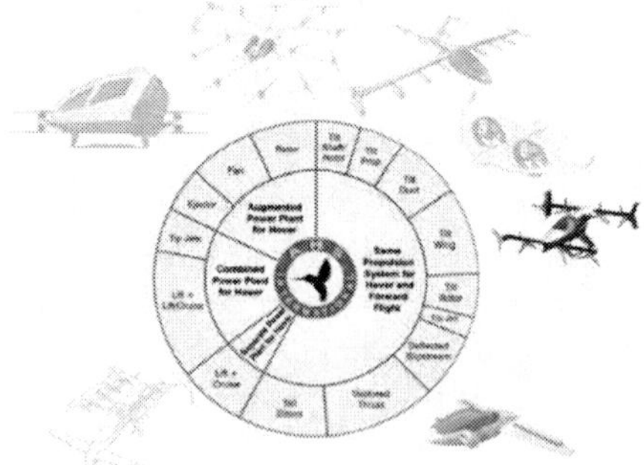

6

Solution

Provide a self-piloted low operating cost aircraft that will enable safe full scale Urban Air Mobility, for both passenger carrying and cargo carrying applications.

Enablers

Direct Operating Cost

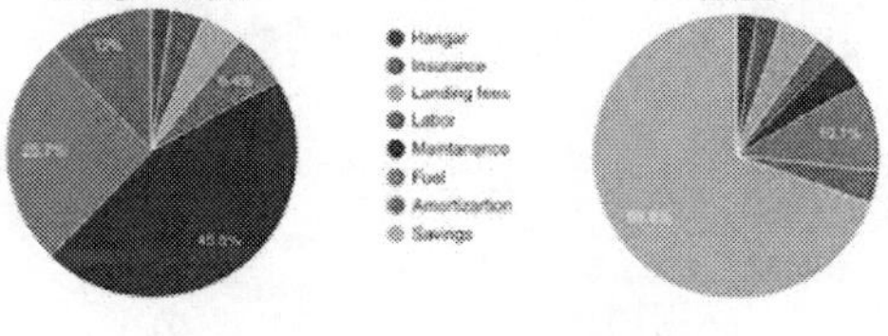

From Porsche 911 Turbo S ...

... to Honda Civic EX

9

Key Features

Smaller size reduces point-to-point travel times, makes the vehicle quieter and enables low-cost composite manufacturing.

The smallest feasible vehicle is a single-passenger transport, and that's good enough — 75% of commuters are single drivers.

Single occupancy requires a high-degree of automation.

Fully self-piloted operation reduces operating cost and enables package delivery and remote sensing.

9

10

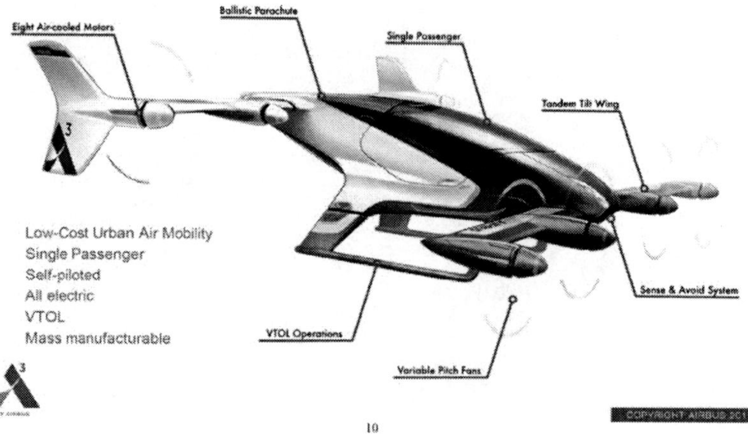

Low-Cost Urban Air Mobility
Single Passenger
Self-piloted
All electric
VTOL
Mass manufacturable

10

11

Dimensions

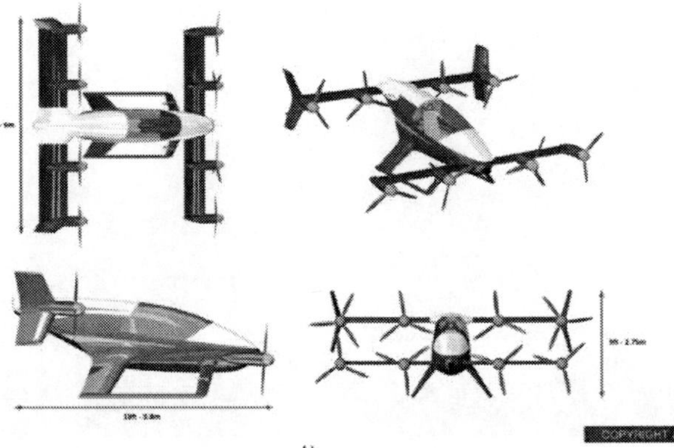

11

12

Accomplishments

Alpha Objectives

- Demonstrate flightworthiness of full scale electric VTOL aircraft concept
- Demonstrate reliable autonomy with sense-and-avoid capability

Accomplishments

- Flown 3 subscale aircraft
- Met with FAA regulators
- Neural network vision system trained
- Tested full scale structural components

12

13

13

14

Timeframe

Each phase (Alpha and Beta) lasts two years

First flights in Q4 of 2017

Productizable demonstrator by 2020

All major systems are in development

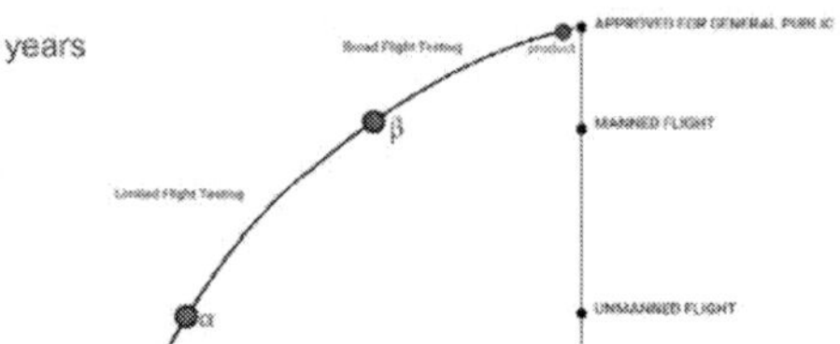

14

AIRBUS : setting the standards [15]

Exceptionally low operating cost

Swappable batteries for quick turn around times

Autonomous platform to improve safety and scalability

Robust vehicle designed to tolerate the harshest of conditions — from extreme temperatures to salt presence

Over forty years of expertise in certification and safety

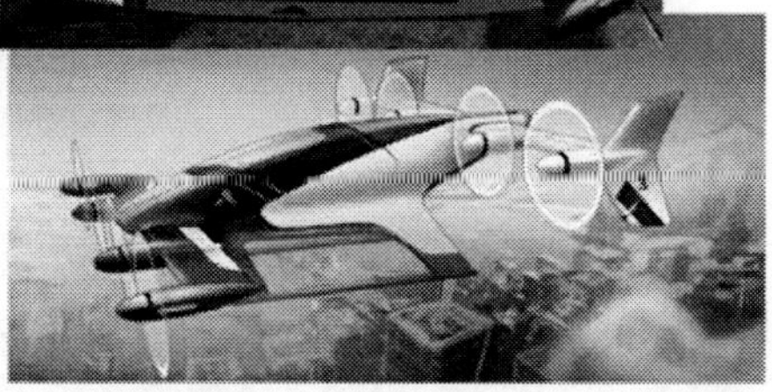

15

Bell Helicopter
A Textron Company
Scott Drennan
Director, Innovation

3

3

4

4

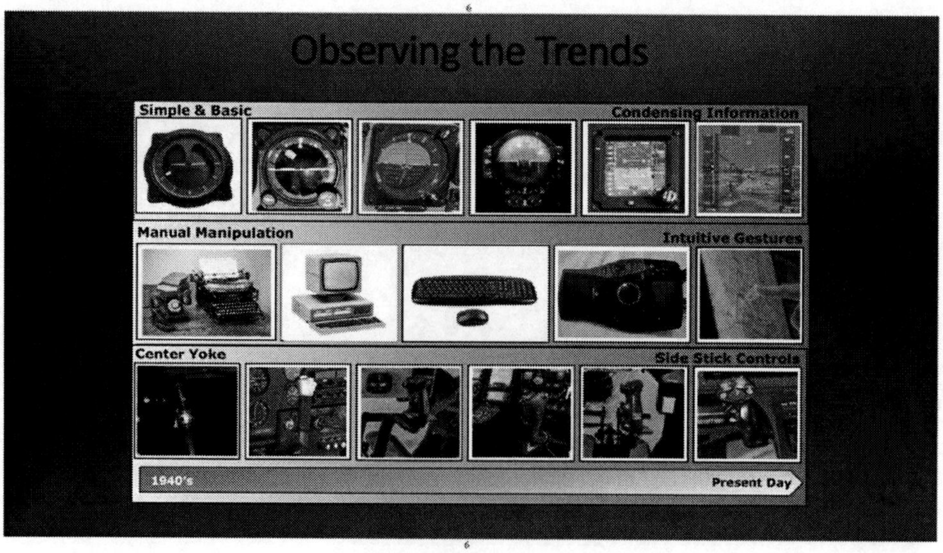
Observing the Trends
Simple & Basic
Condensing Information
Manual Manipulation
Intuitive Gestures
Center Yoke
Side Stick Controls
1940's
Present Day

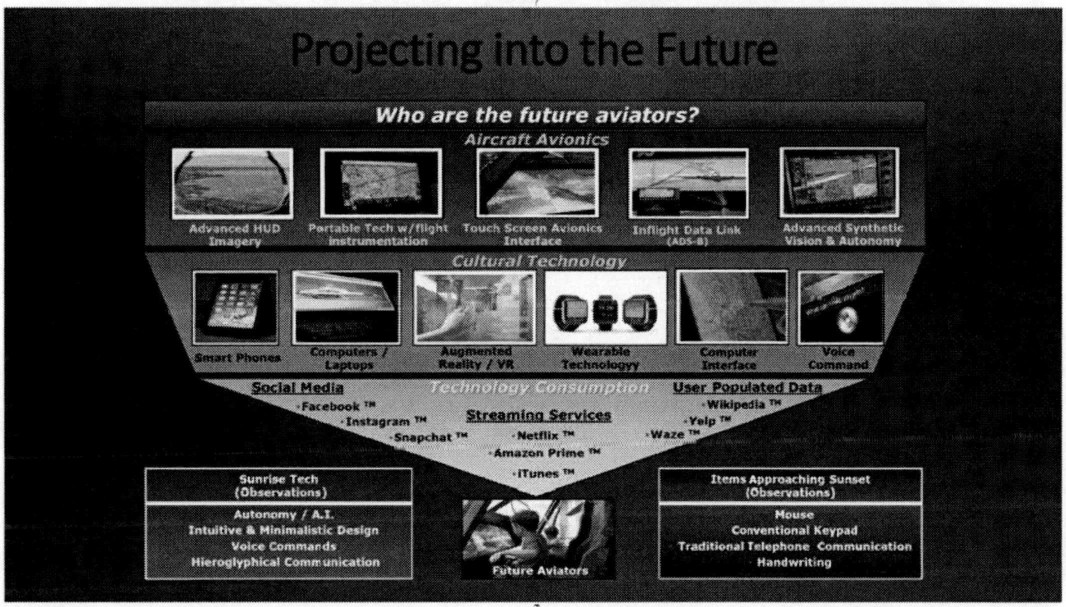
Projecting into the Future
Who are the future aviators?
Aircraft Avionics
Advanced HUD Imagery
Portable Tech w/flight instrumentation
Touch Screen Avionics Interface
Inflight Data Link (ADS-B)
Advanced Synthetic Vision & Autonomy
Cultural Technology
Smart Phones
Computers / Laptops
Augmented Reality / VR
Wearable Technologyy
Computer Interface
Voice Command
Technology Consumption
Social Media
·Facebook ™
·Instagram ™
·Snapchat ™
Streaming Services
·Netflix ™
·Amazon Prime ™
·iTunes ™
User Populated Data
·Wikipedia ™
·Yelp ™
·Waze ™
Sunrise Tech (Observations)
Autonomy / A.I.
Intuitive & Minimalistic Design
Voice Commands
Hieroglyphical Communication
Future Aviators
Items Approaching Sunset (Observations)
Mouse
Conventional Keypad
Traditional Telephone Communication
Handwriting

1

1

2

Video Presentation in
Original e-media

2

Current State of VTOL and other Aircraft

- **Environmental issues are increasingly important in aviation with small improvements being implemented**
- **Basic Aircraft Design and Technology have changed little in decades**
- **Customer needs and expectations are not optimized by legacy technology and currently available products**
- **Performance improvements have been small, but purchase and operating costs continue to rise**

Until Now…

XTI AIRCRAFT

3

3

XTI TriFan 600 Hybrid Electric VTOL

Value, Performance, Technology

Low Purchase Price	Low Operating Costs	High Performance	Green Tech

4 Pillars of TriFan 600

XTI AIRCRAFT

4

4

5

Where We Are Now

- Quality, proven top-level management team in place with detailed plans for successful execution
- Millions invested in concept development and engineering
  - Multiple Patents Approved
  - Transformative Hybrid VTOL Aircraft Performance Predictions
- Financial requirements complete and Multi-faceted capital campaign underway
- Pre-launch marketing campaign
  - Market analysis complete
  - Market attention and positive media
  - Initial Customer reaction strong
- Supplier / partner / vendor conversations underway

XTI AIRCRAFT

5

6

Where We Are Going

- The value and performance of the TriFan Aircraft will be recognized and a solid market will developXTI
- Beyond Executive and Commercial Transport, Medevac and offshore oil and gas, utility – Other potential uses go beyond imagination
  - Operators will invent new ways to use the aircraft
  - Business uses will expand
- Creation of a high value company

XTI AIRCRAFT

6

7

Advances in technologies overcome historical challenges
Result is greater performance and increased safety

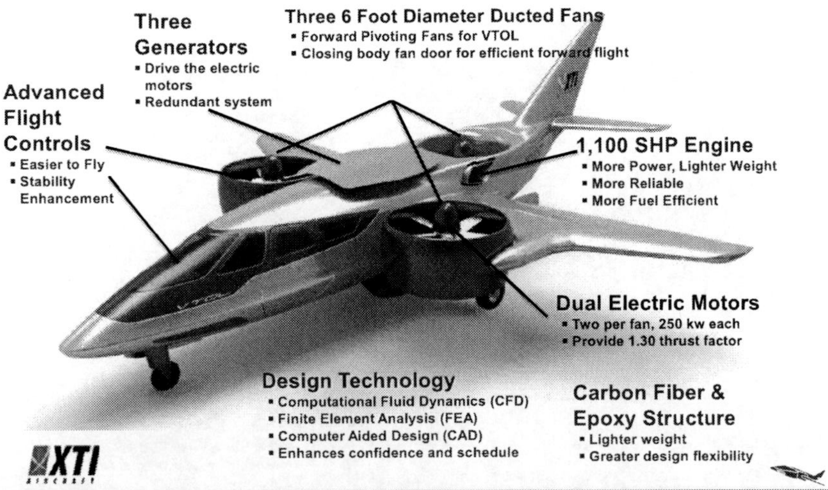

7

8

Impressive Performance

Weights:	
Max Vertical Take-off:	2,404 Kg / 5,300 lbs
Max Short Take-off:	2,858 Kg / 6,300 lbs
Empty Weight:	1,588 Kg / 3,500 lbs
Useful Load:	
Passengers	5 plus 1 Pilot
Vertical Take-off:	816 Kg / 1,800 lbs
Short Take-off:	1,270 Kg / 2,800 lbs
Fuel:	473 Liters / 125 Gallons
Max Cruise Speed:	555 kph / 300 knots
Range:	(@ 300 knots cruise)
Max payload/vertical take-off:	1,060 Km / 660 miles
Pilot only/vertical take-off:	1,240 Km / 770 miles
CTOL Max Range	2,222 Km / 1,200 miles
Performance Features:	
Max cruise altitude:	8,840 Meters / 29,000 ft
Achieve max cruise altitude:	11 minutes

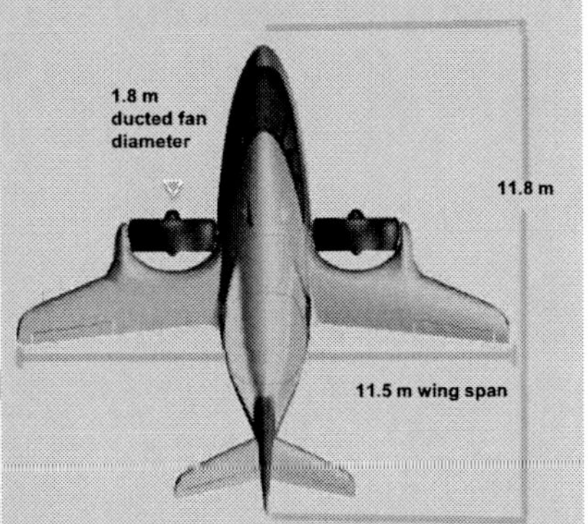

8

9

Executive Cabin Comforts

Interior Cabin Features:
- Interior Cabin Width 70 inches
- Seats 4 passengers
- Baggage 200 lbs
- Cabin Pressure 6.4 PSI
- 68 dbi at cruise altitude

XTI AIRCRAFT

9

9

10

Market Analysis

XTI AIRCRAFT

10

10

11

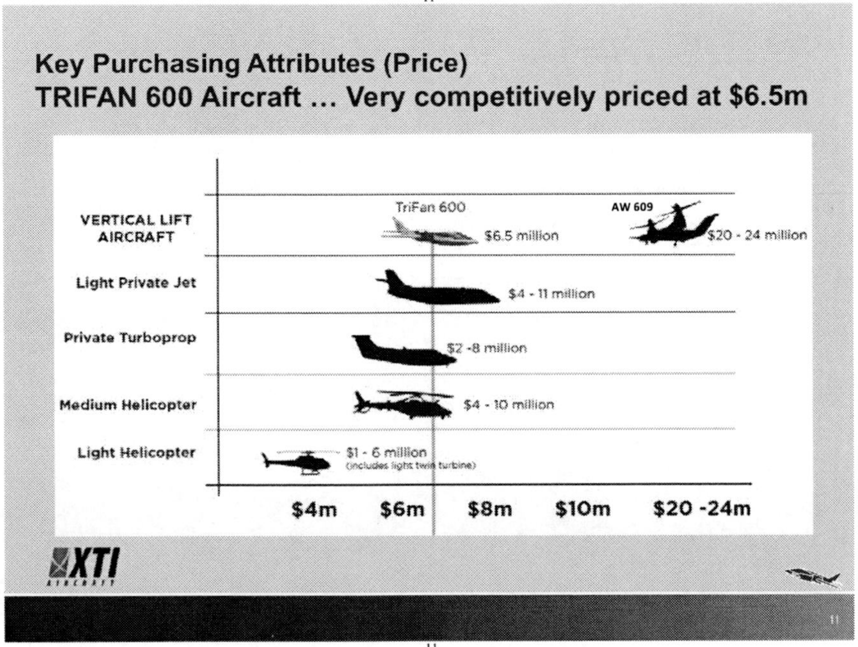
Key Purchasing Attributes (Price)
TRIFAN 600 Aircraft ... Very competitively priced at $6.5m
VERTICAL LIFT AIRCRAFT
TriFan 600
$6.5 million
AW 609
$20 - 24 million
Light Private Jet
$4 - 11 million
Private Turboprop
$2 -8 million
Medium Helicopter
$4 - 10 million
Light Helicopter
$1 - 6 million
(includes light twin turbine)
$4m
$6m
$8m
$10m
$20 -24m
XTI
AIRCRAFT
11

11

12

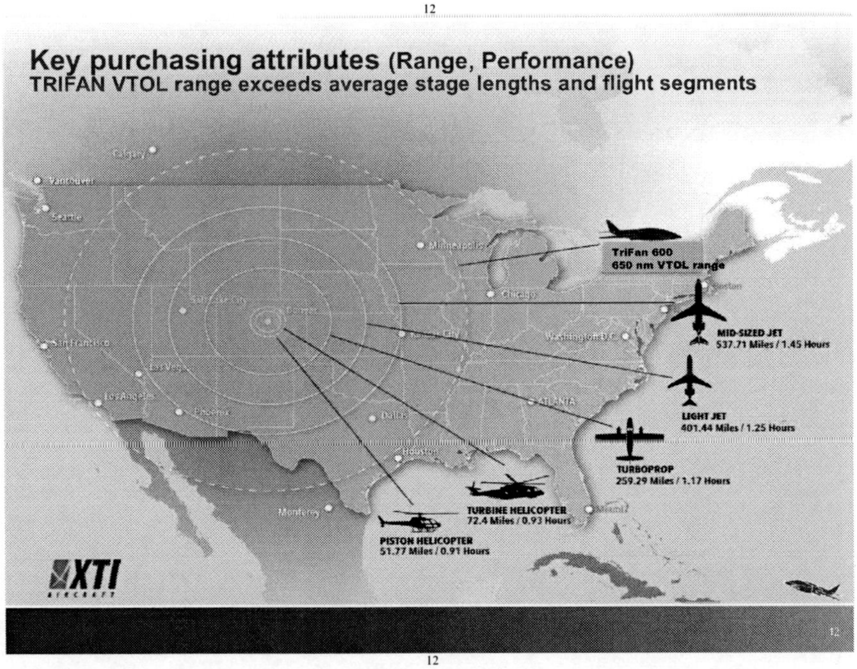
Key purchasing attributes (Range, Performance)
TRIFAN VTOL range exceeds average stage lengths and flight segments
TriFan 600
650 nm VTOL range
MID-SIZED JET
537.71 Miles / 1.45 Hours
LIGHT JET
401.44 Miles / 1.25 Hours
TURBOPROP
259.29 Miles / 1.17 Hours
TURBINE HELICOPTER
72.4 Miles / 0.93 Hours
PISTON HELICOPTER
51.77 Miles / 0.91 Hours
XTI
AIRCRAFT
12

12

13

Commercial Market Demand Summary

XTI TRIFAN 600 Target Customer

- **Airplane and helicopter owners with ability to spend $6-8M on aviation travel solution**
- **Executives of leading companies, High Net Worth individuals**
- **Emergency Medical Services transport providers**
- **Range "sweet spot" of between 250-825 miles**
- **Values time efficiency and convenience**

Market Size

XTI estimates the TRIFAN 600 commercial market size is 40-100 / year

Based on Teal Reports, Grace Dickerson's professional analysis, and management's experience

Only sales to civilian users considered; does not include military or government sales

XTI AIRCRAFT

13

13

14

14

Colin Theodore, NASA

Associate Project Manager, **Revolutionary Vertical Lift Technology Project (RVLT)**

Principal Investigator, **Design Environment for Novel Vertical Lift Vehicles (DELIVER)**

- **Outline of Briefing:**
  - DELIVER overview / objectives
  - Small UAS noise
  - RVLT design and optimization toolsets
  - Urban eVTOL vehicle noise
  - NASA role / key technology areas

0

DELIVER Vision

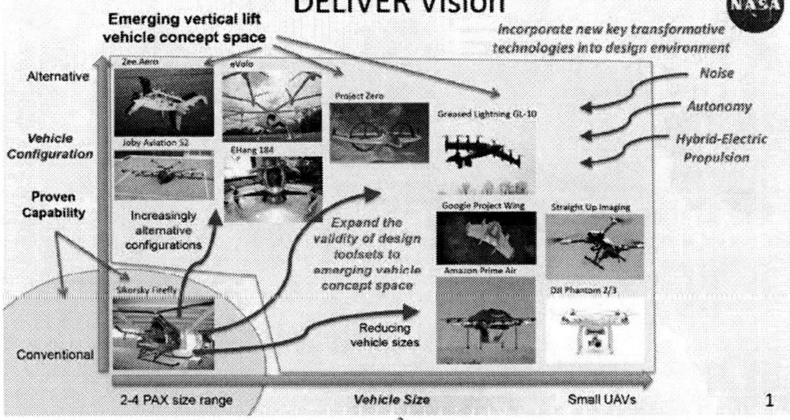

1

3

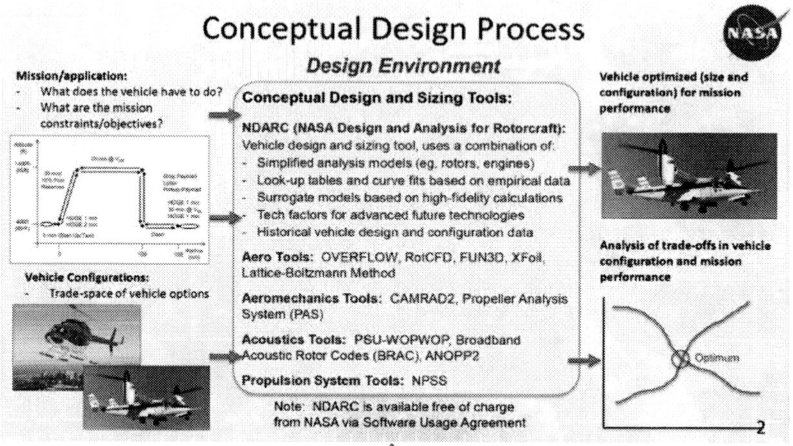
Conceptual Design Process
Design Environment
Mission/application:
- What does the vehicle have to do?
- What are the mission constraints/objectives?
Vehicle Configurations:
- Trade-space of vehicle options
Conceptual Design and Sizing Tools:
NDARC (NASA Design and Analysis for Rotorcraft):
Vehicle design and sizing tool, uses a combination of:
- Simplified analysis models (eg. rotors, engines)
- Look-up tables and curve fits based on empirical data
- Surrogate models based on high-fidelity calculations
- Tech factors for advanced future technologies
- Historical vehicle design and configuration data
Aero Tools: OVERFLOW, RotCFD, FUN3D, XFoil, Lattice-Boltzmann Method
Aeromechanics Tools: CAMRAD2, Propeller Analysis System (PAS)
Acoustics Tools: PSU-WOPWOP, Broadband Acoustic Rotor Codes (BRAC), ANOPP2
Propulsion System Tools: NPSS
Note: NDARC is available free of charge from NASA via Software Usage Agreement
Vehicle optimized (size and configuration) for mission performance
Analysis of trade-offs in vehicle configuration and mission performance
Optimum
2

3

4

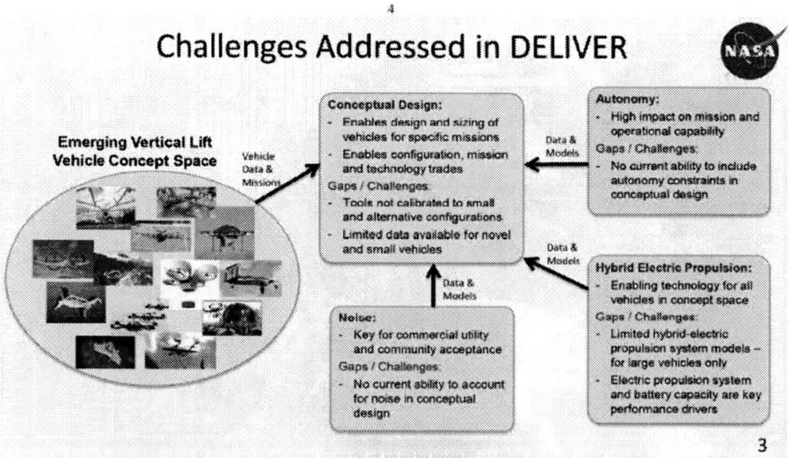
Challenges Addressed in DELIVER
Emerging Vertical Lift Vehicle Concept Space
Vehicle Data & Missions
Conceptual Design:
- Enables design and sizing of vehicles for specific missions
- Enables configuration, mission and technology trades
Gaps / Challenges:
- Tools not calibrated to small and alternative configurations
- Limited data available for novel and small vehicles
Data & Models
Autonomy:
- High impact on mission and operational capability
Gaps / Challenges:
- No current ability to include autonomy constraints in conceptual design
Data & Models
Hybrid Electric Propulsion:
- Enabling technology for all vehicles in concept space
Gaps / Challenges:
- Limited hybrid-electric propulsion system models – for large vehicles only
- Electric propulsion system and battery capacity are key performance drivers
Data & Models
Noise:
- Key for commercial utility and community acceptance
Gaps / Challenges:
- No current ability to account for noise in conceptual design
3

4

5

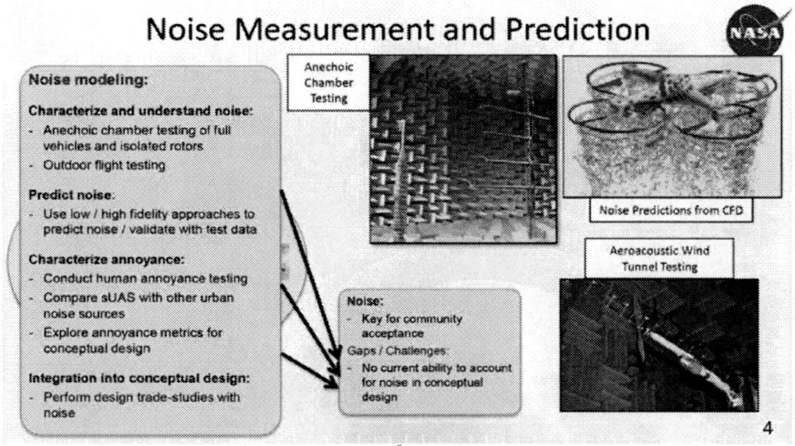
Noise Measurement and Prediction
NASA
Noise modeling:
Characterize and understand noise:
- Anechoic chamber testing of full vehicles and isolated rotors
- Outdoor flight testing
Predict noise:
- Use low / high fidelity approaches to predict noise / validate with test data
Characterize annoyance:
- Conduct human annoyance testing
- Compare sUAS with other urban noise sources
- Explore annoyance metrics for conceptual design
Integration into conceptual design:
- Perform design trade-studies with noise
Anechoic Chamber Testing
Noise Predictions from CFD
Aeroacoustic Wind Tunnel Testing
Noise:
- Key for community acceptance
Gaps / Challenges:
- No current ability to account for noise in conceptual design
4

5

6

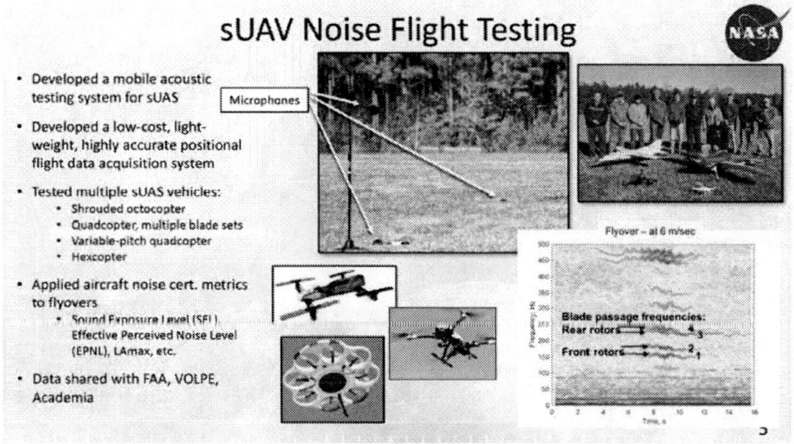
sUAV Noise Flight Testing
NASA
• Developed a mobile acoustic testing system for sUAS
• Developed a low-cost, light-weight, highly accurate positional flight data acquisition system
• Tested multiple sUAS vehicles:
  • Shrouded octocopter
  • Quadcopter, multiple blade sets
  • Variable-pitch quadcopter
  • Hexcopter
• Applied aircraft noise cert. metrics to flyovers
  • Sound Exposure Level (SEL), Effective Perceived Noise Level (EPNL), LAmax, etc.
• Data shared with FAA, VOLPE, Academia
Microphones
Flyover – at 6 m/sec
Blade passage frequencies:
Rear rotors
Front rotors

6

7

Interactional Aerodynamic Noise

- Determine contributions of rotor and rotor-airframe interactions to radiated noise for simple vehicle configurations
- Rotor and airframe support test stand
  - Physically separate from one another
  - Able to vary rotor tip separation distance (Δ)
  - Airframes of constant and variable cross-section (cones) considered
- Results show
  - Small rotor to airframe clearances generate many harmonics of noise
  - Placing rotor ½ rotor radius below airframe, noise is essentially the same as an isolated rotor case

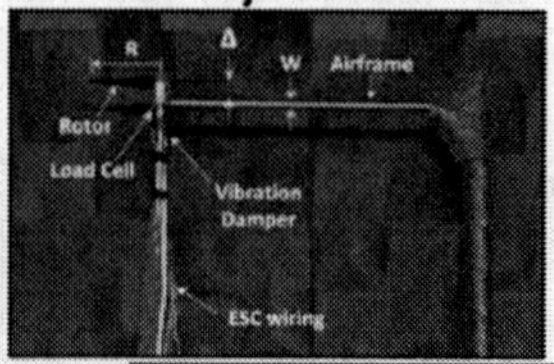

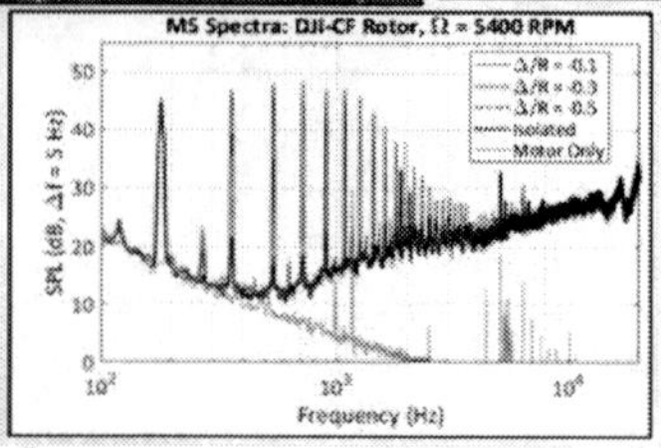

6

7

8

Initial Assessment of Annoyance

Key Objectives:

- Evaluate the level of annoyance endemic to acoustics for multi-copters as compared with other vehicles that make noise
- Determine what characteristics are driving the annoyance for multi-copters, ie., frequency range, multiple frequencies, time variations, speed
- Conduct psychoacoustics experiments in the Exterior Effects Room (EER) at NASA Langley

Analysis so far:

- Significant offset between the groups of recorded road vehicles and sUAS used in this study
- The implication: the common belief that ***'as long as a sUAS is no louder than a delivery truck, operators will not have to worry re: noise for package delivery'*** may not hold true.

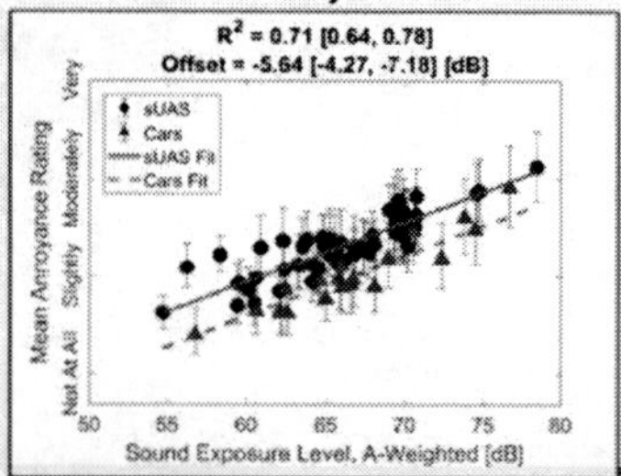

7

8

9

NASA Request for Information (RFI) - Released

- ***Posted March 27, 2017 on FedBizOps*** **www.fbo.gov**
- ***Solicitation #:*** *RFI-VTOL-UAS-2017*
- ***Responses due May 26, 2017***
- ***RFI: Designing Next Generation Vertical Take-Off and Landing (VTOL) Unmanned Aerial Aircraft Systems (UAS)***
  - "The objective of this Request for Information (RFI) is to gather input to inform future NASA Aeronautics Research Mission Directorate (ARMD) research in Vertical Take-off and Landing (VTOL) Unmanned Aircraft Systems (UAS) technology, and to identify major technical barriers to increased use of 5- to 1000-lb VTOL UAS for civil missions."
  - This RFI is specifically targeted at vehicle technologies to enable a new generation of VTOL UAS that are low-noise, low-emission, and significantly more durable, efficient and reliable than current configurations.
  - Responses to this RFI will help to inform the planning, advocacy, prioritization, and execution of research at NASA related to VTOL UAS in the next five years.

9

9

10

RVLT Technical Challenge - MDAO Design Process

Demonstration of an MDAO Process for Vertical Lift Vehicles

Planned Schedule: FY16-18

Objectives:

- Integrate new discipline tools into a design optimization framework
- Demonstrate new MDAO process to generate low-noise, low-emission VTOL conceptual designs

Technical Areas and Approaches:

- Develop integrated discipline modules that operate within a streamlined optimization framework
- Create conceptual designs of VTOL aircraft that meet emission and external noise metrics

Vehicle A: 6 person, 400 nm range

Baseline: Single Main Rotor Helicopter

Configurations Evaluated for Low Emission Design
Coaxial: conventional gas turbine, compressed H2 fuel cell, Li-Ion battery
6-rotor tiltrotor: Li-ion battery, Turbo-electric
2-rotor tiltrotor: variable-speed gas turbine

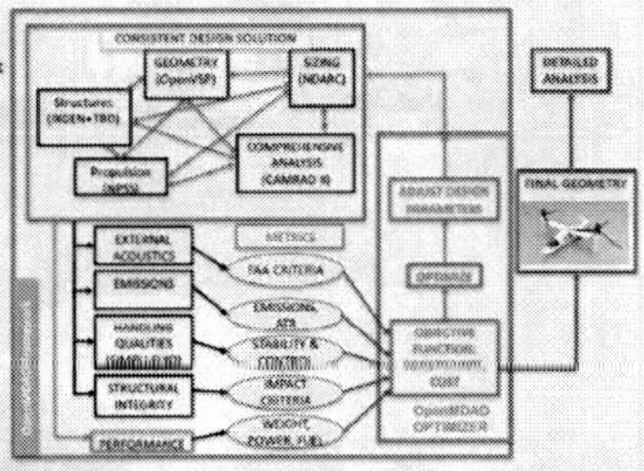

9

10

11

Sample Vehicles to Demonstrate Design Process

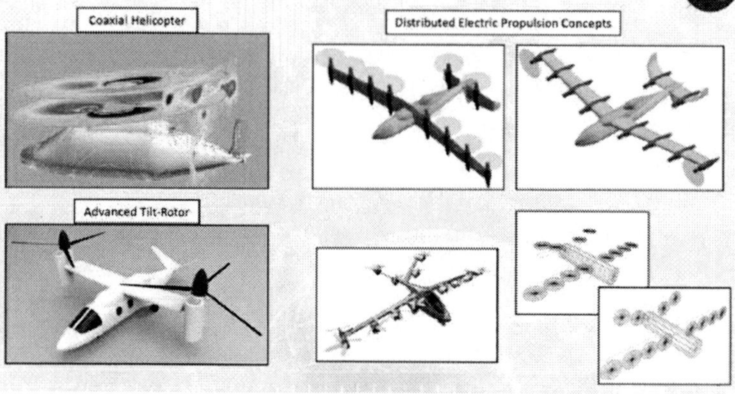

11

12

Acoustic Modeling of Urban eVTOL Systems

Background:

- Human response to aircraft community noise is a complex perception phenomenon that is a function of both acoustic and non-acoustic factors
- Noise goals usually specified in terms of certification metrics, which may not fully reflect acoustic factors related to human response, nor are intended to reflect non-acoustic factors

Approach:

- Develop some notional vehicle designs similar to those being proposed for Urban eVTOL missions
- Use current validated aeroacoustics tools and methods to develop an simplified auralization of these notional vehicles
- Develop scenario to assess noise levels and annoyance of different vehicle designs

NASA eVTOL Multi-Rotor
- 2 passenger, MTOW: 450 kg
- 18 props, 1.8m diameter
- 2-bladed props
- 160 m/s tip speed

NASA eVTOL Tilt-Rotor Concept
- 4 passenger, MTOW: 1250 kg
- 6 props on front wing
- 2 props on tail
- 4-blades props
- 140 m/s tip speed

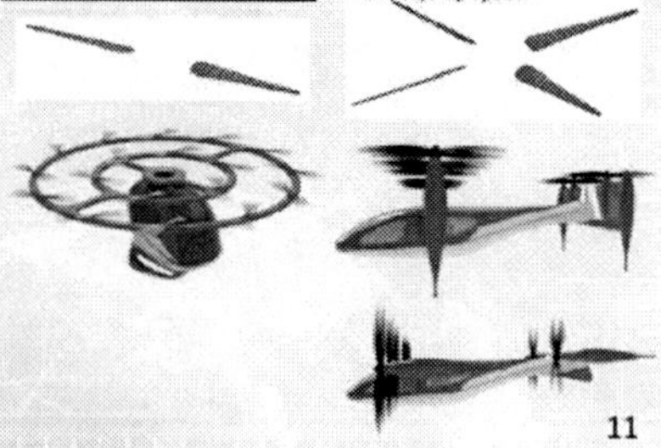

11

12

13

Vertical Take-Off Scenario – Tilt-Wing Vehicle

Acoustic Modeling

- Zero angle of attack
- Rotors RPM – 2% variation (DF)
- Simple model for Rotor Structure Interaction (RSI)

Trajectory

- 10s hold at ground, 2.5 m/s rise to 75m
- Sideline observer at 30m

Next Steps:

- Assess noise and annoyance metrics for use in multi-disciplinary design and analysis tool that include performance, safety, energy consumption, as well as noise

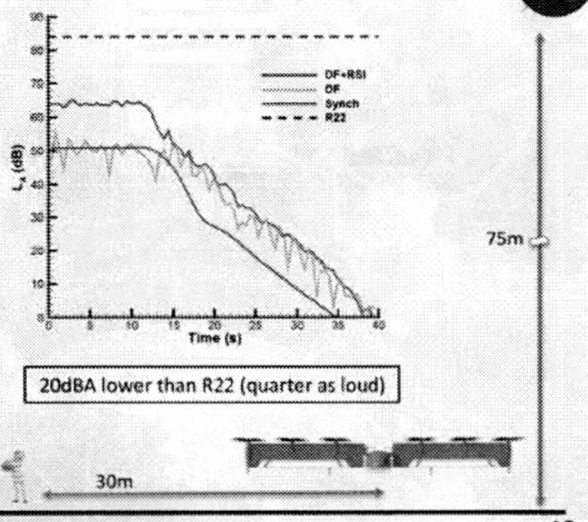

13

13

14

The role of NASA Aeronautics is generally acknowledged to be the national focal point for supporting and supplying long-term research that bolsters the aviation industry and invests in its future.

- *As new configurations and systems are proposed and developed, do we understand enough at the component and the system integration level to model the concepts and conduct meaningful trade studies? What pieces are missing?*
- *From the perspective of a research organization, can we move fast enough and provide value to the community in research results and tools?*
- *What is the best approach for NASA to engage with the rapidly growing number of companies developing advanced VTOL vehicles to encourage, promote, advance technology development to best enable these vehicles? In-house demonstrations? Performing collaborative research? Sponsor technical community gatherings? Performing independent evaluations? Provide data to regulators to assist in developing certification standards? Sponsor Grand Challenges?*
- *What is the best way to collaborate with other Government agencies such as the FAA and DoD to leverage investments?*

14

14

Technology Areas to Address

- *Conceptual design and analysis tools that are applicable to emerging vehicle and propulsion system architectures.*
- *Vehicle noise and footprint on the ground.*
- *Community acceptance, particularly safety and annoyance.*
- *Life cycle carbon footprint, including manufacture, operation and disposal.*
- *Weather-tolerant operations.*
- *Move to autonomous, self-aware vehicles.*
- *Certification and regulation – electric propulsion, autonomous operations.*

15

NASA ON-DEMAND MOBILITY (ODM) OVERVIEW

Transformative Vertical Flight for Urban Mobility Panel

AHS Forum, 11 May 2017

Ken Goodrich, ODM Lead, Safety and Ease of Use, NASA Langley Research Center

ODM Takeaways

Potential, small passenger (or equivalent cargo) aircraft and missions enabled by technology convergence

– "On-demand" access to aviation similar to cars

Exists today in niches, but generally expensive, risky, and adverse community impacts

Can't "scale" meaningfully in current system

NASA conducting & supporting initial, exploratory research on feasibility, scalability, desirability and system concepts – need to focus investments in areas that best leverage industry and FAA. Need your continued feedback

Outline

Overview of what "ODM" means to NASA

Review of NASA-FAA ODM and Emerging Aviation Technologies Workshops

Recommendations to NASA Aero by NASA multi-center ODM planning team

What is Aviation-Based, On-Demand Mobility (ODM)?

- *Transportation* where users have access to immediate and flexible air travel
  - Users dictate trip origin, destination, and timing
  - 1-9 passengers or equivalent payload
- ~2-4x faster than cars & hub-spoke over ~20-500 mile range

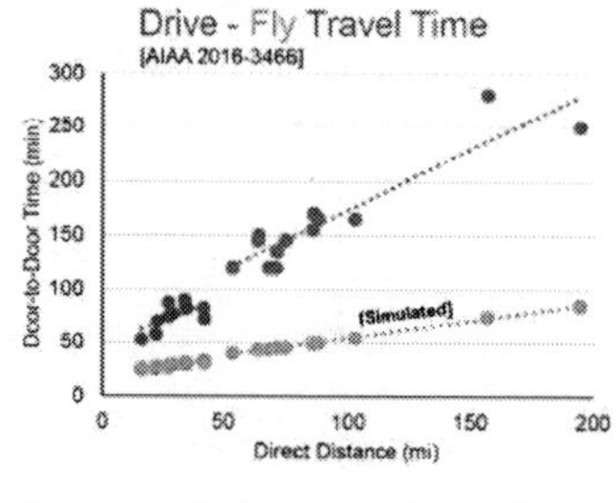

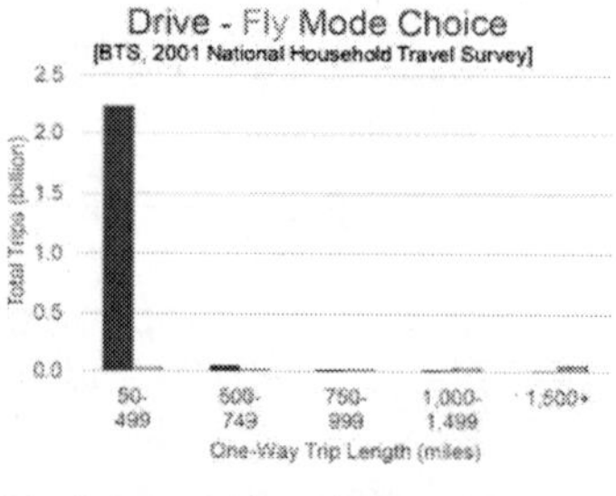

Cars operate **12x more trips of 50+ miles** and **>500x more vehicle** miles than air travel
ODM can bring speed of aviation into ***daily*** lives.

5

What is Aviation-Based, On-Demand Mobility (ODM)?

- *Transportation* where users have access to immediate and flexible air travel
  - Users dictate trip origin, destination, and timing
  - 1-9 passengers or equivalent payload
- ~2-4x faster than cars & hub-spoke over ~20-500 mile range
- ODM is a range of missions, aircraft types, operations enable by integration of Distributed Electric Propulsion (DEP), Vehicle Integration, Vehicle Autonomy, Advanced NAS Operations and Distribution Infrastructure
  - Enables trips that were not time/cost effective with current transport (e.g., CTOL Commuter)
  - Potential to mitigate surface congestion by creating an alternative to car travel (e.g., VTOL Air-Taxi)

6

Why Now? Technology Convergence

Autonomy

- Increase safety (e.g. human error mitigation/avoidance)
- Simplified vehicle operation
- Network connectivity
- High-density airspace operations
- Increase user base 100x

Distributed Electric Propulsion

- Scale-free Propulsion
- Highly Redundant and Reliable
- High power/weight, Quiet
- Robust Operations
- High cruise efficiency concepts

ODM Convergence Drivers

- NASA Strategic Implementation Plan's "Mega-Drivers"
- Industry investment
- FAA receptiveness & potential to improve safety
- Automotive synergy
- Public acceptance

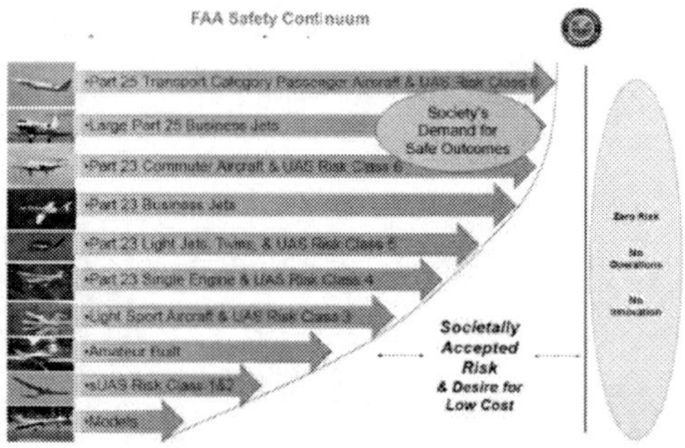

Outline

Overview of "ODM"

Review of NASA-FAA ODM and Emerging Aviation Technologies Workshops and supporting studies

Recommendations to NASA Aero by NASA multi-center ODM planning team

9

Overview of ODM Workshops

- Joint FAA sponsorship (Small Airplane Directorate, ACE-100)
- 5 workshops from July 2015 through September 2016
- Augmented by ARMD-funded ODM studies and tasks
- Goals:
  - Identify potential and necessary investments underpinning ODM and advanced small aircraft
  - Develop a means of assessing/prioritizing investments
  - Identify community of interest

Primary outcomes:

- Community including ~120 worldwide, engaged organizations
- Barriers, metrics, and reference missions
- Systems engineering framework
- Community technology roadmaps from 4 working groups
- Presentation archive: http://www.nianet.org/ODM/roadmap.htm

9

10

Workshop Attendees, Subset

Government
- NASA (all 4 aero centers, HQ), FAA, DARPA, ONR, US Army CASCOM, Combined Arms Support Command, US Army Research Laboratory

Academia
- MIT, Georgia Tech, Embry-Riddle, Purdue, Virginia Tech, NIA, Duke, UIUC, Texas, ODU, WPI, Kansas, Cincinnati, Hampton, others

International
- EASA, JAXA, KARI, ONERA, Clean Sky JU, Flight Research Laboratory

Professional/Trade Organizations
- GAMA, AOPA, AIAA, AHS, SAE International, CAFE Foundation

Operators
- Uber, Cape Air, ImagineAir, Surf Air

Traditional Aircraft OEMs
- Boeing, Airbus Group, Textron (Cessna, Bell Helicopter), Mooney, Piper, Cirrus

Emerging Aircraft OEMs
- A^3 (Airbus), Joby Aviation, Zee.Aero, Aurora Flight Sciences, Terrafugia, Toyota, Tzunum, Elytron Aircraft, Lilium GmbH, E-Volo GmbH, Aero Electric Aircraft Corp, Eviation, Wright Electric, others

Aircraft Subsystems and Avionics
- Garmin, GE Aviation, Rolls-Royce, Continental Motors, Avidyne, Honeywell, BRS, Lycoming Engines, others

Venture Capitalists and Prizes
- Capricorn Investment, Yarbrough Capital, The Gofly Prize, XPRIZE

Consultants
- Josephson Engineering, Gigajoule Jug Consulting, Holmes Consulting, Ellis Industrial Design

~120 organizations participated

10

11

Prioritized Barriers to ODM

Ease of Certification	Affordability	Safety	Ease of Use	Door-to-Door Trip Speed
Average Trip Delay	Community Noise	Ride Quality	Efficiency	Lifecycle Emissions

Prioritized order, but any barrier can limit feasibility, utility, growth

12

Metrics to Assess Progress Toward Overcoming Barriers

Ease of Certification Time/Cost Required	**Affordability** Total Operating Cost/Pax Mile	**Safety** Fatal Accidents per Vehicle Mile	**Ease of Use** Required Operator Training Time & Cost	**Door-to-Door Trip Speed** Equivalent Speed
Average Trip Delay Time	**Community Noise** Perceived Annoyance	**Ride Quality** Passenger Comfort Index	**Efficiency** Energy/Pax Mile	**Lifecycle Emissions** Total Emissions /Pax Mile

Barriers & metrics involve far more than direct, vehicle parameters.

Four Working Groups Formed to Identify Potential Technology Investments

Simplified Vehicle Operations (SVO)

- Integration of automation, on-, off-board human resources for safety, ease-of-use, operational performance/capabilities

Electric Propulsion and Configuration Integration

- Electric and Hybrid Battery/Engine Power Systems
- Overall aircraft design and integration

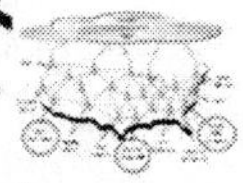

ODM Airspace Integration

- Airspace operations/procedures and air traffic management
- Supporting ATM infrastructure (e.g. secure CNS, FIS)

Manufacturing, Integrated Structures & Community Impact

- Low cost, high quality products at low production volumes
- Lower weight and more capable airframes
- Ultra-low community noise at close proximity
- Failsafe recovery modes
- Icing protection

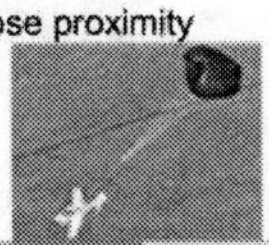

ODM Reference Missions Identified

- Highest relevance to workshop participants
- Provide near- and longer-term opportunities for new technologies, vehicles, operations, & transformational capabilities

CTOL Commuter reference mission

9 pax, ~400 mi, 225+ mph, single-pilot
Moderate industry interest
Tech scale up/down missions:
- 1-4 PAX CTOL air-taxi / advanced GA (X-57)
- 40 PAX regional airliner

VTOL Air-Taxi reference mission

4 pax, ~125 mi, 150+ mph, single-pilot/pilotless
Strong industry interest
Tech scale up/down missions:
- Small public-use UAS (e.g., first responder)
- 9 pax, VTOL commuter

15

ODM System Engineering Framework Identified

Barriers/metrics and reference missions provide a basis for systems engineering framework to

a) guide evaluation of various ODM system concepts and
b) asses potential impacts of alternative technologies
c) manage technology R&D

Systems Engineering Framework

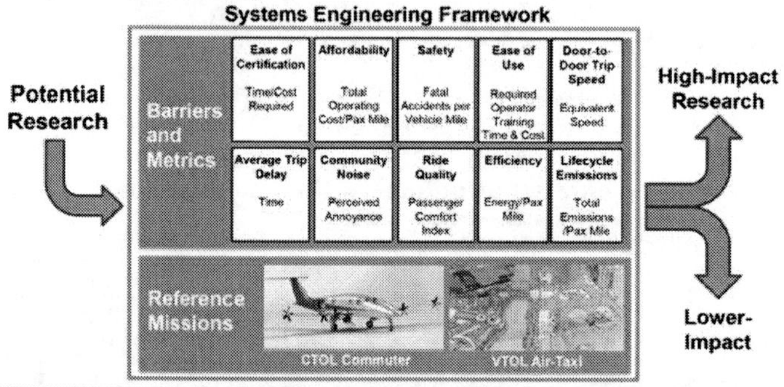

15

16

Technology Working Group Roadmaps to Address Barriers

www.nianet.org/ODM/roadmap.htm

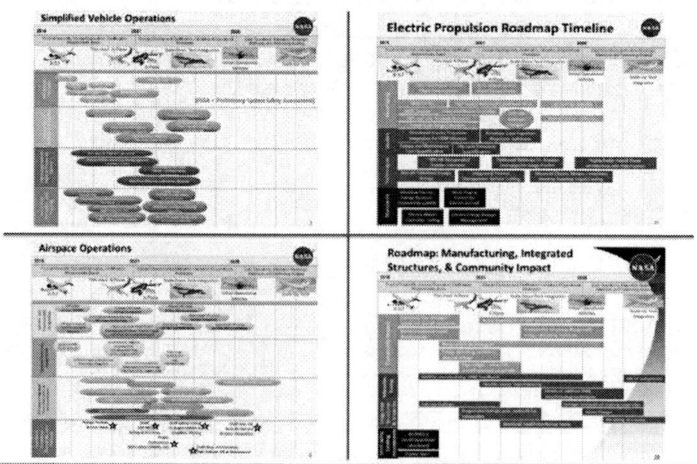

16

17

Outline

Overview of "ODM"

Review of NASA-FAA ODM and Emerging Aviation Technologies Workshops and supporting studies

Recommendations to NASA Aero by NASA multi-center ODM planning team

On-Demand Mobility | 17

17

18

Recommended NASA Role & Early Research: Develop, Validate, & Demonstrate

Primary NASA roles – architect, integrator, technologist, partner

- Develop system concept that's *deployable* & *scalable*
  - *Potential to accommodate ~1 million aircraft peak aircraft aloft nationally vs ~8000 today*
- Establish and build stakeholder partnerships
- Vibrant application space for NASA & external investments
- Integrated consideration of vehicle airworthiness, pilot/operator & automation certification; operations

DEP Enabled Vertiplanes
Safe, quiet, weather-tolerant

Simplified Vehicle Operations
Autonomy and interfaces enabling transition from Experts -> operators -> autonomous
Inner & outer loops; execution & judgment

Urban Aviation System Architecture & NAS Integration
Ground and flight guidelines & procedures for high-capacity, urban aviation

On-Demand Mobility 18

18

19

GAMA PROPOSED DECONSTRUCTION BASED ON FAA PILOT TRAINING AND TESTING AREAS (FAR PART 61)

GAMA

Preflight Preparation | Preflight Procedures | Airport Operations | Takeoff, Landing & Go-Arounds | Performance Maneuvers

Slow Flight & Stalls | Basic Instrument Maneuvers | Navigation | Emergency Operations | Postflight Procedures

Multiengine Operations | Night Operations

EACH AREA HAS:

- Overall objective
- Tasks and task elements
  - Skills
  - Knowledge
  - Risk Management
- Conditions & standards of acceptable performance

19

20

GAMA PROPOSED DECONSTRUCTION BASED ON FAA PILOT TRAINING AND TESTING AREAS (FAR PART 61)

GAMA

Preflight Preparation | Preflight Procedures | Airport Operations | Takeoff, Landing & Go-Arounds | Performance Maneuvers

Slow Flight & Stalls | Basic Instrument Maneuvers | Navigation | Emergency Operations | Postflight Procedures

Multiengine Operations | Night Operations

DECONSTRUCTED PILOT

Basic Airmanship | Navigation | Communicatio n | Terminal Procedures | Takeoff & Landings

Systems Management | Detect & Avoid | Planning & Decision Making | Emergency Procedures

20

21

GAMA PROPOSED DECONSTRUCTION BASED ON FAA PILOT TRAINING AND TESTING AREAS (FAR PART 61)

GAMA

- Preflight Preparation
- Preflight Procedures
- Airport Operations
- Takeoff, Landing & Go-Arounds
- Performance Maneuvers
- Slow Flight & Stalls
- Basic Instrument Maneuvers
- Navigation
- Emergency Operations
- Postflight Procedures
- Multiengine Operations
- Night Operations

DECONSTRUCTED PILOT (AREAS WHERE AUTONOMY CAN BE MORE CAPABLE)

- Basic Airmanship
- Navigation
- Communicatio n
- Terminal Procedures
- Takeoff & Landings
- Emergency Procedures
- Detect & Avoid
- Planning & Decision Making
- Systems Management

21

22

Vibrant Mix of Traditional and Emerging Aviation Operations

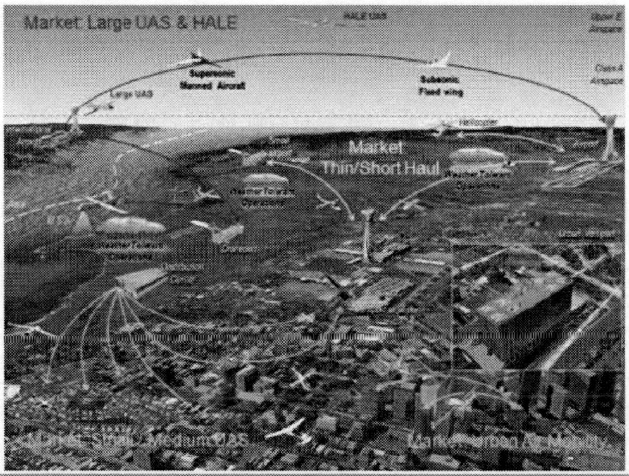

On-Demand Mobility | 22

22

1

Transformation of UAS Technology

Enabling Transformative Transportation Concepts

Presented by: Wes Ryan,
Manager, Programs & Procedures
(Advanced Technology)
FAA Small Airplane Directorate

Date: April 2017

1

2

What is Certification?

- **Acknowledgement that FAA requirements are met for:**
  - Aircraft, Aeronautical Products, Airmen, Mechanics, Controllers, Operators, etc.
- **Gain "Safety Assurance"**
  - Confidence that a product or operation will meet FAA safety expectations to protect the public
  - Our risk-based processes are well-proven

Concept

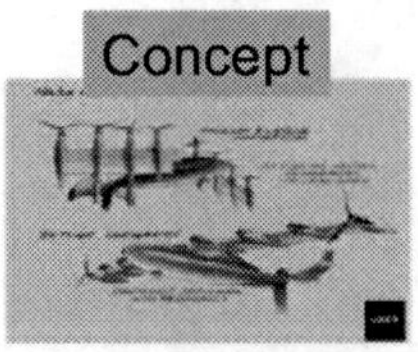

Development

Operation

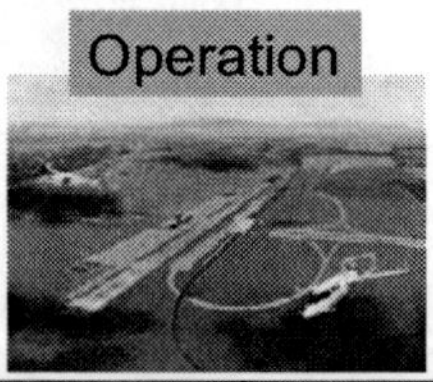

Federal Aviation Administration
www.faa.gov/uas
2

2

3

Safety Assurance

- **Comes from combination of factors**
  - **Airworthiness** – Safe condition for its intended use
  - **Design** – Verify design, engineering, construction, etc. meet applicable requirements in certification basis
  - **Maintenance –** Repair/replace prior to failure
  - **Operations** – Limits sufficient for the acceptable risk
  - **Airspace** – Level of integration, traffic exposure, controller involvement, and equipage
  - **Pilot/Automation** – Trained for aircraft and level of risk
- **Role of Pilot/Automation Key for Safety**
  - Pilot currently provides mitigation – role in the future?

Federal Aviation Administration
www.faa.gov/uas
3

3

4

Transformational Enablers

- **Regulatory improvements for Part 23**
  - Performance-based regulations - top-level safety goals
  - Customizable design/certification requirements through collaborative Means of Compliance
- **Flight path control & automation**
  - Appropriate certification rigor for intended use
- **Electric propulsion/lift systems**
  - Potential safety benefits
- **Operational integration**
  - Equipage for redundancy - emerging sensor fusion

Federal Aviation Administration
www.faa.gov/uas
4

4

5

Flight Path Control & Automation

- **Reduce/replace functions of pilot**
  - **Aviate** – Fly the aircraft – speed, altitude, etc.
  - **Navigate** – Follow intended flight path, respond to air traffic instructions, and avoid hazards.
  - **Communicate** – Interact with ATC primarily verbal
  - **Decision Making** – Managing safe flight & contingency
- **Developing "Assured" Automation**
  - **Bounded Behavior** – Safe laminations of authority and expected outcomes – architecture is key
  - **Fault Tolerant, Fail-Safe** – Continues to function after a failure

Federal Aviation Administration
www.faa.gov/uas 5

5

6

Opportunities for Collaboration

- **UAS safely prototyping technology that will revolutionize flight**
  - Automation & Flight Controls
  - Auto Collision Avoidance
  - Refuse-to-Crash Logic
- **Key to passenger carrying, highly-automated aircraft**
  - Requires early collaboration
  - FAA, NASA, industry, academia, municipalities

Federal Aviation Administration
www.faa.gov/uas 6

6

7

"We're all For Bringing a Silicon Valley Attitude into Aviation, as Long as You Let Us Bring A Little Aviation Attitude into Silicon Valley"

Federal Aviation Administration
www.faa.gov/uas 7

7

8

QUESTIONS?

Federal Aviation Administration
www.faa.gov/uas 8

8

9

Regulatory Improvements

- **Performance-based regulations**
  - Revised Part 23
- **Provide foundation for innovation**
  - Custom methods of compliance
- **Industry standards**
  - Allow for industry to assist in developing requirements
- **Focus on design and operation**
  - Design technology is maturing – batteries, EP systems
  - Need to focus on operational integration – safe automation

Federal Aviation Administration www.faa.gov/uas 9

9

10

Enabling Innovation & Access

Rapid Growth

Open Market & Airspace

Rapidly Evolving Use Cases

Innovative Technology

Low Regulatory Burden

Judicious Integration

Defined CONOP & Mitigations

Standardization

Risk Based Regulations

FAA Safety Goals

Seek Balance to Enable Access

Federal Aviation Administration www.faa.gov/uas 10

10

11

Why Do We Certify?

- **FAA must safely manage the airspace and civil aircraft operations**
  - Title 49 U.S. Code § 40103(a)(1)
- **Manage design/operational risks**
  - Apply resources/rigor based on risk
- **Certification provides "Safety Assurance"**
  - Confidence that a proposed product or action will meet FAA safety expectations to protect the public

11

12

Key Strategic Priorities

Safety: *Enable safe operations of UAS and Transformative Flight Concepts*

Adaptability: *Create an environment where emergent technology can be safely and rapidly introduced into the NAS*

Global Leadership: *Shape global standards and practices for technology through international collaboration*

12

Advances and Applications in Comprehensive Gearbox Aero-Thermodynamics and Tribology Simulation

Sean McIntyre
Research Associate
The Pennsylvania State University
Applied Research Laboratory
University Park, PA, USA

Robert Kunz
Professor
The Pennsylvania State University
Department of Mechanical Engineering
University Park, PA, USA

ABSTRACT

The current status of an ongoing effort to develop a comprehensive gearbox aero-thermodynamics and tribology simulation tool, named PSULOL, applicable to both well-lubricated and loss-of-lubrication operation is presented. PSULOL employs a multi-scale approach, wherein various physical effects including: meshing tribology, convection heat transfer within the system and to the environment, high-frequency thermodynamic effects induced by the gear meshing frequency, and the long-time response of the overall gearbox temperature to a net imbalance of heat generation and transfer to the environment are simulated separately and coupled with one another iteratively through appropriate boundary and initial conditions. As established in 2014, the first-generation version of PSULOL was the first computational fluid dynamics-based (CFD-based) tool coupled with an all-lubrication regime tribology model capable of simulating transient loss-of-lubrication failure of high-speed gearboxes. This first version was built on an in-house research CFD code, NPHASE-PSU, and relied on a number of physical simplifications, particularly with regard to the geometric configuration of the housing, and to the effects of the coupling between multiphase flow within the system (retained lubricant dynamics) and meshing tribology. Here, present efforts underway to increase the physical fidelity and ease-of-use of PSULOL are outlined, and progress toward simulating physically-realistic gearbox configurations is shown. This includes modeling the disperse multiphase flow of lubricant droplets and film within the system, and transitioning from NPHASE-PSU, an in-house research CFD code, to Star-CCM+, a commercial code that is more user-friendly and features additional geometry-handling capabilities.

INTRODUCTION

Over the past five years, an effort has been underway at Penn State to develop a coupled aero-thermodynamics and tribology simulation capability for loss-of-lubrication operation of high-speed gearboxes (Ref. 1-7). This computational framework, called PSULOL, employs a multi-scale simulation approach wherein the salient physical effects and their associated space and time scales were identified, and are simulated separately (Ref. 1, 2, 7). The important physical effects identified are: gear tooth meshing tribology, high-frequency aero-thermodynamic effects induced by the gear meshing frequency and rotation rate, multiphase convection within and outside the gearbox, and the long-time response of the temperature distribution in the gearbox to net imbalance of heat generation and removal.

As it was not known *a priori* exactly what numerical simulation capabilities would be required by PSULOL, the first-generation version was implemented in NPHASE-PSU, an in-house finite volume CFD code with multiphase flow and conjugate heat transfer capabilities (Ref. 8). Use of an in-house code was beneficial for the early efforts, allowing uninhibited access to modify and add to the functionality of the numerical method where necessary.

A number of physical and geometric simplifications were made in developing the first-generation version of PSULOL. To reduce the computational cost of the initial tests, the flow between the gears and shrouding within gearboxes was modeled as cyclic-symmetric, so that the CFD simulation of the convection process within the gearbox could consider only a single tooth sector. NPHASE-PSU requires face-matched grids in the fluid and solid domains for conjugate heat transfer simulation. As a result, the housing had to be modeled as a cylinder with end plates with mass and external surface area approximately matching that of the real gearbox housing to conform to the cyclic-symmetric treatment of the internal convection process (Ref. 1, 2, 7).

Presented at the AHS International 73rd Annual Forum & Technology Display, Fort Worth, Texas, USA, May 9-11, 2017.

Initially, PSULOL was applied to simulation of loss-of-lubrication experiments in the 3.5" pitch diameter spur gear rig at NASA Glenn Research Center (Ref. 9, 10, 11). The decision to model spur gears, rather than gears with more complicated tooth shapes such as spiral bevel and helical allowed the initial all-lubrication-regime tribology model to consider line contact, rather than elliptical patch contact (Ref. 2-6).

Physically, the physics of gear meshing tribology and the multiphase aero-thermodynamics of a gearbox are coupled with one another through the temperature and lubricant film thickness on the gear flanks as they mesh, and through the resulting meshing frictional heat flux. The tooth temperature affects the lubricant viscosity, and more importantly under loss-of-lubrication operation, determines the survival or failure of the molecular-scale non-metallic surfaces layers that form on the gear teeth and prevent scuffing. To date, the all-lubrication-regime tribology model accepts the incoming tooth temperature as an input from the aero-thermodynamics simulation of the gearbox and returns the heat generation by meshing friction, but the dependence of the friction on the incoming lubricant film thickness is modeled empirically entirely within the tribology model.

Here, we describe efforts currently underway to enhance the predictive capability of PSULOL by improving the physical fidelity of the underlying simulations and models. The main goals of the effort described here are to:

1. Enable simulation with a geometrically-representative housing configuration,
2. Simulate, rather than model, the multiphase flow within the gearbox and its coupling with the gear meshing tribology.

A detailed description of the aero-thermodynamics simulation approach for the first-generation PSULOL framework, as well as a thorough discussion of the content in the multiphase flow modeling section below can be found in (Ref. 1). The all-lubrication-regime tribology model is described in detail in (Ref. 3, 6).

MULTIPHASE FLOW MODELING

In high-speed gearboxes, such as those used in helicopters, lubricant is supplied to the meshing gear teeth directly by high-pressure jets, since the pitch-line velocity (linear speed) of the gear teeth is far too high for dip lubrication as is used in low-speed applications. The multiphase flow of air and lubricant within a high-speed jet lubricated gearbox features continuous jets of lubricant, dispersed lubricant droplets of varying size suspended in air, and wall-bounded lubricant films on all surfaces—both rotating and stationary. Multiphase flow effects within high-speed gearboxes have important ramifications on the performance of these systems under both normal and loss-of-lubrication operation.

The most important effects of multiphase flows within gearboxes, indeed the very reasons that lubricant is used in them at all, are reducing the friction between contacting solid surfaces in relative motion (e.g. gear teeth and bearing components), and removing the heat generated by friction within these moving contacts. Under normal operation, these functions are achieved by the forced circulation of cooled lubricating oil through lubrication jets. Under loss-of-lubrication, these are achieved passively, and with severely reduced effectiveness, through the multiphase flow dynamics of the deposition of dispersed airborne droplets to wall-bounded films and the breakup of these wall-bounded films into droplets due to shear stress. Figure 1 provides a notional sketch of the deposition of airborne dispersed droplets to a continuous wall-bounded film, along with the effects of droplet splashing and bouncing, and Figure 2 depicts the breakup of a wall-bounded film into droplets due to shear instability (Kelvin-Helmholtz) brought about by the film-adjacent boundary layer in the air. Because the lubricant jets break up almost immediately into droplets, and impinge directly on the gear teeth, it is reasonable to consider dispersed droplets and continuous wall-bounded films as the only relevant small-scale configurations of the heterogeneous mixture of lubricant and air within high-speed gearboxes.

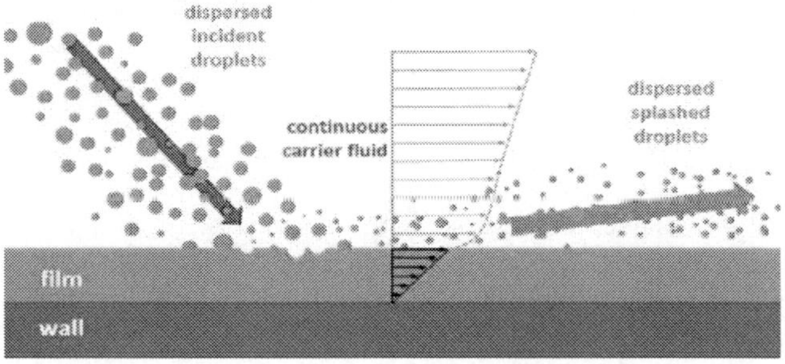

Figure 1: Droplet-laden flow impinging on a wall-film with deposition, bouncing of droplets off of the film, and splashing of secondary droplets formed from deposition

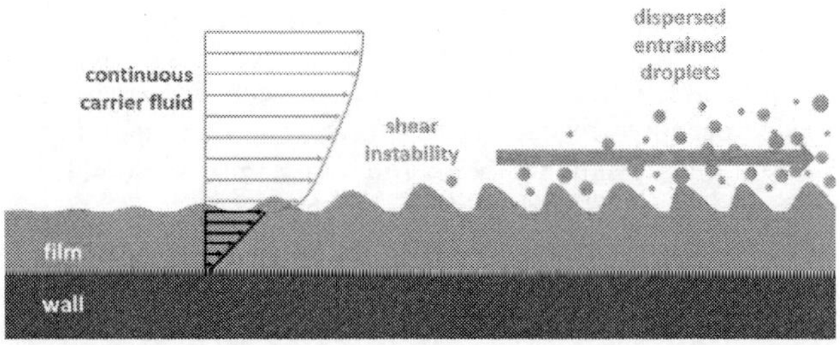

Figure 2: Breakup of a wall-film into dispersed droplets by carrier fluid shear (Kelvin-Helmholtz instability)

CFD methods for heterogeneous multiphase flows fall into four categories based on the choice of a Lagrangian or Eulerian description of motion, and on whether an interface-resolved or averaged formulation of the equations of motion is employed. With an Eulerian description of motion, the degrees of freedom are associated with points in a spatial grid of control volumes, while the degrees of freedom are associated with points that move *with the fluid* for a Lagrangian formulation. Interface-resolved means that all scales of interface between the distinct fluid species are fully resolved. Examples of Eulerian interface-resolved multiphase formulations are the volume-of-fluid (VOF) method of Hirt and Nichols (Ref. 12), and the level-set method of Sussman (Ref. 13). Within an Eulerian framework, interface-resolved multiphase CFD requires multiple computational cells to resolve each individual droplet, film, etc. The interface is smeared over several cells to avoid spurious oscillations that arise due to discontinuities in numerical solutions.

Interface-resolved Lagrangian methods for sprays track each individual droplet, but typically still employ an Eulerian formulation for the continuous background (carrier) fluid species. Each droplet can be treated as a point mass, and the interactions between the droplets and the background flow (e.g. drag and added mass), as well as with other droplets (e.g. breakup and coalescence), can be modeled. Thus, the control volume grid for the continuous Eulerian fluid species (air) need not fully resolve each feature of the Lagrangian dispersed species (droplets). When the number of droplets in a flow of interest is relatively small, this relaxation of the grid spacing requirement can make interface-resolved Lagrangian simulation less computationally expensive than interface-resolved Eulerian simulation such as VOF.

When it is computationally feasible, interface-resolved simulation is preferable, since the interfacial transfer of mass, momentum, and energy can be directly simulated rather than modeled. Most heterogeneous multiphase flows of engineering interest, and indeed those that arise in jet-lubricated high-speed gearboxes, contain far too many droplets or bubbles for interface-resolved simulation to be feasible. Instead, averaged multiphase equations of motion are developed based on the expected density of droplets, bubbles, etc. per unit volume at a given point. Such formulations are developed by multiplying the local instantaneous balance laws for mass, momentum, and energy by an indicator field that is either unity or zero, demarcating the space occupied by each fluid species, and then applying an averaging technique (e.g. time, volume, ensemble, Boltzmann, etc.). In the resulting averaged equations, the expected value of the indicator function for a fluid species is called the volume fraction of that species; it takes on values between zero and one. Upon averaging, the local instantaneous interfacial compatibility conditions (e.g. no phase change, no slip, and isothermal) lead to averaged interfacial transfer terms that must be modeled (e.g. Stokes drag, added mass, droplet lift force, etc.). Ishii and Hibiki (Ref. 15), as well as Drew and Passman (Ref. 16), and Berry (Ref. 17), all give good theoretical background for the development of the averaged multi-material equations of motion. As interface-resolved multiphase gearbox simulation is not feasible, we employ an ensemble-averaged formulation.

With an ensemble-averaged formulation of the equations of motion selected, there remains a choice between Eulerian and Lagrangian descriptions of motion. Both have some aspects that make them desirable for this problem, and some that make them undesirable. With an Eulerian description of motion, it is possible to formulate an implicit temporal discretization of the equations of motion, which allows large or even infinite (steady flow) time-steps to be considered, where a Lagrangian description of motion intrinsically requires an unsteady simulation (even for a steady flow), and is generally limited to explicit temporal discretization, which comes with very restrictive limitations on the allowable time-step size. On the other hand, it is much easier to formulate droplet splashing and bouncing when using a Lagrangian description of droplet motion (e.g. LANL KIVA (Ref. 14)) than when using an Eulerian formulation. We have decided that the possibility of steady solutions and longer time-steps afforded by the Eulerian description of motion outweighs the added difficulty of modeling the transfer of mass, momentum, and energy between droplets and wall-films in an Eulerian framework.

Wall-Bounded Film Modeling

Wall-bounded films can be modeled either as surface fields or as volumetric fields, with some algorithmic considerations to ensure that such films remain attached to walls if a volumetric approach is employed. In a three-dimensional CFD simulation, the wall-film equations of motion would only be solved on a set of two-dimensional faces corresponding to wall surfaces if a surface-based field is employed. Treating the wall-film as a surface-based field is a more elegant approach in some ways. First, there is no need to implement algorithmic tricks to force the film to remain wall-bounded. Perhaps more importantly, a surface-based treatment is able to accommodate wall-bounded films with thickness ranging from zero to larger than the wall-normal grid spacing.

The interfacial compatibility condition (no-slip, isothermal, etc.) between the wall-bounded film and continuous carrier fluid is generally enforced at the wall boundary of the CFD mesh, so that there is a geometric error in the location of the enforcement of the interfacial compatibility conditions that is equal to the film thickness. Generally, wall-bounded films are quite thin, so that this inconsistency does not lead to noteworthy accuracy degradation. Some difficulty does arise when applying such surface-based wall film treatment to incompressible flow in enclosed domains, since the volume of the domain does not decrease as droplet (volumetric) mass is deposited to the surface film field, requiring either a nonphysical decrease in density or increase in mass. As long as there isn't a whole lot of droplet mass getting transferred to the film, this is often acceptable physically, but limits the convergence of global mass balance. Implementation of a surface-based field within a volume-based CFD framework can be quite complicated.

If the wall-bounded films will *always* be thick relative to the well-normal grid spacing, then a VOF-like technique can be used to directly simulate the film species as a volumetric field. Surface tension can be modeled using, for instance, the continuum surface force (CSF) of Brackbill (Ref. 18), so that the interfacial physics that lead to effects like film breakup due to shear can directly captured. As the physics that confine these films to the wall are directly simulated, no special treatment is required to enforce wall confinement. On the other hand, a VOF-like treatment for wall-films requires that these films always be large relative to the wall-normal spacing, and additionally that the wall-tangential spacing be small relative to the film thickness whenever the breakup of films under shear needs to be captured.

When the thickness of a wall-bounded film is smaller than the wall-normal grid spacing, such approaches fail, and the effects of the wall-film have to be accommodated differently. If a volumetric field is used to represent wall-bounded films with thickness less than the wall-normal grid spacing, then the configuration of the film and carrier fluid species within wall-adjacent cells must be constructed from the volume fraction using the knowledge that the film is wall-bounded. Figure 3 shows the configuration of wall-film (blue) and air (white) in a single wall-adjacent computational cell. Here, the film height is $hf = \alpha_f V_P / A_{wall}$, where α_f is the volume fraction of the film species in cell P, V_P is the volume of cell P, and A_{wall} is the area of face s, which corresponds to the wall.

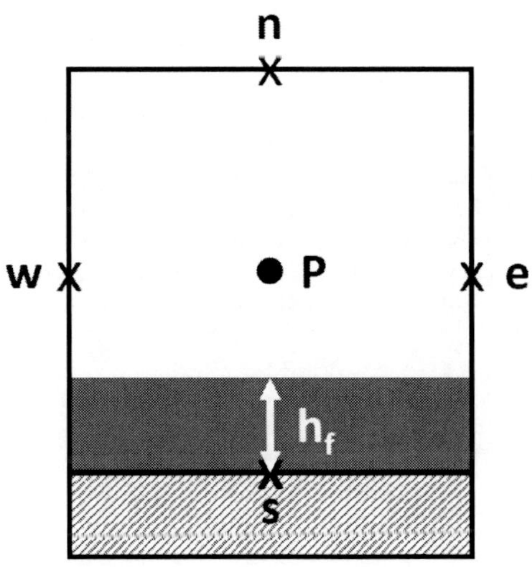

Figure 3: Single two-dimensional wall-adjacent control volume (cell) with wall-adjacent film height $h_f = \alpha_f V_P/A_{wall}$

With a volume-based field for wall-bounded films, the free-stream value of the film volume fraction (away from the wall) must be initialized to some negligibly small but nonzero value to ensure that the equations of motion remain defined throughout the domain. The volume fraction of wall-film becomes large in wall-adjacent cells by virtue of mass transfer via droplet deposition, but some additional treatment is necessary to ensure that the film mass in wall-adjacent cells remains confined to the wall. This can be accomplished either by introducing a wall-normal body force to offset any other wall-normal forces experienced by the wall-film field, or by modifying the mass fluxes across cell faces to be nonzero only if a face connects two wall-adjacent cells. Figure 4 shows several computational cells in the vicinity of a reverse wedge geometry with wall-films. If the flux-modification wall-film confinement technique is used, then the cell faces marked green are left open for film field mass flux, while the film mass flux is set to zero on the faces marked red. Note that the faces at the back end are closed off, so that film mass builds up at the back of the trailing edge until the film thickness gets large enough that the rate of film breakup equals the rate of film inflow plus droplet deposition.

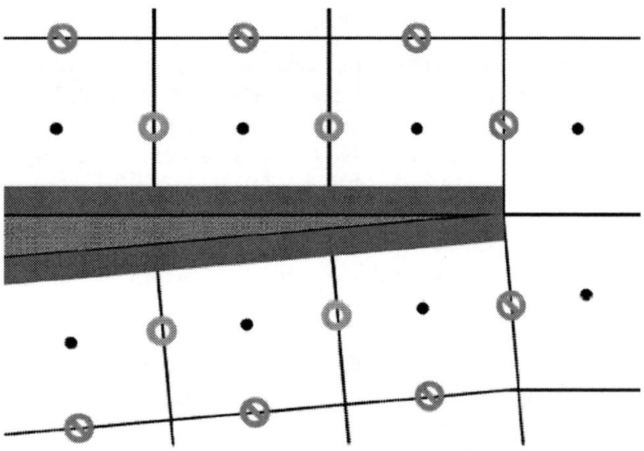

Figure 4: Control volumes in a notional computational grid at a trailing edge with wall-film treatment. Faces open to film flux are marked green, while those that are closed are marked red.

We have chosen to model the film field as a volume-based field, rather than a surface-based one due to the simplified implementation within a volume-based CFD code, and to the mass conservation difficulty that arises with a surface-based film field in enclosed domains, such as the interior of a sealed gearbox. Additionally, we have selected the flux modification approach to wall-film retention rather than the introduction of a body force, since the latter cannot force the wall-boundedness constraint exactly, while the flux-modification approach can.

Wall-bounded films tend to break up due to shear instability when their Weber number, defined as $We = \rho v^2 h / \sigma$ with σ the surface tension, exceeds 17. In high-speed gearboxes, the shear rate (and therefore shear stress) in the air next to gears and shrouding is sufficiently high, that the maximum film thickness before the Weber number exceeds 17 is of the order of 0.1 mm. Generally, the wall-adjacent spacing in the CFD grid will be larger than this value for high-speed gearbox simulations, but care should be taken to verify that the predicted wall-film thickness is smaller than the wall-normal grid spacing with this wall-film model.

Droplet—Wall-Film Interaction

Having established an averaged Eulerian multiphase CFD method that can accommodate continuous air, and lubricant in the form of droplets and wall-bounded films, as well as a model for the mutual drag between the air and wall-films, we still require models for the interaction of droplets and wall-films. There are two primary types of droplet—wall-film interaction in high-speed gearboxes, shown in Figure 1 and Figure 2. Figure 1 shows droplets impinging on a wall-film, with some amount of droplet mass splashed or bounced back with some normal and tangential velocity. Figure 2 shows the breakup of a wall-film into droplets due to shear instability.

Mass transfer from wall-film to droplet fields due to film breakup is simple to accommodate, as it simply re-identifies film mass, momentum, and energy that has been sheared off as now being droplet mass. Thus, the mass transfer rate is simply the rate at which film mass is sheared off. The momentum transfer rate is equal to the mass transfer rate time the film surface velocity, and the energy transfer rate is equal to the mass transfer rate times the specific internal energy of the film. Modeling the interfacial transfer of mass, momentum, and energy between droplets and wall-films for droplet deposition with splashing and bouncing, as in Figure 1, is less simple since there must be some mutual force to generate the wall-normal velocity of the splashed and bounced droplets. In such a case, there are also two distinct velocities for droplets: the average velocity of incident droplets, and the average velocity of splashed and bounced ones. Thus, some physics could be missed if the splashed and bounced droplet mass is returned to the same droplet field that is impinging on the wall-film. On the other hand, if a separate droplet field is used to accept splashed droplet mass, respecting the distinct velocities of incident and splashed droplets, a closure problem arises if these droplets are then allowed to deposit to film with splashing, so there must be some gradual re-grouping of splashed droplet mass into incident fields once drag with the air reduces the velocity discrepancy between splashed and non-splashed droplets.

Clearly, the simplest bookkeeping option to accommodate droplet deposition with splashing and bouncing is to simply return the splashed droplet mass to the incident field, and hope that there is not a great reduction in physical fidelity. This approach could be made more physically realistic by including the mutual force between the wall-film and droplets that accounts for the additional wall-normal momentum of the splashed and bounced droplets. Alternatively, the splashed and bounced droplet mass can be transferred to a separate receptor field that has a miniscule free-stream volume fraction, so that the splashed and bounced droplets have no effect whatsoever on the incident ones. Using a separate receptor field for the splashed and rebounded droplets requires development of a method to gradually return this splashed droplet mass to the original droplet field in order to avoid having to account for secondary, tertiary, etc. impingement of splashed droplets onto wall-film.

To investigate whether the simpler approach of returning rebounded droplet mass to the incident droplet field could prove adequate, a two-dimensional single-cell test case was devised with a configuration as shown in Figure 3. The north and west faces were treated as fixed-velocity inlets, the south face was a wall with a wall-bounded film field, and the east face was a fixed-pressure outlet. The incoming flow angle relative to the wall normal direction was varied from zero to ninety degrees (impinging directly onto the wall through flow entirely in through the west face and out the east face). To eliminate dependence on the segregated SIMPLEC-based numerical algorithm, the fully-nonlinear discretized equations of motion were solved using Powell's hybrid method via the fsolve routine of SciPy (Ref. 19), which is a front-end for MINPACK's hybrd routine (Ref. 20).

If there is no splashing or bouncing of droplets, then all of the wall-normal mass flux of incoming droplets should get deposited to the film. That is, the mass transfer rate from the incident droplets to the film is equal to $\rho_d \, \alpha_d \, (\mathbf{v}_d \bullet \mathbf{A}_s)$, where ρ_d is the droplet species density (e.g. ~1000 k/m^3 for water), α_d is the droplet volume fraction, $\mathbf{v}_d$ is the droplet velocity, and $\mathbf{A}_s$ is the outward area vector for the south face. When some fraction of the incident droplet mass is splashed or rebounded, it is imparted with wall-normal velocity *away* from the wall, so that any secondary impingement of such rebounded droplets would not be expected to occur in the same cell. Thus, if ϕ is a number from zero to one representing the fraction of incident droplet mass that is splashed or rebounded, one would expect the collection efficiency β (the ratio of droplet mass deposited to film to the incoming wall-normal droplet mass flux) to equal $1 - \phi$.

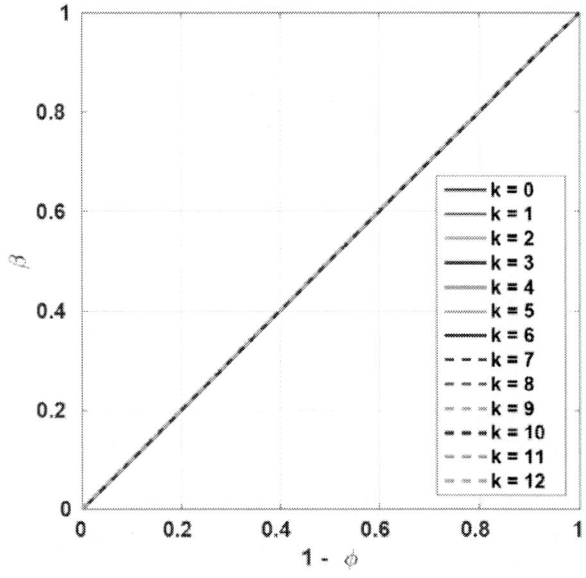

Figure 5: Collection efficiency vs. prescribed collection efficiency in a single wall-adjacent cell with incoming flow angles from 0 (wall-normal) to $\pi/2$ (wall-tangential) using a separate receptor field to accept non-deposited incident droplet mass

Figure 5, Figure 6, and Figure 7 show the resulting collection efficiency β compared to its prescribed value 1-ϕ at wall-normal flow angles equal to $k\pi/24$, with k=0 wall-normal and k=12 entirely along the wall. We see in Figure 5 that using a separate receptor field to accept the splashed and rebounded droplet mass gives the desired result of $\beta = 1-\phi$ for all flow angles and all values of ϕ, while returning the mass to the same field as the incident droplets only approaches the desired result in the limiting case of wall-tangential flow, when there is very little deposition to begin with. As such, we conclude that splashed and rebounded droplet mass *must* be accepted by a field other than the incident one. Otherwise, droplet mass non-physically accumulates in wall-adjacent cells until the collection efficiency is nearly unity regardless of the fraction of incoming droplet mass that rebounds.

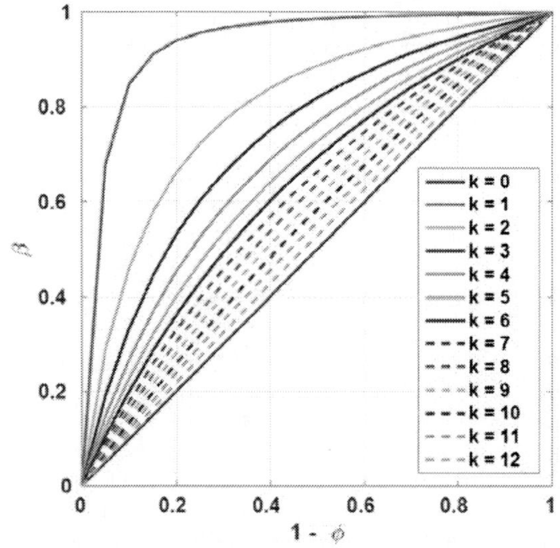

Figure 6: Collection efficiency vs. prescribed value, returning the non-deposited incident droplet mass to the incident droplet field without applying a restoring force

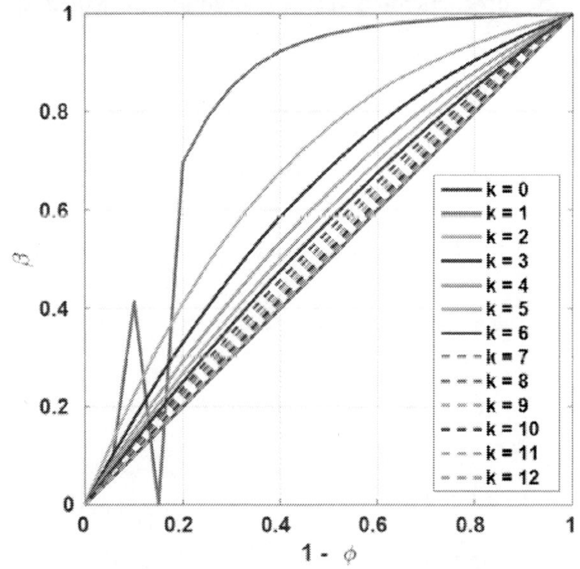

Figure 7: Collection efficiency vs. prescribed value, returning the non-deposited incident droplet mass to the incident droplet field with a restoring force applied to offset the excess incident momentum

Modeling Droplet Splashing and Bouncing

So far, we have identified an ensemble-averaged Eulerian multiphase CFD approach as the best option for simulation of high-speed gearbox aero-thermodynamics, and we have established that as separate receptor field should be used to accept splashed and rebounded droplet mass, rather than returning the mass to the same incident droplet field that is impinging on wall-bounded film. Now, we still require models for the fraction of incident droplet mass that is splashed or rebounded, as well as the size distribution of the splashed and rebounded droplets, and their wall-normal and wall-tangential velocity components.

LEWICE 3.0 is a simulation tool developed at the NASA Glenn Research Center for simulation of icing on airfoils. It employs a transient Lagrangian formulation for the droplet dynamics, and can accept CFD or potential flow simulation results for the air flow field. The droplet deposition model used in LEWICE 3.0 includes the effects of splashing and rebounding of droplets, and has been extensively validated against experimental measurements of collection efficiency on airfoil sections in the NASA Glenn Icing Research Tunnel (Ref. 21). Because there is currently no validation data available for droplet deposition, splashing, and bouncing within the context of high-speed gearboxes, the LEWICE 3.0 models for splashing and rebound have been converted from an interface-resolved Lagrangian formulation to an ensemble-averaged Eulerian one and implemented for use in high-speed gearbox simulation. Figure 8 compares collection efficiency results vs. non-dimensional distance from the leading edge for a NASA MS-317 airfoil in droplet-laden flow with a mean volumetric diameter of 92 micron. Experimental results from the NASA Glenn Icing Research tunnel are compared to

LEWICE 3.0 results as well as the averaged Eulerian model implemented in NPHASE-PSU with and without splashing and bouncing models active. We see that both NPHASE and LEWICE 3.0 greatly over-predict the collection efficiency without the splashing and bouncing models active, but match experimental results reasonably well with both turned on. Examining the behavior of NPHASE-PSU with only the splashing model compared to with both splashing and bouncing models, it is apparent that the splashing model affects the result most in the vicinity of the leading edge (s/c = 0), where the flow is approximately normal to the wall, while the droplet bouncing model has a greater effect further from the leading edge, where the flow is more wall-aligned.

SIMULATION OF INCREASINGLY COMPLEX GEARBOX CONFIGURATIONS

PSULOL is a CFD-based multi-scale tool for the simulation of coupled aero-thermodynamics and tribology in high-speed gearboxes under loss-of-lubrication conditions. It is the first comprehensive simulation tool for analyzing gearbox loss-of-lubrication performance to include a

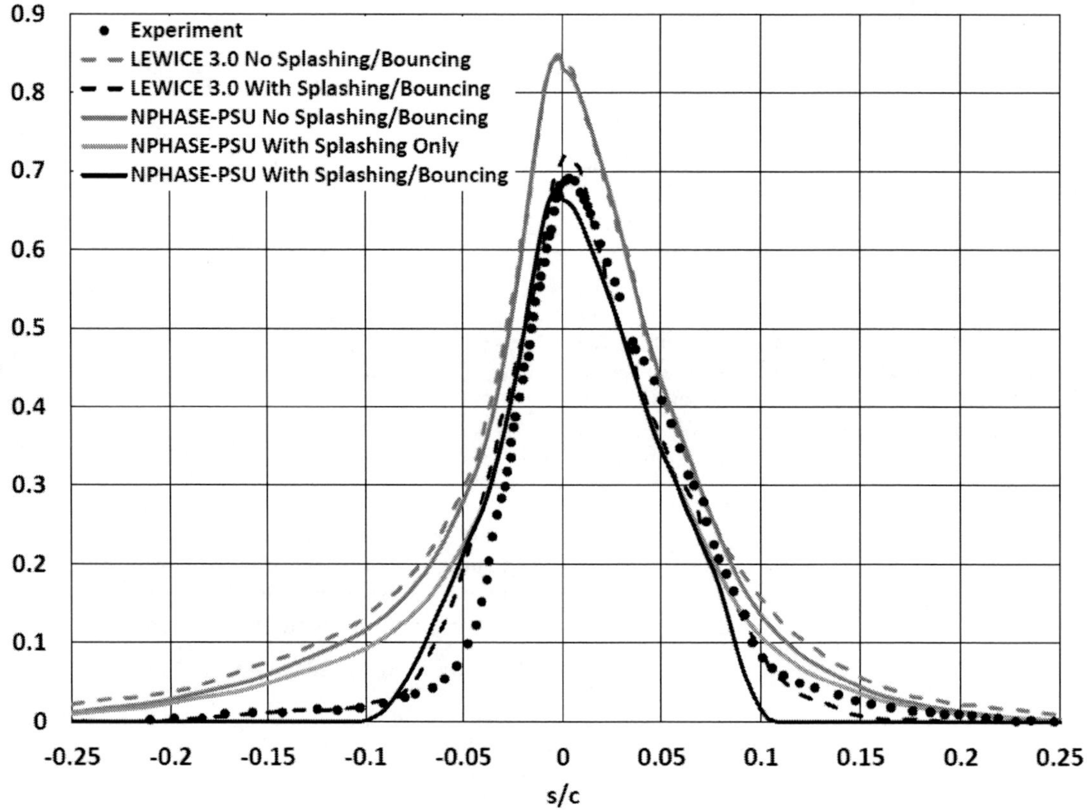

Figure 8: Comparison of collection efficiency results vs. nondimensional distance from the leading edge on a NASA MS-317 airfoil from icing tunnel tests at NASA Glenn Research Center, and numerical simulation using LEWICE 3.0 and NPHASE-PSU with and without droplet splashing and bouncing models

tribology model for all lubrication regimes from elasto-hydrodynamic lubrication (EHL) through boundary lubrication and scuffing failure. The first-generation computational framework for PSULOL was built on NPHASE-PSU, an in-house finite-volume CFD code with Eulerian multiphase flow and conjugate heat transfer capabilities. The level of access to the underlying source code afforded by using an in-house research code was instrumental in establishing the first-generation version of PSULOL, as it made it possible to modify every aspect of the code's functionality. After the baseline multi-scale gearbox model was established, it was observed that the features required to implement such a multi-scale model are also available within the context of commercial CFD codes that provide an application programming interface (API) to transfer field functions between the CFD code and a separate user-coded tribology model.

In order to hasten the establishment and assessment of the initial loss-of-lubrication simulation capability, a number of physical and geometric simplifications were made in developing the first-generation PSULOL capability. Simulations of the NASA Glenn Research Center loss-of-lubrication tests performed using PSULOL demonstrated that the selected modeling approach was capable of producing phenomenologically similar results to what was observed in experiments with a reasonable computational cost (Ref. 1, 2, 7). These initial shakedown runs either directly simulated (e.g. finite-volume CFD and heat transfer) or modeled all of the physical effects that were identified as significant in determining the onset of failure by severe scuffing. Much of the strong coupling between the aero-thermodynamics and tribology was captured directly, with the conjugate heat transfer simulation of the gearbox providing the meshing tribology model the surface temperature of the meshing gear teeth, and the tribology model returning the meshing frictional heat flux to the aero-thermodynamics module.

The physics that determine the thickness of the lubricant film and load support by solid contact in a non-conformal contact (e.g. gears, cams, and roller bearings) as a function of the lubricant properties, amount of lubricant present, material properties and roughness of the contacting surfaces, temperature, speed, etc. are complicated, and development of accurate physics-based models has proven elusive. The significant mixed lubrication regime, where the contact load is supported substantially by both solid contact (boundary lubrication) and hydrodynamic pressure generation (Stokes flow/Reynolds equation), is particularly challenging since simple statistical theories based on isolated asperity contact and an otherwise uninterrupted lubricant film are valid only in the light mixed lubrication regime, where there is minimal solid contact. Because there are currently no validated models available for the load share by solid contact in the significant and deep mixed lubrication regimes, the solid contact load fraction $\varphi_s \in [0, 1]$ was initially modeled as a decaying exponential function in time,

$$\phi_s = c_{eq}\left[1 - \exp\left(\frac{-t}{\tau}\right)\right]. \quad 1)$$

Here, $c_{eq} \in [0, 1]$ represents the final equilibrium level of lubricant starvation, and the time constant τ defines the rate at which the solid contact load share approaches c_{eq}. This assumes that the rate of lubricant loss is proportional to the amount of remaining lubricant in excess of its final (possibly zero) equilibrium value, and that the relationship between load share by solid contact and the excess remaining lubricant is nearly inverse linear between fully-lubricated and fully-starved operation. The modeled value of ϕ_s was used in the significant mixed lubrication model to interpolate between physics-based models for the light mixed and boundary lubrication regimes, with some special treatment included to accommodate boundary film failure.

Having made such simplifications to the coupling between multiphase flow and the tribology model, it was reasonable to make other geometric and modeling simplifications to decrease computational cost and model complexity without affecting the overall accuracy. These included treating the entire housing and support structure as a cylindrical shell with the same thermal mass and similar external surface area in the heat transfer simulation, treating the internal convection as cyclic-symmetric, ignoring the non-cyclic-symmetric aerodynamics where the gear teeth mesh, and ignoring multiphase windage effects, since the mass loading of lubricant under loss-of-lubrication is minimal and the multiphase--tribology coupling was already accommodated in the model for ϕ_s.

Relevant Physics and Scales

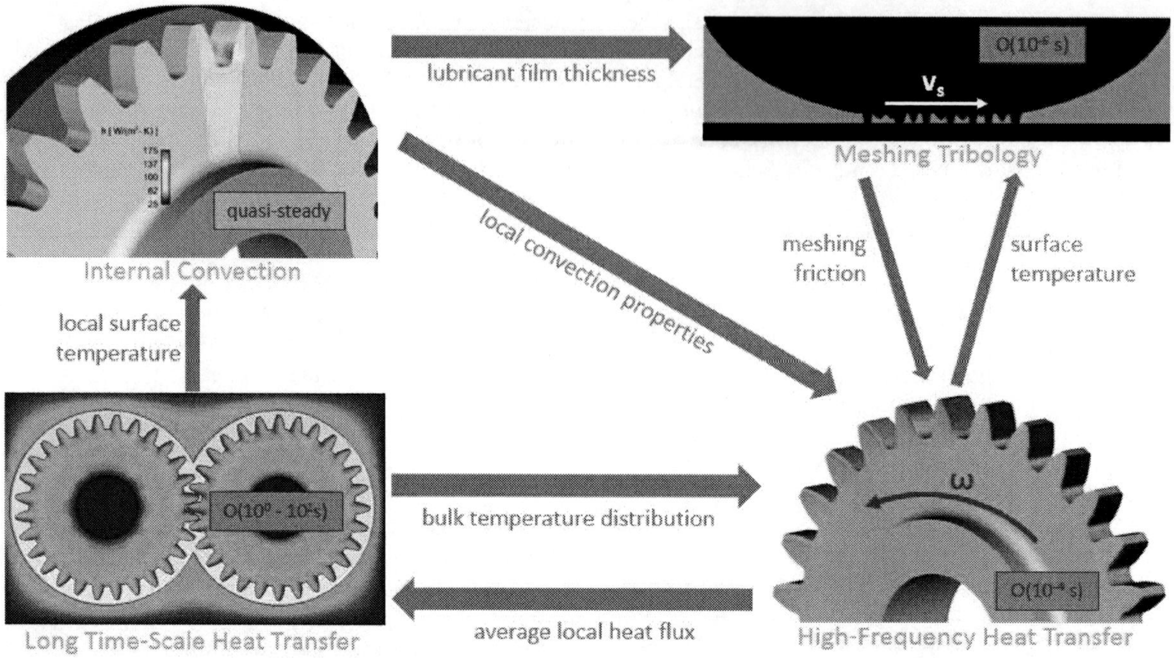

Figure 9: Physical effects and scales of high-speed gearbox loss-of-lubrication operation, along with the mechanisms that couple them with one another

As described in detail in previous publications (Ref. 1, 2, 7), the salient physical effects governing gearbox loss-of-lubrication performance are coupled and span a very wide range of space and time scales. The smallest relevant scale is that of the tribology of the meshing gear teeth, which takes place on a spatial scale of $O(10^{-4}$ m) and a time scale of $O(10^{-6}$ s). The gear meshing frequency and rotation rate define another important time-scale of high-frequency aero-thermodynamics associated with the displacement of fluid by the meshing gear teeth, and with the heating and cooling cycle gear teeth experience every revolution as they heat rapidly due to meshing friction and cool gradually throughout the rest of the revolution. For a typical high-speed gearbox, such high-frequency thermodynamic processes operate on a time scale of $O(10^{-4}$ s) with a spatial scale similar to the tribology-relevant scale of $O(10^{-4}$ m) for the smallest-scale aero-themodynamic features (e.g. recirculation zones, sharp temperature gradients, etc.). Given the high speed of rotation of rotorcraft gearboxes (tens of thousands of rpm for some components), it is necessarily the case that the high-frequency component of the aero-thermodynamics be *very* nearly time-periodic. For instance, if the temperature of a gear tooth coming into mesh increments by any meaningful amount over the course of a single rotation (i.e. at the same point in the rotation), then thermal failure would occur immediately, as the temperature of that tooth would increase by that amount hundreds or thousands of times within a second.

There is another scale associated with the response of the average temperature distribution to the net imbalance of heat generation and heat removal through conduction to the support structure and convection to the environment (filtering out the once-per-revolution periodic variation). A relevant time scale for this long-time or bulk thermal response is $O(1$ s). The enclosed mixture of air and lubricant within a high-speed gearbox must have a low mass loading of lubricant (or the windage/churning loss would be enormous), so the heat capacity of this multiphase mixture is negligible compared to that of the gears, shafts, and housing. As such, the multiphase convection process within a gearbox at any given time can be considered quasi-steady, determining the convection heat flux from the hot gears to the cooler housing and support structure in response to a temperature difference between the two.

Figure 9 provides a graphical description of these four identified physical effects and scales, and the mechanisms by which they are coupled across a very wide range of scales. Considering these separate physical effects and their coupling, a multi-scale simulation algorithm was devised wherein each of the scales (tribology, high-frequency heat transfer, convection within the gearbox, and long time scale heat transfer) is simulated separately, and coupling is achieved through averaged conjugate boundary conditions and initial conditions during one time step. Figure 10 shows the flow of information between the various simulation modules within one outer (long time scale) time step of the first-generation simulation framework.

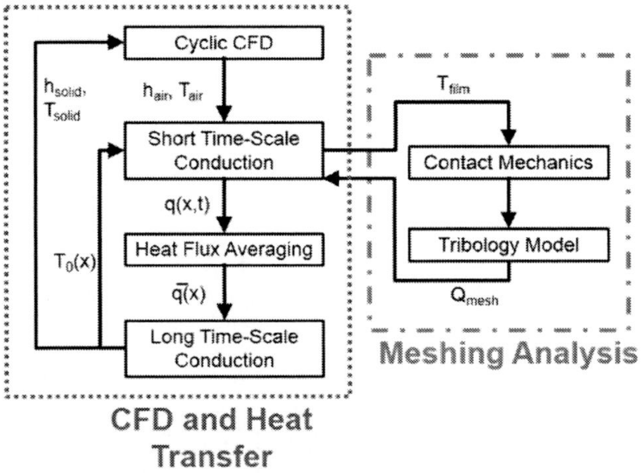

Figure 10: Simulation modules and their interaction within one outer time-step of the first-generation PSULOL simulation framework

sector. Since the high-frequency heat transfer process is not effectively cyclic-symmetric, this sub-simulation was carried out on the entire gear and tightly coupled with the meshing tribology model via incoming tooth surface temperature and meshing friction. The surface heat flux was averaged in time and cyclically onto a single-tooth reference sector, and used as a fixed-flux boundary condition in advancing the temperature distribution throughout the gear in the long time-scale heat transfer module, which treated the gears as cyclic-symmetric since the periodic temperature variations each gear revolution are filtered out.

As it was implemented in NPHASE-PSU, the conjugate heat transfer framework required face-matched grids between the solid and fluid domains for conjugate heat transfer. Because there is no clear way to map the cyclic-symmetric heat flux from the internal multiphase flow into the housing (internal convection) to a non-cyclic-symmetric housing configuration, the housing was treated as a cylinder around the gear with thermal mass and external surface area approximately equal to that of the real housing. In loss-of-lubrication simulations, the temperature of gears themselves is of the most interest, as this is where failure occurs. As long as the thermal response of the cylindrical housing

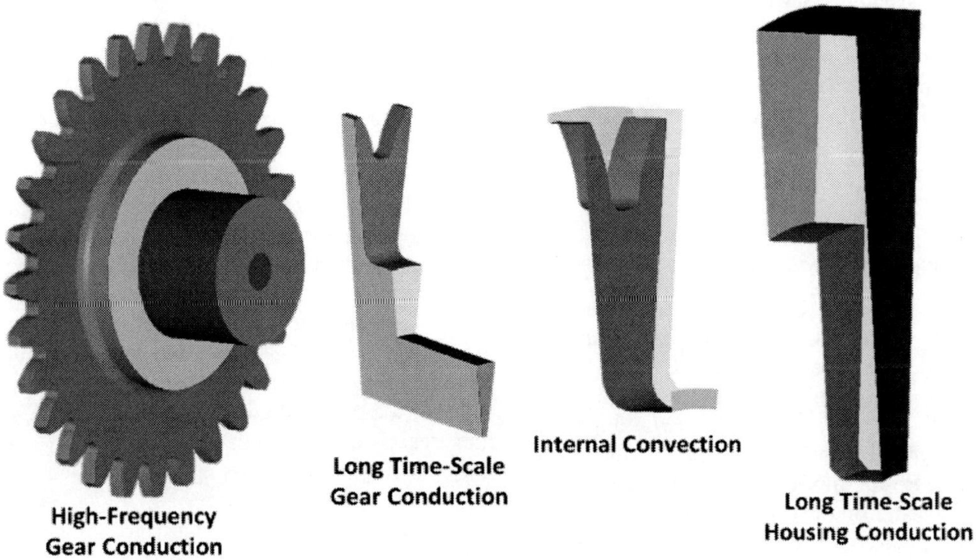

Figure 11: Computational domains used in sub-simulation modules of the first-generation PSULOL framework with cyclic-symmetric internal convection and housing

approximation matches that of the actual housing, there will be no loss of accuracy in the thermal response in the gears. On the other hand, the physics of lubricant retention and re-distribution within gearboxes are very much non-cyclic-symmetric, with lubricant collecting in features specifically designed to retain and passively re-distribute it under the action of fluid dynamic shear to maintain lubricant film on the gear teeth under loss-of-lubrication.

Currently, all multiphase flow effects are lumped into the model for the time-dependent behavior of the load share

Limitations of the first-generation framework

The first generation PSULOL framework (Ref. 1, 2, 7) employed a cyclic-symmetric approximation of the enclosed flow of air and lubricant around the enshrouded gears. Figure 11 shows the computational domains used in the CFD and heat transfer sub-simulation modules in the original PSULOL framework. High-frequency heat transfer was simulated only for the gears themselves, with heat transfer coefficient and near-wall flow temperature distribution given by making n_{teeth} copies of the results from the cyclic-symmetric internal convection simulation of a single tooth between solid contact and hydrodynamic pressure given by equation (1), so that it is not yet possible to directly simulate the impact of changes to lubricant retention features on loss-of-lubrication performance. To improve the predictive capability of PSULOL, an effort is underway to replace this *ad hoc* model for the time-dependent behavior of the load sharing with a model that accepts the incoming lubricant film thickness as an input from fully three-dimensional multiphase CFD simulation of the flow within the gearbox.

Carrying out such a fully three-dimensional multiphase CFD simulation of the flow within the gearbox, along with conjugate heat transfer and simulation of convection to the environment, within NPHASE-PSU would require a large amount of code development to accommodate some of the geometric challenges of gearbox simulation. In particular, methods to handle the close gaps and dynamic contact between gear teeth as they mesh, as well as a capacity to map surface values between non-face-matched grids would be required. While these could have been implemented within NPHASE-PSU, the cost of such code development would have been large, and the process of case setup, running, and post-processing would have remained challenging for the end user. Since it already has close-gap and dynamic contact capabilities with overset meshes, can interpolate between non-conformal and non-face-matched grids, and provides sufficient means of coupling with user-code (the tribology model), we decided to implement the second-generation PSULOL framework within the commercial CFD package Star-CCM+, rather than implement the necessary geometry-handling capabilities within NPHASE-PSU.

Fully 3-D CFD & heat transfer approach

Two approaches to the gearbox aero-thermodynamics simulation have been identified that can accommodate the meshing gears, conjugate heat transfer (CHT), and a geometrically accurate representation of the housing. The first, which we call the *transient multi-scale* approach, employs a moving-grid overset approach for the fully-coupled multiphase aero-thermodynamics, tribology model, and conjugate heat transfer on the high-frequency time scale. This fully-unsteady, fully-coupled sub-simulation is run for several system rotations, and the local instantaneous heat flux on all solid surfaces is time averaged to remove the time-periodic variations that arise due to rotation. The local time-averaged heat flux is then used in a long time scale conduction analysis with a time step on the order of seconds to update the temperature distribution within the gearbox. Therefore, each outer time step consists of an unsteady moving-grid overset sub-simulation of the coupled multiphase aero-thermodynamics, conjugate heat transfer, and tribology models on a short time scale, and a conduction simulation using the averaged heat flux distribution from the short time-scale module to update the bulk temperature distribution.

The second identified approach to simulating gearbox aero-thermodynamics and CHT with a geometrically representative housing is called the *frozen rotor* approach. With this method, overset grids are still used, but there is no grid motion. Rather, moving wall boundary conditions are applied on all rotating components such as the gears and shafts, and stationary wall boundary conditions are applied on stationary components such as the housing. A frozen rotor simulation is very similar to using a mixing plane in turbomachinery calculations where regions associated with rotors are solved in the relative frame and those associated with stators are solved in the stationary frame. With the mixing plane approach, mixing is formally enforced by averaging, while with the frozen rotor approach, it is assumed that numerical and physical diffusion generate sufficient mixing to prevent non-physical fluid dynamic interactions between stationary and rotating components.

The transient multi-scale approach is certainly more physically accurate. It is capable of capturing all of the aero-thermodynamic effects including the transient displacement of the air and lubricant mixture by the gear teeth as they mesh, and could be considered an exact treatment of the multi-scale problem in that all short time scale effects are simulated in a time-accurate coupled fashion and the time-periodic variation is formally filtered out by time-averaging for use by the long time scale analysis.

While its accuracy is appealing, the computational cost of the transient multi-scale approach is daunting. One would imagine that a typical transient loss-of-lubrication simulation would take several hundred or maybe several thousand outer time steps of order one second. If the high-frequency heat transfer module requires unsteady moving-grid overset simulation of the full gearbox for several revolutions, then each outer time-step will take several hours to complete so that a full simulation with this approach would take several months to complete. This is in the ballpark of what today's state-of-the-art simulations take to complete, and it is within the realm of possibility to carry out gearbox loss-of-lubrication simulations using this approach.

On the other hand, the frozen rotor approach is a factor of $O(100)$ less computationally costly than the transient multi-scale approach. When using a frozen rotor approach for a CHT simulation, cyclic averaging should be performed on the temperature, heat transfer coefficient, and heat flux (including the meshing friction, so that it must be applied to all teeth) at the fluid-solid interface on rotating components. If this averaging is not performed, some level of non-physical variation in the temperature distribution around the rotating components may arise due to (physical) asymmetry in the flow field when the in-place topology is used. Such asymmetry does not arise in reality since rotation acts to average out the effects of non-symmetric flow features on rotating components. The frozen rotor method does lose the physics of displacement of air and lubricant by the gear teeth as they mesh, but this likely plays a secondary role in the overall thermal balance in the system.

In light of the substantial computational cost savings of the frozen rotor method, we have selected this approach for the second-generation PSULOL framework. The frozen rotor approach necessarily simulates the fluid dynamics on the long time scale, filtering out time-periodic effects that arise from rotation. The heat transfer coefficients and near-wall flow temperature could still be taken from a frozen-rotor simulation and given to a rotating components-only conduction analysis of the high-frequency heat transfer. In this case, the second-generation PSULOL algorithm would be identical to the first-generation one, but with the full housing and support structure instead of a cylindrical approximation. While simulating the NASA Glenn 3.5" spur gear tests, it was observed that the temperature rise across mesh (and therefore the magnitude of the time-periodic temperature variation) was never more than several degrees except after the onset of severe scuffing and system failure. Thus, at least for that configuration, the high-frequency temperature variation can be safely ignored. In the present work, the frozen rotor aero-thermodynamics simulation is coupled directly with long time scale conduction analysis in the gears, housing, and support structure. These are solved simultaneously each outer time-step as a CHT problem, and coupled directly with the tribology model.

Tribology coupling approach

One major advantage of using NPHASE-PSU in developing the first-generation PSULOL framework was the direct access to modify the code, and manipulate the parallelism of any added routine. This allowed the tribology model to be compiled directly into NPHASE-PSU. In general, the cell faces corresponding to the gear tooth flanks (meshing surfaces) will not belong to the same parallel process. So, global gather operations must be carried out to get the required properties from the aero-thermodynamics simulation for the tribology model, and global scatter operations are necessary to disseminate the meshing friction information to all processes owning gear tooth flank faces. Implementation of this was straightforward in NPHASE-PSU. The necessary calls to MPI_Gather and MPI_Scatter were simply added, and the tribology model was run only on the head process (processor 0). Figure 12 provides a picture of the flow of information between the parallel processes and the tribology model in the NPHASE-PSU-based framework.

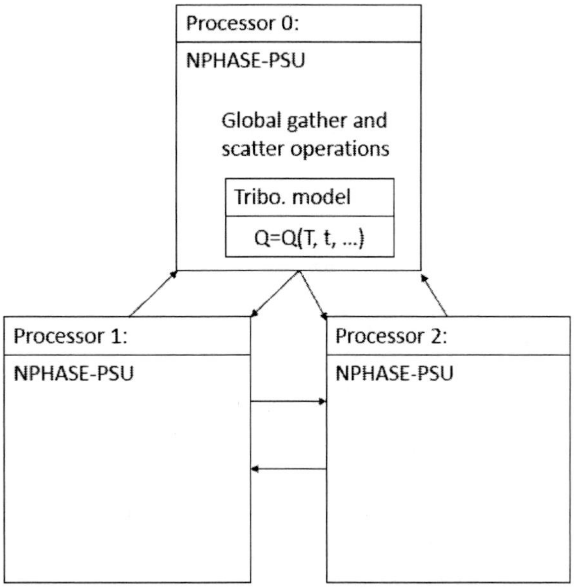

Figure 12: Tribology model implementation and parallelism in the NPHASE-PSU version of PSULOL

While Star-CCM+ is certainly desirable for its additional geometry-handling capabilities, robustness, and ease-of-use, the switch from NPHASE-PSU to Star-CCM+ greatly limits our options for coupling it with the gear meshing tribology model. Coupling the two requires passing the temperature field and eventually the lubricant film status on the gear teeth from Star-CCM+ to the tribology model, and then returning the heat flux distribution. The two viable options for coupling with user code are creating a user-coded boundary condition, and creating a co-simulation.

User-coded boundary conditions are C/C++/Fortran routines that the user writes, which must include special code to interact with the way Star-CCM+ stores data and communicates with user code. User-coded boundary condition routines are run only on processors owning faces of that boundary type, and do not know about boundary faces on other processors. Figure 13 shows the flow of information between the processes and the tribology model with the tribology model implemented as a user-coded boundary condition. Inter-processor communication occurs only in the fluid dynamics and heat transfer solution and the tribology model is run separately on each processor based only on processor-local data.

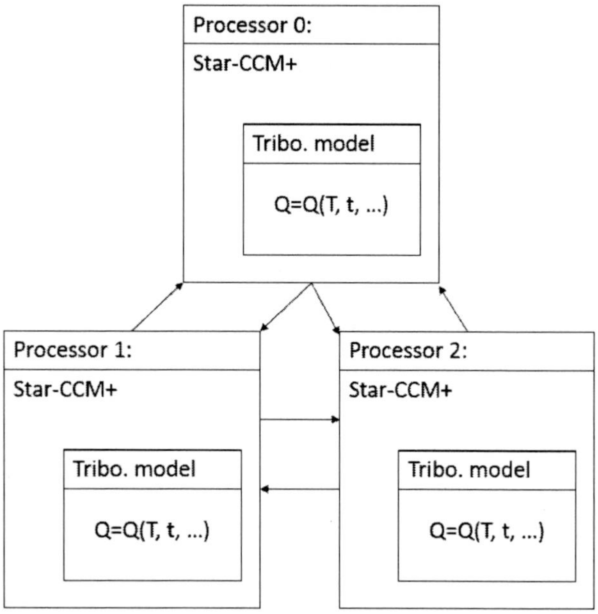

Figure 13: Tribology model implementation as a user-coded boundary condition in Star-CCM+

The second way of coupling Star-CCM+ with the user-coded tribology model is to implement it as a co-simulation. In contrast to the user-coded boundary condition, where user code is run entirely within Star-CCM+, a co-simulation runs completely separately and the two exchange data using functions provided by Star-CCM+ with its C/C++/Fortran API. All quantities of interest are mapped to a two-dimensional grid of the boundary surfaces of interest (in this case, all gear tooth flanks). This two-dimensional grid is stored both in Star-CCM+ and the user code with a consistent face ordering scheme. Scalar fields (e.g. temperature) on this communication surface are passed back and forth as arrays of floating point values. Star-CCM+ handles all of its own parallelism internally, globally interpolating field values to the communication grid across all processors. Coupling with the co-simulation can either be loose (every time-step or at fixed time intervals) or tight (every nonlinear solver iteration). Figure 14 depicts the flow of information between Star-CCM+ and the tribology model co-simulation within this framework.

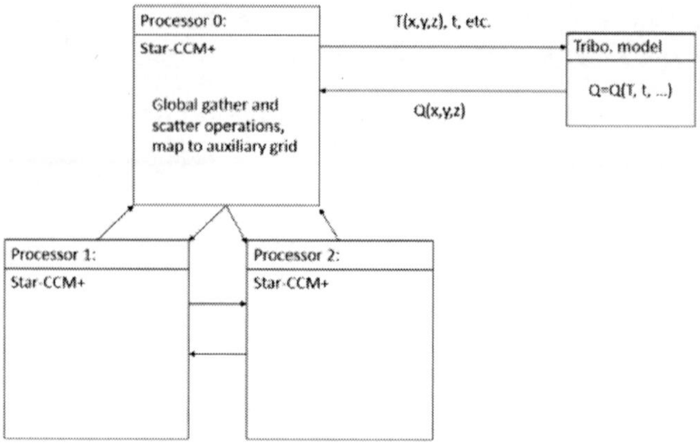

Figure 14: Tribology model implementation as a co-simulation coupled with Star-CCM+

The tribology model was implemented both as a user-coded boundary condition and as a co-simulation. With the frozen rotor approach, the cyclically-averaged temperature distribution on the tooth faces is required and the meshing friction heat flux is applied to all of the teeth. As a result, the tribology model is simulating the same exact problem on each processor which owns tooth flank faces. This duplication of effort is not necessarily a deal-breaker since all processors still have to wait for input from the tribology model with the co-simulation approach even if the effort is not duplicated. On the other hand, it was found that user-coded boundary conditions are difficult to debug because they are encapsulated entirely within Star-CCM+, making it particularly challenging to track down, for instance, which sub-routine throws a floating point exception signal. User-coded boundary conditions are also called every nonlinear solver iteration (potentially many times per time step). The dependence of meshing friction on temperature is not incredibly strong, so it is not necessary to update the meshing friction every time the temperature is updated during the nonlinear solution iterations in a single time-step. While the computational cost of the tribology model is low compared to CFD, it is high enough that calling it every single nonlinear solve slowed down the simulation progression substantially. In the interest of improved performance and easier debugging, we have selected to use the co-simulation implementation of the meshing friction tribology model.

PRESENT STATUS

Current efforts are being directed toward simulation of loss-of-lubrication tests carried out in the Penn State University Gear Research Institute's (PSU-ARL/GRI) 3.5" four-square test rig, which has an identical test section to the NASA Glenn facility (Ref. 9, 10, 11). These simulations will be carried out using the second-generation implementation of PSULOL in Star-CCM+. All of the parts of the Star-CCM+ based implementation are in place and functioning and demonstration cases have been run to verify their functionality. Computational grids have been generated for the full housing, gears, shafts, and room around the gearbox, but these still require some additional refinements to improve performance of the overset grid hole cutting algorithm in some tight regions within the gearbox.

Conjugate heat transfer with overset

To evaluate the selected frozen rotor overset grid approach for simulation of aero-thermodynamics and conjugate heat transfer within and outside of a gearbox, a simplified configuration was simulated using the 3.5" gears from the PSU-ARL/GRI with a greatly simplified housing geometry. Figure 15 shows the configuration of this simplified geometry. There is no internal shrouding, and an inlet has been added to the lower left and an outlet to the upper right to accelerate convergence of the flow field simulation. Though this configuration is geometrically much simpler than the full gearbox, shrouding, gears, and shafts, it is a suitable test case since the major challenges to this simulation are associated with accommodating the close gaps and contact between the meshing gear teeth.

Figure 15: Geometric configuration of the simplified housing and gears for testing the frozen rotor overset approach to aero-thermodynamics and conduction

Cells at the edge of an overset grid that are not on a physical boundary of the computational domain are called fringe cells. Coupling between overset grids is accomplished by finding a nearby donor cell in the other grid, which gets identified as a neighboring degree-of-freedom. Fringe cells for which no adequate donor cell can be found are called orphan cells and substantially degrade solution quality and numerical stability. If there are more than one or two orphan cells nearby, the numerical solution often fails. With finite volume and finite difference discretization, fringe cells cannot also be donor cells (this is not necessarily the case with overset finite element discretization). As a result, wall boundaries in two different overset grids must be separated from one another by several (about four) cells to avoid introducing orphans and ruining the numerical solution.

When two surfaces on different overset grids come into physical contact (e.g. meshing gear teeth) or get closer than the near-wall grid spacing, the introduction of orphans and associated solution degradation is unavoidable unless some modification is made to the overset grid method. In the case of meshing gear teeth, flow features in the immediate vicinity of contact or near-contact between two gear teeth do not have much effect on the overall flow or heat transfer characteristics. So, an overset grid method for simulating gearbox aero-thermodynamics need not physically resolve what is happening in these sub-millimeter scale regions. Rather, it simply needs to avoid generating orphans.

Star-CCM+ has recently introduced a new overset mesh interface, called the ZeroGap overset interface, that is intended to accommodate overset grids with contact and under-resolved gaps that would otherwise generate orphans. This is accomplished by simply cutting out any cells in under-resolved gaps between nearby walls of two different overset grids, and applying wall boundary conditions on all of the faces of the active fringe cells bordering these new holes to prevent non-physical mass loss/flow leakage. While this approach certainly does not give accurate results in the immediate vicinity of the ZeroGap treatment, as mentioned above, we're not concerned with the actual flow there, but simply with avoidance of solution failure due to orphan cells.

To test the performance of the ZeroGap overset treatment within the context of conjugate heat transfer simulation, a simulation of the simplified gearbox geometry was performed with a background overset grid for the housing and one overset grid for each of the two gears. Because they are in contact, a ZeroGap overset interface was used between the two gear meshes, while standard overset interfaces were used between each gear mesh and the background mesh. Figure 16 through Figure 18 show a slice at the axial mid-plane of the overset fluid domain grids used in this simulation. Figure 19 shows both the fluid and solid (heat transfer) grids in the vicinity of the meshing gear teeth. Conjugate heat transfer is applied anywhere the surface of the gear heat transfer grid meets a fluid domain cell that hasn't been cut out by the ZeroGap logic, and adiabatic (no heat flux) boundary conditions are applied on gear surface faces where the neighboring fluid cell was cut out.

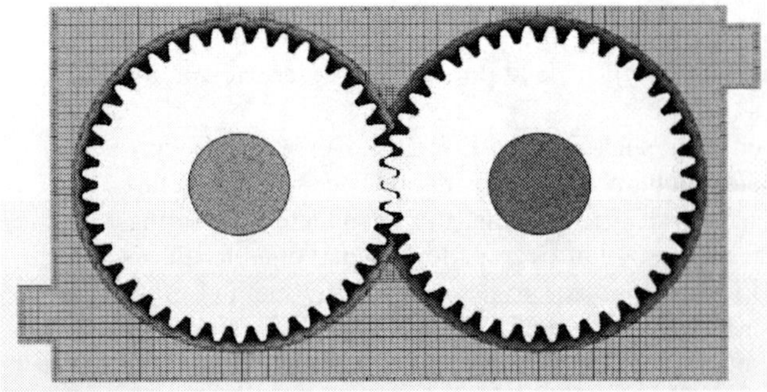

Figure 16: Fluid domain overset grids in simplified 3.5" pitch diameter spur gearbox test case

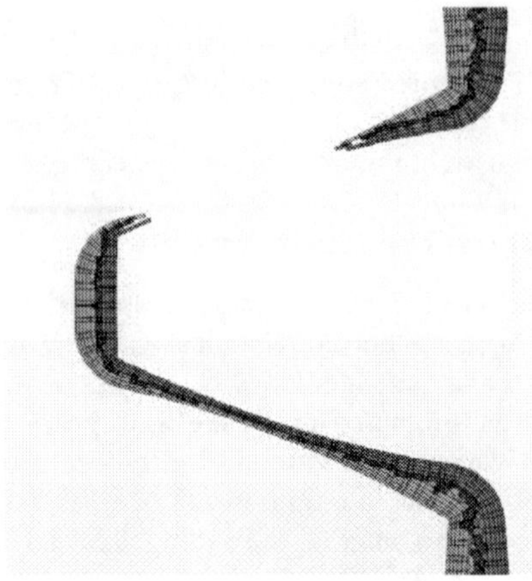

Figure 18: Overset fluid domain grids for both gears at tooth contact showing the hole cut out by the ZeroGap overset treatment at the contact location

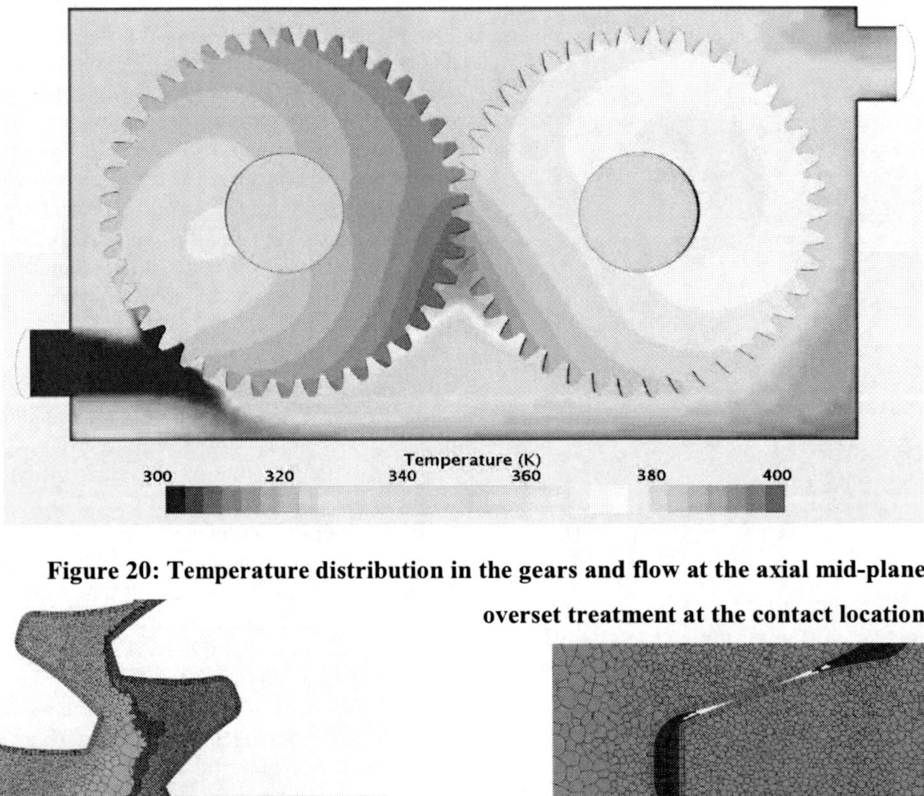

Figure 20: Temperature distribution in the gears and flow at the axial mid-plane

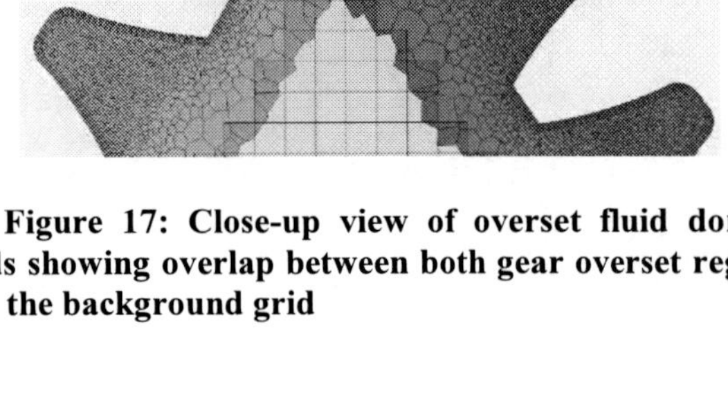

Figure 17: Close-up view of overset fluid domain grids showing overlap between both gear overset regions and the background grid

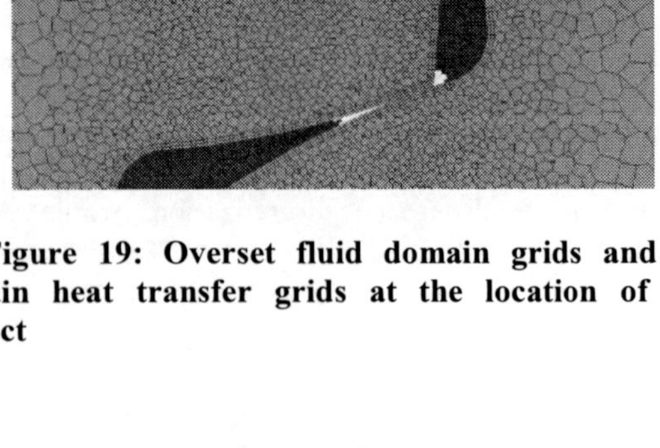

Figure 19: Overset fluid domain grids and solid domain heat transfer grids at the location of tooth contact

Using the overset grids shown in Figure 16 through Figure 19, a frozen rotor conjugate heat transfer was carried out with a constant meshing friction heat flux of 300 W applied evenly over all of the gear tooth flanks (marked red in Figure 15). Because cyclic averaging of the conjugate heat transfer on rotating components has not yet been implemented, some asymmetry arises in the temperature distribution in the gears. This asymmetry is much greater here than it would be in a real gearbox since 300 K air is being fed in through the inlet and impinging directly on the left gear, which would not be the case in a real gearbox configuration. Figure 21 provides flow streamlines colored by temperature for the same simulation.

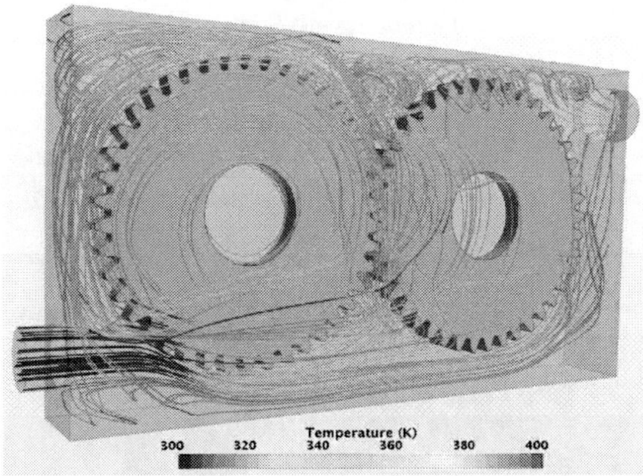

Figure 21: Flow streamlines colored by temperature

With this successful outcome of this frozen rotor overset gearbox aero-thermodynamics and conjugate heat transfer test simulation, it is expected that the selected simulation approach will be viable for fully-resolved simulations with a geometrically-representative housing configuration.

Tribology model co-simulation

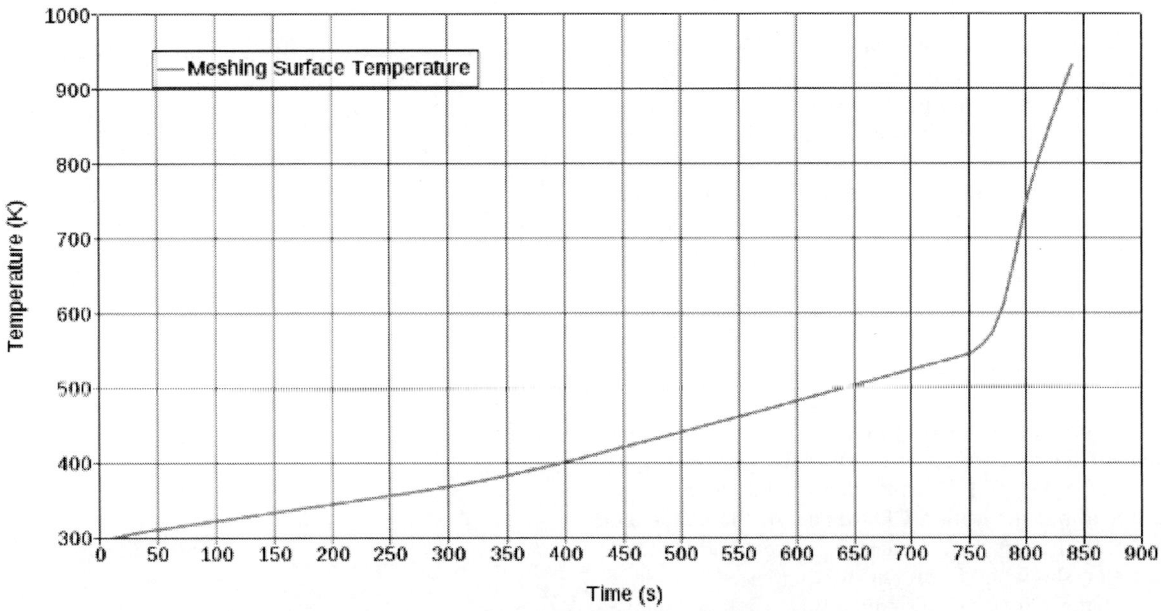

Figure 22: Time trace of average "meshing" surface temperature from tribology model test case

The line contact tribology model presented in (Ref. 2-6) was implemented in C++ and coupled with Star-CCM+ using the co-simulation API. To verify the functionality of the tribology model co-simulation coincident with a conjugate heat transfer boundary condition in the CFD analysis, a simple test case was devised with a 10 cm steel cube (similar mass to a 3.5" gear and portion of shaft) in flow. The windward face of the cube was treated as a meshing gear flank, and the tribology model with coupled with the CFD analysis through the surface temperature and meshing friction. Figure 22 shows the computational domain and grid resolution on the surface of the block, as well as the temperature distribution on and near the block surface, and the axial velocity distribution in a plane at the span-wise midsection at the end of the simulation (after scuffing failure is predicted by the tribology model). Figure 23 provides the time history of the average temperature over the front "meshing" surface of this block. The general behavior of this test case agrees with previous results, and demonstrates that co-simulation implementation of the tribology model functions as it should.

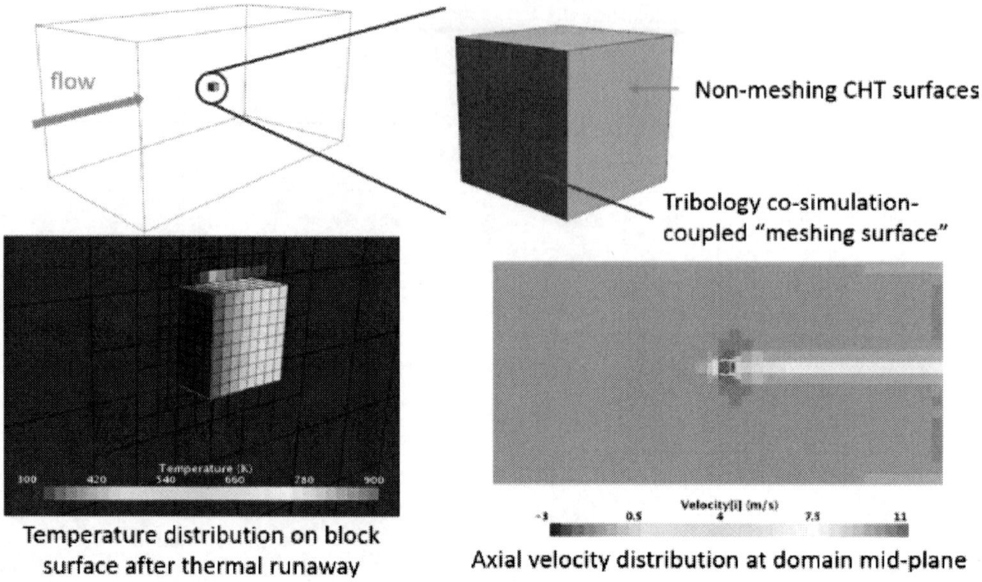

Figure 23: Tribology model co-simulation test case configuration

Full system simulation

With the required parts of the Star-CCM+ based second-generation PSULOL framework implemented and tested, we are in the process of setting up a loss-of-lubrication simulation of the PSU-ARL/GRI 3.5" pitch diameter spur gear tests with the full housing and shafts, as well as conjugate heat transfer with a CFD model of the ventilated room in which the test stand resides. Fully-featured CAD of this test stand, shown in Figure 24, was imported into Star-CCM+, and the built-in CAD editor was used to remove complicated features such as bolts and bolt holes that would require a great deal of grid to physically resolve and have little to no effect on heat transfer in the system. The defeatured CAD model is shown in Figure 25, and a detail view of the test gears, shafts, and internal shrouding after defeaturing is provided in Figure 26.

Figure 24: Gears, shafts, and inner shroud after defeaturing

Overset computational grids were generated for the background fluid domain (housing exterior and interior), rotating regions around both gears and shafts within the housing, and rotating regions around both shafts outside the housing. The level of grid resolution and complexity of the configuration of the gears, shafts, and shrouding within the housing is shown in Figure 27. The total cell count is currently $O(10^8)$ between all fluid and solid domains, which is similar to the cell count in current state-of-the-art large eddy simulations. The grids shown here still require some modification to improve the performance of the overset hole cutting in some regions within the housing where the housing and internal shrouding CAD has very tight gaps, but are nearly ready. Once the overset grids are complete, full-system simulations will be carried out using the currently-implemented tribology model, followed by simulations using a refined tribology model for elliptic contact patches that is presently under development.

Figure 25: Defeatured CAD model of the PSU-ARL/GRI 3.5" pitch diameter spur gear test rig for use in aero-thermodynamics and heat transfer simulation

Figure 26: Fully-featured CAD of the PSU-ARL/GRI 3.5" pitch diameter spur gear test rig after import into Star-CCM+

SUMMARY

Current progress in an ongoing effort at Penn State to develop a design-relevant simulation framework for high-speed gearbox aerothermodynamics and tribology under loss-of-lubrication operation has been described. The second-generation simulation framework described here is based on the first-generation framework described in (Ref. 1, 2, 7). While the first-generation version of PSULOL was implemented within the context of an in-house finite volume

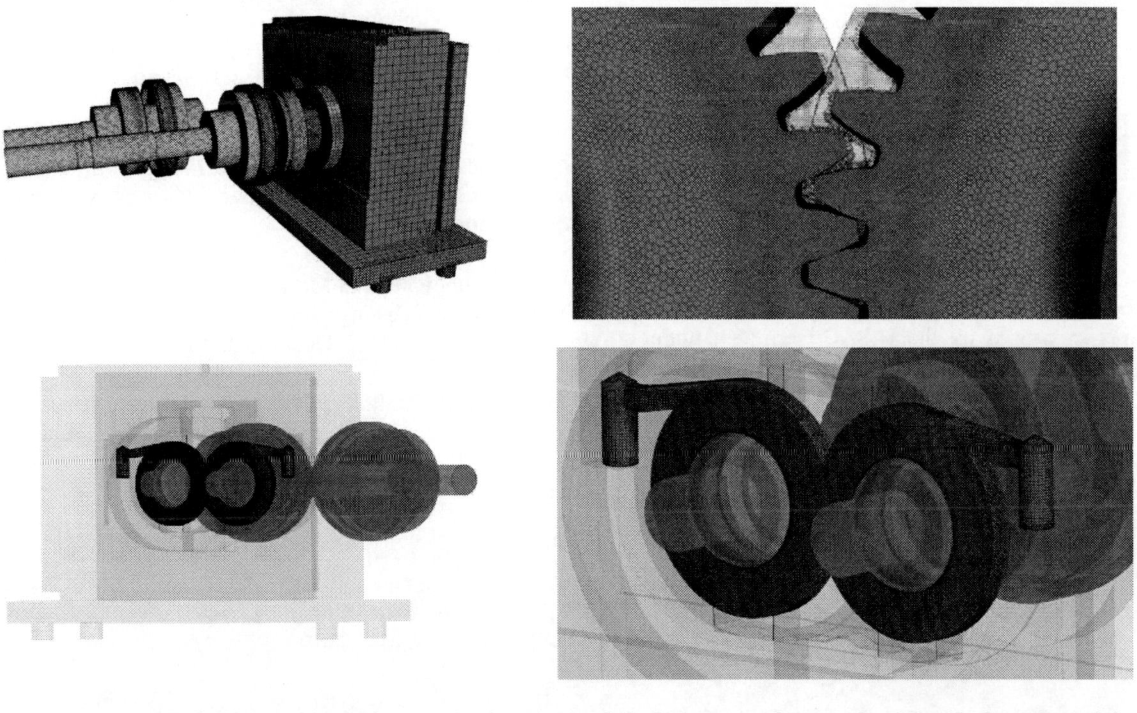

Figure 27: Surface grids on the exterior housing and shafts, gear surfaces and axial mid-plane slice of the overset fluid domain grid around each gear, and the inner surface of the internal shrouding

CFD code, NPHASE-PSU, which features conjugate heat transfer and multiphase flow capabilities, the second-generation version is implemented in Star-CCM+, a commercial finite-volume code with many of the same features as NPHASE-PSU, but with improved geometry-handling capability and robustness, and greater ease-of-use for case setup and processing.

Focus of the present effort is on improving approximations that were made in developing the first-generation effort. In particular, the housing, shrouds, and support structure had been previously modeled as a cylinder with end caps to facilitate the use of cyclic-symmetric CFD simulation of the multiphase flow within gearboxes. The housing and support structure is now represented in a geometrically-accurate fashion, and the fully three-dimensional flow and conjugate heat transfer within and outside the gearbox is now accommodated. The previously-developed all lubrication regime line contact tribology model (Ref. 2-6) has been implemented as a Star-CCM+ co-simulation and successfully coupled with the second-generation aero-thermodynamics model, and the second generation framework has been tested and verified to function on simplified cases.

Grid generation and overset interface setup is nearly complete for the first loss-of-lubrication simulation of a full gearbox with complete geometrically-accurate housing (the PSU-ARL/GRI 3.5" spur gear rig), and these will be carried out soon. Following completion of this, an enhanced elliptical contact patch tribology model will replace the line contact tribology model currently in use, and efforts will be directed toward coupling the tribology model with numerical simulation of the lubricant distribution within the gearbox, whereas all coupling between the lubricant distribution within the gearbox and gear meshing friction are currently modeled empirically within the tribology model.

Author contact: Sean McIntyre smm5164@psu.edu, Robert Kunz rfk102@engr.psu.edu

ACKNOWLEDGMENTS

The authors are grateful to the U.S. Army's National Rotorcraft Technology Center program for co-funding this research. Effort sponsored by the U.S. Government under Other Transaction number W15QKN-10-9-0003 between Vertical Lift Consortium, Inc. and the Government. The U.S. Government is authorized to reproduce and distribute reprints for Governmental purposes notwithstanding any copyright notation thereon. The views and conclusions contained herein are those of the authors and should not be interpreted as necessarily representing the official policies or endorsements, either expressed or implied, of the U.S. Government.

REFERENCES

[1]McIntyre, S., "A Multi-Scale Model for Geared Transmission Aero-Thermodynamics," The Pennsylvania State University, 2016.

[2]McIntyre, S., Yu. Q., Kunz, R., Chang, L., and Bill, R., "A Computational System Model for Gearbox Loss-of-Lubrication," *70th Annual American Helicopter Society Forum*, American Helicopter Society, May 2014.

[3]Yu, Q., "Development of a Gear Meshing Tribology Model with Application to Loss-of-Lubrication Analysis," The Pennsylvania State University, 2016.

[4]Yu, Q., Chang, L., and Yeau-Ren, J., "A Mathematical Model for Rolling/Sliding Line Contacts in Boundary and Near Boundary Lubrication," *Proceedings of the Institution of Mechanical Engineers, Part J: Journal of Engineering Tribology*, 2015.

[5]Yu, Q., McIntyre, S., Chang, L., Bill, R., and Kunz, R., "A Boundary Lubrication Model Including Surface-Film Failure for Gear Contact Analysis under Loss-of-Lubrication Condition," *Proceedings of the Annual Forum of the Society of Tribologists and Lubrication Engineers*, May 2014.

[6]Yu, Q., McIntyre, S., Kunz, R., Chang, L., and Bill, R, "A Physics-Based Tribology Model and Parametric Assessments of Loss-of-Lubrication using Reduced-Order System Level Analysis," in *72nd Annual American Helicopter Society Forum*, West Palm Beach, 2016.

[7]Yu, Q., McIntyre, S., Kunz, R., Saribay, Z., and Bill, R., "A Multi-Scale Computational Framework for Modeling the Thermo-Fluid Dynamics of High Speed Gearbox Systems," *69th Annual American Helicopter Society Forum*, May 2013.

[8]Kunz, R. F., *NPHASE-PSU, V3.1 Theory and User's Manual*, The Pennsylvania State University Applied Research Laboratory, 2006.

[9]Handschuh, R. F. and Morales, W., "Lubrication System Failure Baseline Testing on an Aerospace Quality Gear Mesh," Tech. Rep., DTIC Document, 2000.

[10]Handschuh, R., Polly, J., and Morales, W., "Gear Mesh Loss-of-Lubrication Experiments and Analytical Simulation," Tech. Rep. NASA/TM-2011-217106, NASA Glenn Research Center, 2011.

[11]Handschuh, R. F., "Thermal Behavior of Aerospace Spur Gears in Normal and Loss-of-Lubrication Conditions," *71st Annual American Helicopter Society Forum*, American Helicopter Society, May 2015.

[12]Hirt, C. W. and Nichols, B. D., "Volume of fluid

(VOF) method for the dynamics of free boundaries," *Journal of Computational Physics*, Vol. 39, No. 1, 1981, pp. 201—225.

[13]Sussman, M., Smereka, P., and Osher, S., "A level set approach for computing solutions to incompressible two-phase flow," *Journal of Computational Physics*, Vol. 114, No. 1, 1994, pp. 146—159.

[14]O'Rourke, P. J. and Amsden, A., "A Spray/Wall Interaction Submodel for the KIVA-3 Wall Film Model," Tech. Rep., SAE Technical Paper, 2000.

[15]Ishii, M. and Hibiki, T., *Thermo-Fluid Dynamics of Two-Phase Flow*, Springer Science, 2nd ed., 2011.

[16]Drew, D. A., and Passman, S. L., *Theory of multicomponent fluids*, Vol. 135, Springer Science & Business Media, 2006.

[17]Berry, R. A., *Ensemble Averaged Conservation Equations for Multiphase, Multi-Component, and Multi-Material Flows*, Idaho, National Laboratory, 2003.

[18]Brackbill, J., Kothe, D. B., and Zemach, C. "A continuum method for modeling surface tension," *Journal of Computational Physics*, Vol. 100, No. 2, 1192, pp. 335—354.

[19]Jones, E. Oliphant, T. and Peterson, P., "SciPy: Open source scientific tools for Python," 2014.

[20]Moré, J. J., Sorensen, D. C., Hillstrom, K. E., and Garbow, B. S., " The MINPACK project," *Sources and Development of Mathematical Software*, 1984. Pp. 88—111.

[21]Wright W.B., "Validation results for LEWICE 3.0," 2005.

TH-1H Finite Element Model Validation and Flight Loads Development

Jeffery Brenna
Staff Aerospace Engineer
MERC
Warner Robins, GA, USA

Robert McGinty
Chief Scientist
MERC
Warner Robins, GA, USA

Nick Hatcher
Senior Mechanical Engineer
MERC
Warner Robins, GA, USA

Mary W. Schleider. P.E.
Chief, System Engineering
SOF/PR RW, USAF
Warner Robins, GA, USA

Gregory T. Wood
Aerospace Engineer
Rotary Wing ASIP Manager
SOF/PR RW, USAF
Warner Robins, GA, USA

ABSTRACT

This paper presents the testing and analysis methods used in estimation of TH-1H flight loads and validation of a finite element model using data gathered during a static crane lift and flight strain survey. Analysis methods and results are emphasized here as most testing methods and data have been shared in previous forums. This is followed by a discussion of the resulting computed load values and their verification through comparisons to operational parameters such as aircraft weight and vertical acceleration. It was found that computed loads and moments compare well to flight conditions. For example, main rotor lift force correlates well to the product of aircraft weight and vertical acceleration (Nz). Preliminary results also show that the computed static and dynamic loads lead to strain predictions that have a good overall trend when compared to measured values, but more work is needed to improved correlations with individual sensors. Further development of these loads will allow the H-1 program office to use them to better address structural repairs, flight severities, and risk analyses. Accurately predicted aircraft response at the sensors from both static and dynamic inputs will enable the development of stress spectra to be used for durability and damage tolerance analyses and service life development.

NOTATION

AHS – American Helicopter Society

ASIP – Aircraft Structural Integrity Program

DADTA – Durability and Damage Tolerance Analysis

EFH – Equivalent Flight Hours

FEM – Finite Element Model

G – Unit of Gravity Load

L/ESS – Loads / Environment Spectra Survey

MERC – Mercer Engineering Research Center

Nz – Aircraft Vertical Acceleration in G's

PITS – Point in the Sky

USAF – United States Air Force

W – Aircraft Weight

INTRODUCTION

The United States Air Force (USAF) is seeking to quantify the service life of its TH-1H Huey II helicopter fleet. The TH-1H is a recently modified UH-1H that serves as a training platform for USAF helicopter pilots. Mercer Engineering Research Center (MERC) is working several Aircraft Structural Integrity Program (ASIP) tasks to establish the service life and to manage the fleet. Figure 1 shows the roadmap for service life development, with completed tasks checked and the current task circled in red.

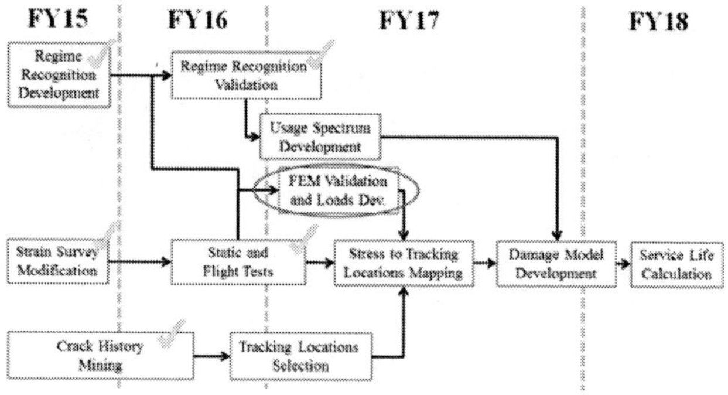

Figure 1. TH-1H Service Life Development

ASIP initiatives required for loads development and finite element model (FEM) validation included FEM development, usage data from developed regime recognition software in conjunction with a Loads / Environmental Spectra Survey (L/ESS), and strain gage and accelerometer

---

Presented at the AHS International 73rd Annual Forum & Technology Display, Fort Worth, Texas, USA, May 9-11, 2017.

data from a flight strain survey. This paper focuses on FEM validation, with background of the testing and analysis used for flight loads development for the TH-1H helicopter. The analysis is followed by a discussion of the resulting computed load values and their verification through comparisons to operational parameters such as aircraft weight and vertical acceleration (Nz).

TESTING AND ANALYSIS

ASIP initiatives that were required for loads derivation included creation and validation of a detailed global finite element model (FEM), a collection of usage data from the Loads / Environment Spectra Survey (L/ESS), and strain gage and accelerometer data from a flight strain survey. These tools and data sets were combined to develop a sensitivity matrix relating forces and moments on the aircraft to strains and accelerations at locations where sensors were placed in the flight strain survey. The sensitivity matrix was then inverted and used with the experimental data to compute forces on the aircraft for any point in the sky (PITS) and corresponding flight regime. In addition, statistical tests were incorporated in the analyses to identify and exclude statistically insignificant loads and to help identify and remove occasional outlying experimental data points.

FINITE ELEMENT MODEL

MERC developed the TH-1H Nastran FEM under previous ASIP tasks (Figure 2). Each structural part is modeled as a separate body and connections between parts are modeled with multipoint constraint equations. Most components are meshed with 2-D shell elements with a nominal mesh size of 0.5 inches on structural beams and stiffeners and 1 inch on skin panels. Areas immediately surrounding the strain gage locations from the strain survey are refined to 0.125 inch elements. The model contains 7,500,000 degrees of freedom, 1,250,000 nodes and 1,300,000 elements.

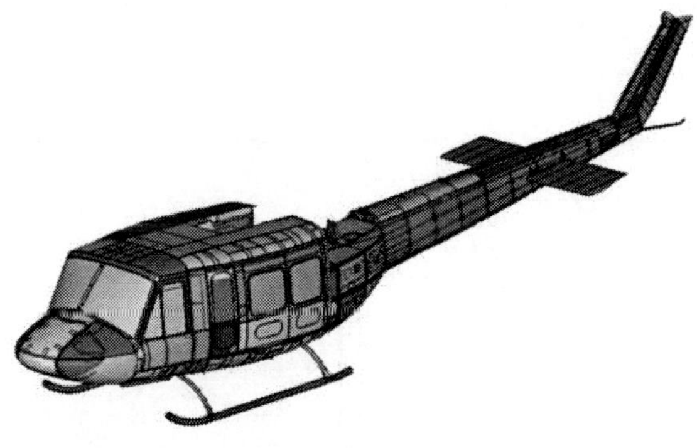

Figure 2. TH-1H Finite Element Model

FLIGHT STRAIN SURVEY

A flight strain survey was conducted, with over 200 hours of operational usage and 10 flight hours of prescribed maneuvers recorded. The TH-1H was instrumented with 59 uniaxial strain gages, two uniaxial accelerometers and six triaxial accelerometers. The strain gages were placed along the main load paths in the cabin and tailboom. Three of the strain gages were placed on aluminum blocks in various sections of the aircraft to monitor thermal drift and electrical noise. A map of the instrumentation for the strain survey is shown in Figure 3.

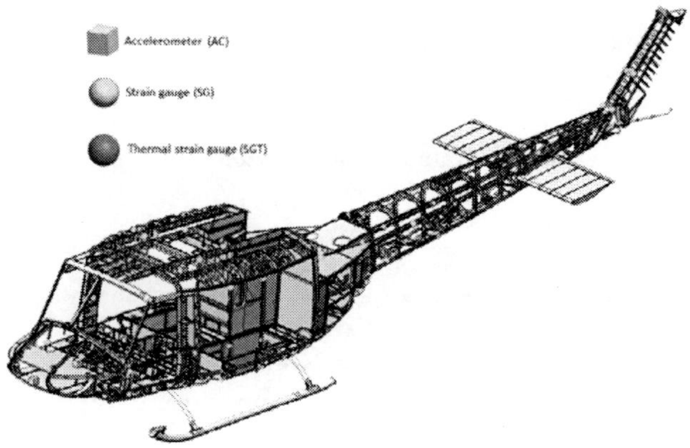

Figure 3. Sensor Map for the Flight Strain Survey

STATIC FLIGHT LOADS DEVELOPMENT METHODOLOGY

The approach used to determine flight loads closely follows the analysis method presented in Ref [1], but incorporates additional steps to properly account for extra complexities as described in Ref [2]. The analysis method begins by proposing that the strain, ε_i, at any location depends linearly on the input forces according to Equation 1, where F_j is the force (or moment) component value at the main rotor, tail rotor, elevator, etc, A_{ij} is a sensitivity coefficient relating the force to the strain value, and the products of the two are summed over the N forces and moments to obtain the strain in the ith gage, ε_i.

$$\sum_{j=1}^{N} A_{ij} F_j = \varepsilon_i \tag{1}$$

The A_{ij} terms are obtained from the FEM by successively applying 1 lb forces and 1 in-lb moments, one at a time, at all considered locations (Figure 4) while recording the resulting predicted strains at each strain gage location. The sensitivity coefficient, A_{ij}, is then equal to the value of the predicted strain at each ith gage for each jth 1 lb force (or 1 in-lb moment).

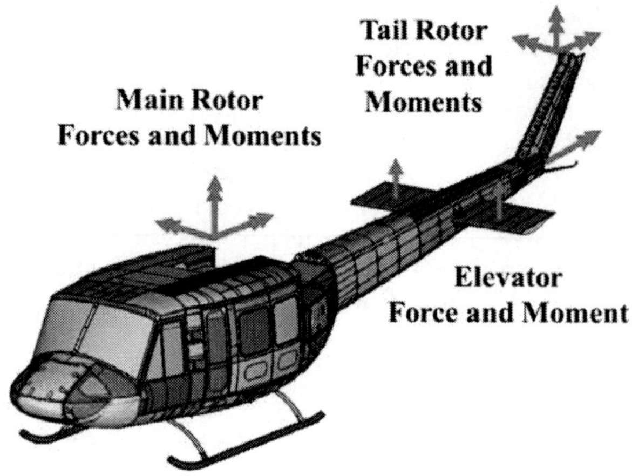

Figure 4. Unit Load Locations

The measured strain values from the gages are used for ε_i, but they must be updated to account for the fact that the aircraft is deformed by its own weight at the time of their application. This correction is made by using the FEM to predict the strain at each gage subject to 1G gravity load, ε_i^{1G}, and updating Equation (1) as follows

$$\sum_{j=1}^{N} A_{ij} F_j = \varepsilon_i + \varepsilon_i^{1G} \quad (2)$$

Equation (2) is written for each strain gage, which leads to more equations than unknown forces, F_j. Therefore, multivariable regression techniques are used to obtain the force values. This process is completed every 0.926 seconds, the time to obtain five complete revolutions of the main rotor, and is defined as one PITS. This definition is necessary to prevent side lobes from occurring in the FFTs used later for dynamic loads development; side lobes would occur if a PITS contained a partial cycle of the main rotor.

FEM VALIDATION

Each load type, F_j, is applied to the FEM to obtain the predicted strains at the strain gage locations for the predicted flight load set at that PITS. The set of predicted strains is compared to the measured strains to obtain an R^2 to quantify the correlation of predicted to measured strains. The results are used to identify locations of poor agreement between measured and predicted values and audit appropriate regions of the structure in the FEM.

Corrections to the model are made, and the updated structure is then used to develop a new flight loads set, creating an iterative process. This is repeated at several different regimes (level flight, pull out, turns, hover, etc.) until acceptable correlations for all gages are obtained. Once this is complete, loads are developed for the entirety of the strain survey, and the same analysis is repeated at a higher level, flight by flight.

DYNAMIC FLIGHT LOADS DEVELOPMENT

Dynamic flight loads development uses the same methodology as the static loads, but predicts dynamic component vibrations based on accelerometer and strain gage oscillations.

Modal analyses are conducted on the FEM to determine sensor output for various input vibrations. The same regression analysis used for static loads is used to predict the dynamic loads on the aircraft. The resulting predicted accelerometer and strain gage vibration amplitudes and phase lags are compared to measured data and used in the same sense as static loads to improve the FEM. This step is performed at each frequency of interest: Main rotor (1st, 2nd, 3rd, 4th, and 8th harmonics), tail rotor (1st and 2nd harmonics), tail rotor drive shaft (1st harmonic), and engine output shaft (1st harmonic).

ANALYSIS RESULTS

STATIC CRANE LIFT

A static crane lift was performed during the strain survey. This lift involved hoisting the aircraft off the ground by lifting only by the main rotor mast. This provided a known load case with which to validate the FEM. The results for the static crane lift are shown in Figure 5. The FEM predictions for all strain gages are within the 10% correlation bands. Because this is the only known load case available to apply to the model, it was anticipated that the strain predictions would be more accurate than those for estimated flight loads.

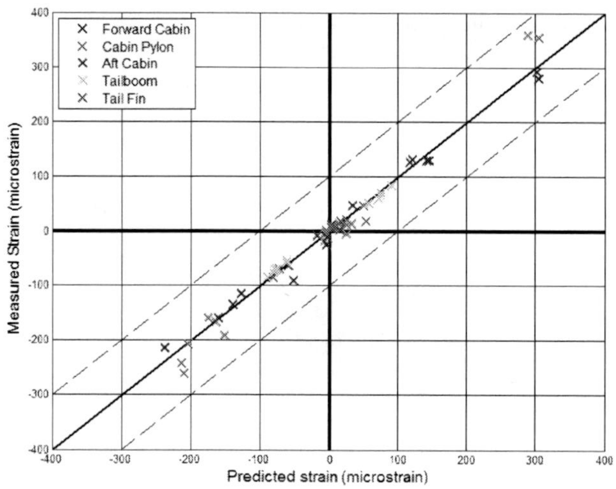

Figure 5. Static Crane Lift Results

STATIC LOADS DEVELOPMENT

Results of the static calculations to date are promising. The computed loads have been used with the FEM to predict strains and compare them to measured values. The resulting average R^2 across all ~20,000 PITS is 0.805.

Figure 6 shows the set of forces and moments [values not shown] found to be statistically significant for a particular climbing left turn. This load case resulted in an R^2 of 0.87. A comparison of the measured and predicted strains for the same predicted loads is shown in Figure 7. This shows a good overall trend, with most falling within the 10% correlation bands defined by Ref [3] for structure not considered safety-of-flight; it also shows that several gages require further investigation.

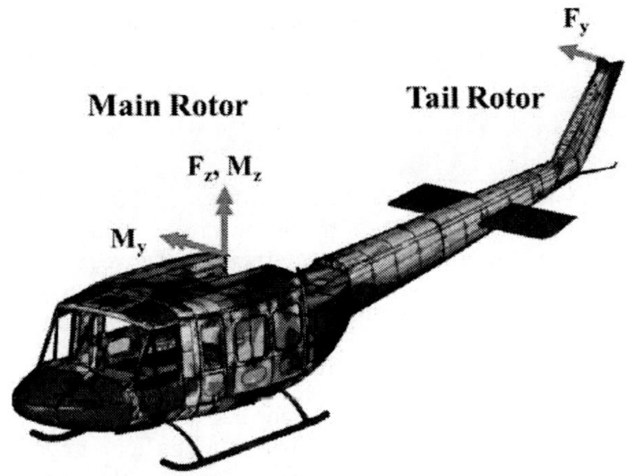

Figure 6. Predicted Loads and Stresses on the TH-1H FEM for a Climbing Left Turn Maneuver

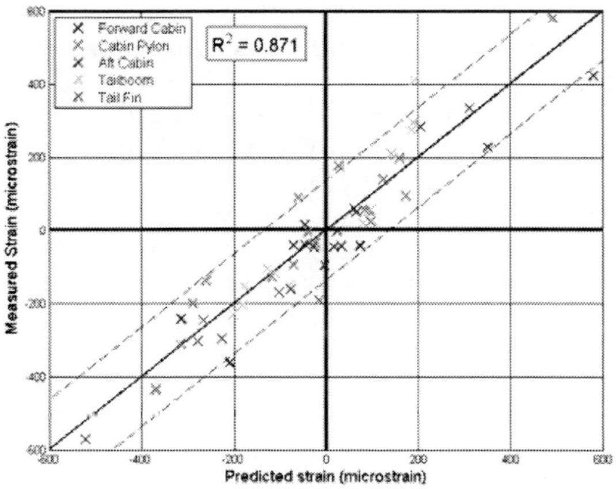

Figure 7. Measured and Predicted Strains for a Climbing Left Turn Maneuver

Figure 8 shows a histogram of R^2 values for the strain survey flight data processed through the time of this reporting. The R^2 values represent regression coefficients between the measured strain values and those predicted by the FEM when the computed loads are applied to it. It shows that 95% of the reliable recorded flight data has an R^2 value between 0.65 and 0.95.

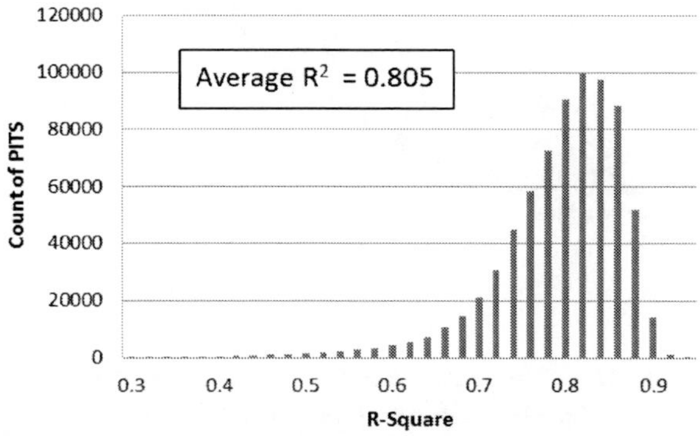

Figure 8. Distribution of R^2 Values

The effects of the 1G gravity loads at the time of strain gage application, ε_i^{1G}, for a subset of flights can be seen in Figure 9. It compares the R^2 values from Figure 8 (in blue) to the values obtained when not accounting for the effects of gravity (in red). The low magnitude of increase in R^2 values is caused by the small strains experienced at most gages under the 1G condition on the ground relative to those experienced in flight.

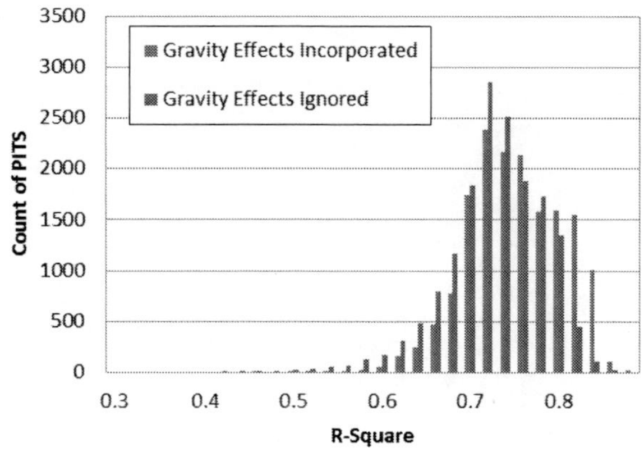

Figure 9. R^2 Distribution with and without 1G Gravity Ground Load Consideration

Computed forces have been found to vary as expected with flight regimes. For example, main rotor lift forces are higher for high-G pull-ups than for straight-level flight, which are in turn higher than for dives and push-overs. The computed lift forces have been found to trend well with (Nz * W), the product of Nz and aircraft weight (W), as shown in Figure 10. The data points in the figure are colored by Nz value and show that this accounts for large variations in the main rotor lift force.

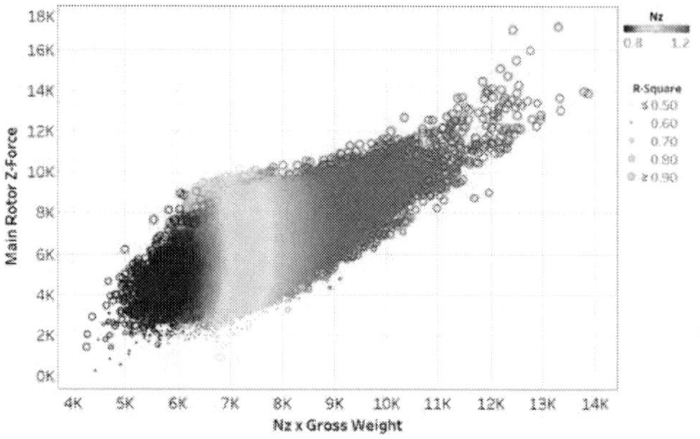

Figure 10. Main Rotor Z-Force Correlation with Vertical Acceleration and Gross Weight

Figure 11 shows the range of forces and moments computed for the TH-1H. The second and third quartiles and average load are shown in the center box, with the whiskers representing the absolute maximum and minimum predicted values. The top two bars show the Y and Z component moments on the main rotor. The remaining two bars represent the forces on the main and tail rotors. Other forces, such as elevator loads, were rarely statistically significant.

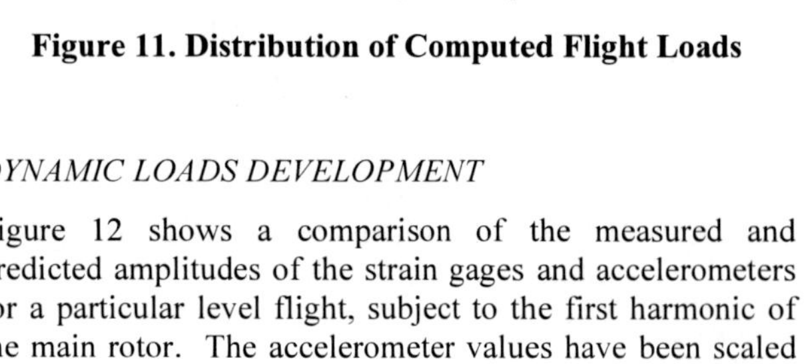

Figure 11. Distribution of Computed Flight Loads

DYNAMIC LOADS DEVELOPMENT

Figure 12 shows a comparison of the measured and predicted amplitudes of the strain gages and accelerometers for a particular level flight, subject to the first harmonic of the main rotor. The accelerometer values have been scaled up by a factor of 250, so that the strain gage and accelerometer data both have a role in loads development without an overbearing weight due to a difference in units.

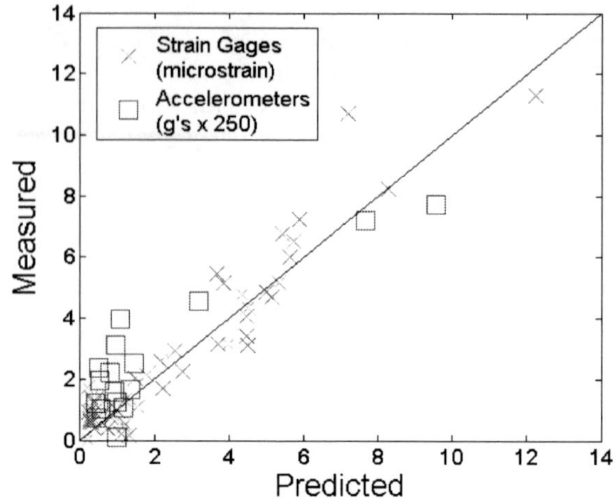

Figure 12. Measured and Predicted Data of Vibration Amplitudes for a Level Flight Maneuver, First Harmonic of the Main Rotor Frequency

This particular maneuver resulted in an R^2 of 0.92 for the strain gages, and an R^2 of 0.68 for the accelerometers. The R^2 values represent regression coefficients between the measured strain and acceleration amplitudes and those predicted by the FEM when the input vibrations are applied to it.

It is important to note there are seven times as many strain gages as accelerometers. Not only is it more difficult to get a stronger correlation with fewer samples, but the scale factor for the accelerometers is set to provide the best overall correlation without reducing the strain gage correlation. These preliminary results provide the data needed to show where the FEM requires further review.

Combining the results of the static and dynamic analyses, the complete response of the sensors can be predicted. The predicted aircraft response to both static and dynamic input will enable the development of stress spectra to be used for durability and damage tolerance analyses (DADTA), equivalent flight hour (EFH) accumulation, and service life development. The global FEM itself will confidently be utilized in global or local structural analyses as a result of loads development processes.

Once the FEM is validated and static and dynamic predictions are improved, the resulting loads will make it possible to better support service life extension efforts for helicopters in their new configurations and flight usage regimes as opposed to the original design specifications. The new load sets will also permit the H-1 program office to better address structural repairs and risk analyses for the fleet.

SUMMARY

This paper has presented the testing and analysis methods being used in the estimation of TH-1H flight loads and validation of the global finite element model using data gathered during a flight strain survey. Analysis methods and results were emphasized here because most testing methods and data were shared in previous forums. Forces and moments are being developed for 20,000 PITS of the strain survey at the time of this reporting and will be used for structural analyses, repair decisions, etc.

It was found that the predicted strains by the FEM trend well with measured values for both static and dynamic flight loads, but further effort is required to improve individual sensor correlations. Confidence in the computed force and moment values is also gained by observing the trend between the static loads and flight parameters such as aircraft weight and vertical acceleration. After further refinement, these loads will be used to develop stress spectra for use in DADTA, EFH, and service life calculations.

FUTURE WORK

Work is ongoing to improve the predicted strains and loads in the TH-1H aircraft through further review of the FEM and flight strain survey data. Additional considerations will be added in the future, including supplementary load types, fuel burn through flight, and ground loading conditions. For the dynamic analysis, other frequency inputs will be investigated to improve the predicted dynamic response. All of these considerations will help to improve future structural analyses with the global FEM, accuracy of predicted loads, and spectrum development for future service life calculations.

REFERENCES

[1]Fujimoto W.W., "Regression-Based Loads Correlation: A Tool for Improving the Accuracy of Damage Tolerance Analyses," The 1999 USAF Aircraft Structural Integrity Program Conference, Nov 30 – Dec 2, 1999.

[2]McGinty R., Wood G., Brenna J., Lamb S., "Determination of Flight Loads for the HH-60G Pave Hawk Helicopter," American Helicopter Society 71st Annual Forum, Virginia Beach, VA, May 2015.

[3]Babish C., Reams R., Day K., "Guidance on Correlating Finite Element Models to Measurements from Structural Ground Tests," United States Air Force Structures Bulletin EN-SB-11-001, 24 June 2011.

Pilot Evaluation of a Real-Time Free Wake Model in a Navy Rotorcraft Fleet Trainer

Jeffrey D. Keller
Senior Associate
Continuum Dynamics, Inc.
Ewing NJ, USA

Daniel A. Wachspress
Senior Associate
Continuum Dynamics, Inc.
Ewing NJ, USA

Jacques C. Hoffler
Senior Engineer
NAVAIR Flight Vehicle Modeling and Simulation Branch
Patuxent River, MD, USA

ABSTRACT

Accurate modeling of rotorcraft aerodynamics and induced flow fields in real-time is important for high fidelity simulation and training. The challenge is heightened for naval rotorcraft flight trainers that must accurately represent the complex aerodynamic interactions between the aircraft and disturbances from the shipboard environment. To address this challenge, a high-fidelity free-wake induced velocity software module that can model coupled rotor wake/ ship airwake interactions has been developed, integrated, validated, and evaluated through analysis and piloted simulations. Use of the free wake model also has allowed vibration cues to be generated that reflect the underlying operating state of the rotor system as opposed to a simulation effect. The free wake induced velocity model was incorporated into a naval rotorcraft fleet trainer and compared with the baseline model during a pilot evaluation session. This paper describes the execution and results of this pilot evaluation.

INTRODUCTION

Pilot training in a high fidelity, motion-base trainer provides a cost-effective, safe method to rehearse normal flight operations and emergency procedures. To adequately represent the aircraft, it is necessary to accurately model the aircraft flight dynamics, as well as replicate the cockpit, flight control characteristics, visual environment, and motion/tactile cues. For rotorcraft simulation, modeling the aerodynamics and induced flow field is of particular importance given the nature and complexities of the rotor aerodynamics and interactions. Modeling rotorcraft flight dynamics for training applications is challenging due to unique flight conditions such as low speed maneuvering flight near the ground, high speed blade stall, and launch and recovery from air-capable ships, as well as hazardous situations such as autorotation, brownout landings, vortex ring state, loss of tail rotor effectiveness, confined area operations and landing on ship decks in high sea states. For a rotorcraft simulator to be an effective training tool, aerodynamic and flight dynamics models must properly characterize the complex operating environment of helicopters in these situations, a precondition paramount for generating proper visual/motion and vibratory cues to the pilot in the virtual simulation environment.

Presented at the AHS International 73rd Annual Forum & Technology Display, Fort Worth, Texas, USA, May 9-11, 2017.

From the perspective of flight dynamics simulation, naval rotorcraft operations are more challenging to model due to the complex aerodynamic environment near the landing spot of a ship. To this end, advances have been made in computational fluid dynamics (CFD) models of the ship airwake aerodynamic disturbances (e.g., Refs. 1-3). Results from these offline CFD analyses have been compiled into ship airwake databases to expand the realism of naval rotorcraft flight dynamics simulators (e.g., Refs. 4-6). For pilot training applications, several modeling approximations are required to achieve real-time operation. Specifically, the rotor induced velocity is represented using low-order finite-state inflow models (i.e., Pitt-Peters dynamic inflow theory) with the ship airwake disturbance (linearly) superimposed to determine the rotor airloads and aircraft dynamic response. In this approach, the interactions between the rotor wake and ship airwake are not directly captured. While research studies have been performed examining the coupled rotorcraft-ship airwake problem (e.g., Refs. 7-9), none of these studies have focused on modeling these effects in a real-time environment with the pilot in-the-loop. Therefore, modeling the rotor wake/ship airwake interaction in the real-time pilot training environment was a central goal of the current research and development effort.

Previous work described in the literature (Refs. 10-12) has focused on the application of real-time free wake models to enhance rotorcraft flight dynamics simulation and pilot training applications. Higher-fidelity physics-based free wake methods are well suited for modeling aircraft ownship interference effects, aircraft-to-aircraft interactions, and phenomena that cannot be captured by simpler inflow models, e.g., vertical descent/vortex ring state (VRS). Free vortex wake elements also have been used to represent both rotor and ship airwake velocity fields, providing a means to capture rotor/airwake interactions for real-time simulation. As part of this previous research, the real-time free wake model has been validated with flight test data for vertical descent and shipboard operations (Ref. 11). Extensive model tuning was not needed for the free wake model, which allows aerodynamic model development to proceed with less dependence on flight test data. Thus, real-time free wake methods potentially can reduce time and cost associated with aircraft flight model development, as well as provide more representative simulations during aircraft preliminary design studies.

The current research and development has led to a software module for integration with an existing flight dynamics simulation architecture, which has been tested in several pilot evaluation sessions (Ref. 11). A common observation during these previous test activities is the importance of vibration cueing. In many rotorcraft trainers, vibration cues are provided using seat shakers and/or motion platforms as a simulation effect that is scheduled (tuned) based on pilot feedback during trainer development. Another application discussed in this paper is the use of a free wake model to provide improved vibration cueing for pilot training. To this end, the free wake software module was integrated with a naval rotorcraft fleet trainer with higher fidelity visual and motion cues. An additional pilot evaluation was performed in the fleet trainer, and results are described in this paper.

The improved physical fidelity of free wake models can provide accurate simulation of rotorcraft flight dynamics, in particular for compound aircraft with multiple rotating and fixed-wing thrust and propulsive components, without extensive wind tunnel or flight testing a priori. A subject of this paper, however, is to assess if an existing trainer, already tuned to match flight test data, can be further improved when the baseline inflow model is replaced with a real-time free wake model. The fleet trainer pilot evaluation provided data to this end. Following an overview of the real-time free wake model, as well as its application to vibration cueing, the fleet trainer test approach and results are presented.

ROTOR FREE WAKE INDUCED VELOCITY MODEL OVERVIEW

Background

The importance of accurate modeling of the rotor induced velocity is well established in many aspects of rotorcraft aeromechanics analysis. For flight dynamics simulation, the importance of inflow models for determining the rotor dynamic response, hub loads, and aerodynamic interference effects is also well established (Ref. 13). Dynamic inflow or finite-state inflow models (Refs. 14-16) are commonly included in the modeling framework of real-time rotorcraft flight dynamics simulations. This class of inflow models benefits from robustness and computational efficiency and has found widespread application in real-time rotorcraft simulation. Empirical and semi-empirical extensions have been implemented to improve these inflow models to capture flight dynamics response characteristics. Tuning of the rotorcraft aeromechanics and inflow models is an essential part of real-time flight dynamics model development as part of fielding a simulator/trainer system. Organizations responsible for simulator development adopt different and unique approaches for model development and tuning (e.g., Refs. 17, 18). It is desirable to seek alternative methods to reduce the effort required, which is one primary motivation for the research and development described in this paper.

Free-vortex wake induced velocity methods, in contrast to conventional approaches described above, model the spatial and time-varying evolution of shed and trailing vorticity from the rotor blades to determine the wake-induced flow field. The resulting induced velocity field can be evaluated at the rotor blades to provide feedback to the rotor airloads and/or away from the rotor plane to evaluate interference or downwash/outwash effects. Free wake models are commonly used as part of rotorcraft comprehensive analyses and have been applied successfully to the prediction of a broad range of rotorcraft aerodynamics and flight mechanics metrics (Refs. 19, 20). Historically, free wake methods have been confined to offline engineering analysis due to computational complexity. Research and development preceding the work described in this paper have led to a rotor free-wake induced velocity software module that is suitable for real-time flight dynamics simulation applications (Refs. 12, 21). This free-vortex wake module has been coupled into academic, commercial and military flight simulation software and shown to provide comparable and improved aerodynamic predictions relative to conventional modeling methods (Refs. 22-24).

CHARM Free Wake / Panel Model

The Comprehensive Hierarchical Aeromechanics Rotorcraft Model (CHARM) is a comprehensive analysis with a free wake model that has been implemented in the rotorcraft flight dynamics simulations described in this paper. Description and validation of the CHARM model has been documented in the literature (Refs. 10, 12, 20, 21). An overview of several aspects of the model relevant to the present study is provided below.

The CHARM free wake model represents the full span vortex sheet that trails from a rotor blade using a Constant Vorticity Contour (CVC) model. This approach represents vorticity components that arise from both spanwise and azimuthal variations in the bound circulation. The CVC wake model is implemented using multiple equal and constant strength curved vortex elements released from the rotor blade (see

Figure 1). Reduced order hierarchical fast vortex methods can be used to greatly reduce the computation time when necessary for real-time applications (Ref. 25).

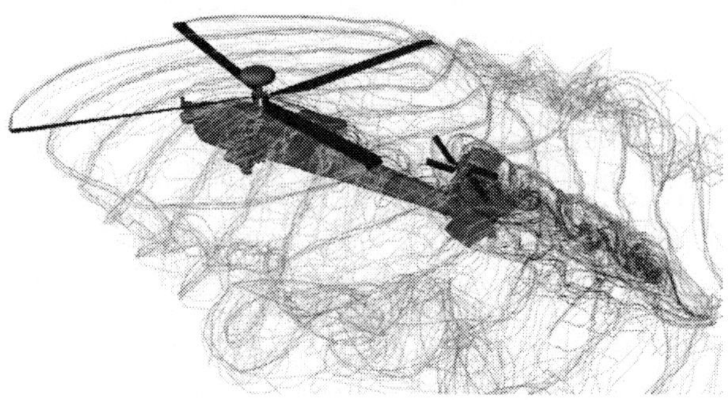

Figure 1. CHARM CVC Wake Model for the AH-64 Main and Tail Rotors

The wake-induced velocity field, which determines the free wake geometry and provides feedback to the rotorcraft aerodynamics loads, is determined from the numerical summation of each vortex element based on the Biot-Savart law. The Biot-Savart summation includes a vortex core model for each vortex element to prevent numerical singularities. The CVC model formulation does not preset the core model radius but instead determines the initial core size based on the blade circulation distribution. This approach concentrates vortex filaments in regions of high vorticity (e.g., near the blade tip) with smaller core radii. Downstream from the rotor blade, the full-span wake is consolidated into either a vortex pair (varying strength root and tip filaments) or a single tip vortex. When performing real-time free-vortex wake modeling, the number of vortex wake elements is typically reduced by transitioning from the full-span, near-wake model to a single tip filament far-wake model within the first rotor revolution. Model settings and internal algorithms have been developed to maximize the fidelity obtained in real-time applications (Refs. 11, 12, 26).

The CHARM module also contains a Hierarchical Fast Panel method that is fully-coupled with the rotor wake solution. The CHARM panel method originally was developed to model rotor-fuselage interactional effects on the free wake geometry and induced velocity field. The panel method permits the effects of general surface structures (e.g., ship deck, hangar, and superstructure) to be introduced into the flow field solution. In this application, the method uses source panel elements with source strengths determined to enforce no normal flow at each panel control point. The CHARM panel solution method has been developed to be computationally efficient and accurate, making use of hierarchical fast multipole methods for evaluation of vortex and source panel far field effects. In the current application, the panel method was used to model the ship landing deck, hangar wall and immediate upstream superstructure (see Figure 2), limiting the model resolution (i.e., number of panels) to support real-time landing operations. The effect of the ocean floor is modeled using an image system, which increases the number of source vortex elements by a factor of two. The panel model supports ship motion and the image system supports sloped terrain, though this latter feature is not required in the context of the current paper.

Validation of the CHARM ground effect and panel models is provided in (Refs. 20, 26). Reference 10 provides additional details of the CHARM model in support of the present research and development work. Figure 2 shows an image of the fully-coupled solution, with the CHARM Module modeling the main rotor, rotor wake, ship deck and ocean floor as described above and the ship airwake as described in the next section of this paper.

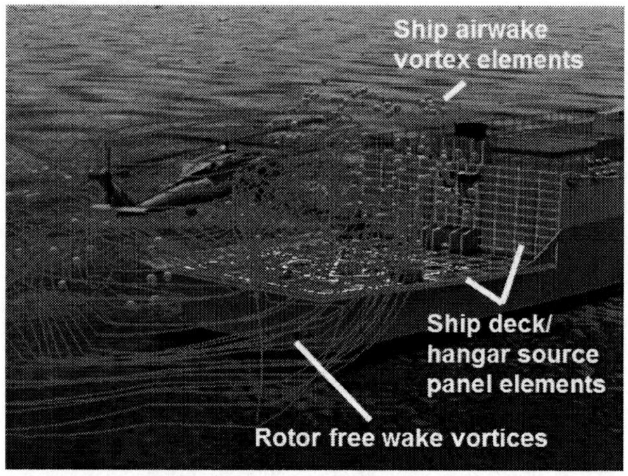

Figure 2. CHARM Model of Rotorcraft Shipboard Operations

Coupled Rotor Wake / Ship Airwake Model

Modeling the interactions between the rotor wake geometry and an aerodynamic disturbance that is characterized by a spatially and time-varying wind field (e.g., the ship airwake) has been a primary focus of this research and development. As discussed previously, the common approach used in rotorcraft flight simulation adds the effects of the ship airwake disturbance, typically determined from offline CFD analysis of a ship configuration, with the rotor inflow determined from dynamic/finite state inflow models. This approach neglects interactional effects between the rotor wake solution and aerodynamic disturbance wind field. Enhancement of the CHARM model to account for these interactions has been performed in research and development preceding the work described in this paper (Refs. 10, 11). A summary of the coupled rotor wake / ship airwake model is provided below.

Nonlinear interactional effects between the rotor wake and ship airwake occur from two primary wake distortion mechanisms. These wake distortion mechanisms provide feedback to the rotorcraft flight dynamics response through the wake induced velocity at the rotor and airframe. The first wake distortion phenomena is the distortion of the rotor wake due to spatial gradients in the wind field. For example, the flow around a structure (ship superstructure) results in regions of "velocity deficit" (blockage effects). The effect of this interaction is a distortion of the rotor wake structure relative to normal operation in "clear air". This effect is modeled in CHARM by accounting for the spatial and time-varying wind field effects in resolving the local flow field vector at each vortex element in the free wake solution. Additional details of this effect are provided in Refs. 10 and 11.

The second rotor wake / ship airwake distortion phenomena is the effect of the rotor wake on the ship airwake. As the magnitude of the rotor downwash increases, due to low speed flight and larger aircraft gross weight/ thrust, there will be a range of operational conditions where the rotor induced velocity will be of similar magnitude to the airwake disturbance, which is proportional to the relative wind over deck velocity. Under these conditions, the approximation of linear superimposition of rotor inflow and ship airwake disturbance can be questioned.

A rotor wake / ship airwake interaction model was developed and implemented in CHARM. The interaction model uses an existing ship airwake dataset that defines the spatial and time-varying flow field without the effect of the rotorcraft, such as an existing airwake database used in real-time rotorcraft simulation (e.g., Refs. 4-6). The ship airwake flow field is converted into an equivalent vortex particle representation. This conversion is performed as a pre-processing step during model initialization and thus incurs no additional CPU time during the piloted simulation. Vortex elements representing the ship airwake flow field are introduced into the CHARM computational domain based on the pre-processed airwake data. Once inserted into the computational domain, the ship airwake and rotor wake vortex elements evolve together under the influence of each other. The flow field velocity at each vortex element and at the rotor blades is determined from a Biot-Savart summation over all vortex elements, including image elements that account for the presence of the ocean floor. Since airwake source data typically are specified for a finite time duration, the airwake equivalent vortex solution is repeated. No time-based blending of the solution is implemented as any discontinuity is smoothed by the induced velocity summation.

Additional details of the CHARM coupled rotor wake / ship airwake model are provided in Ref. 10, which includes numerical results to verify the airwake equivalent vortex element method. Additional validation of the fully-coupled model through both comparison with flight test data and preliminary pilot evaluations is provided in Refs. 10 and 11. The remainder of this paper will focus on enhancements made to the model subsequent to these prior papers associated with acceleration cueing and a description of the execution and findings of the final pilot evaluation.

Flight Dynamics Simulation Integration

The CHARM induced velocity module was integrated with a generic blade element rotor model as an element of the rotorcraft flight dynamics simulation. This integration is performed using the NAVAIR Flight Vehicle Modeling and Simulation (FVMS) models library. Additional details can be found in the literature (Refs. 10, 11). The FVMS blade element model, which includes provisions for switching between CHARM and conventional dynamic inflow models, was used to formulate the flight model for the naval rotorcraft fleet trainer evaluation. This generic rotor model also was integrated with a generic helicopter simulation, designated the ExampleHelo (ExHel) simulation (Ref. 27). The latter model was used for engineering studies, such as the vibration cueing study described below, prior to the fleet trainer evaluation.

VIBRATION CUEING

Prior to the development and testing work described in this paper, several pilot evaluations were conducted within the NAVAIR Manned Flight Simulator (MFS) Multi-Mission Helicopter (MMH) facility at Patuxent River, MD (Refs. 10, 11). One comment that was consistent throughout all previous pilot testing was the desire for vibration cues to provide an indication of transition between flight regimes (e.g., hover-to-forward flight translational lift). In particular, comments from one pilot evaluation session suggested that vibration cues could aid the assessment of VRS entry since vibratory loads increase as the fully-developed VRS state is approached. To illustrate this behavior, analysis of a vertical descent/VRS entry simulation case was performed in which an open loop collective input was applied to mimic pilot inputs in conjunction with an autopilot (pilot model) to minimize roll, pitch, and heading excursions (Ref. 11). Time histories of the collective input, descent speed, and vertical acceleration (load factor n_z) are shown in Figure 3 for one simulation case. During the initiation of the descent, the vertical rate approximately tracks the collective input. The descent rate, however, continues to increase when the collective stick is held fixed approximately 35 seconds into the simulation. At this point, large changes in the vertical load factor can be observed. These characteristics are indications of fully-developed VRS conditions, which also were confirmed using wake visualization. When the collective is rapidly increased at 40 seconds, the descent rate does not appreciably decrease, another known characteristic of VRS.

The relationship between flight condition and vibration level can be demonstrated by spectral analysis of the vertical acceleration. Figure 4 plots the autospectra (PSD) for the simulation data from Figure 3 determined for multiple 10-second windows. These windows correspond to hover (0-10 sec), vertical descent (10-20 sec, 20-30 sec), VRS entry (30-40 sec), and established VRS (40-50 sec, 50-60 sec) phases of the simulation. It can be seen that as VRS conditions are established and maintained, the vertical acceleration power increases by two orders of magnitude, corresponding to an increase in the vibration amplitude of one order of magnitude. Although the vibration response magnitude has not been validated, the relative change can be used to drive an appropriate pilot cue (e.g., motion seat/shaker), which is derived directly from the physical fidelity of the model.

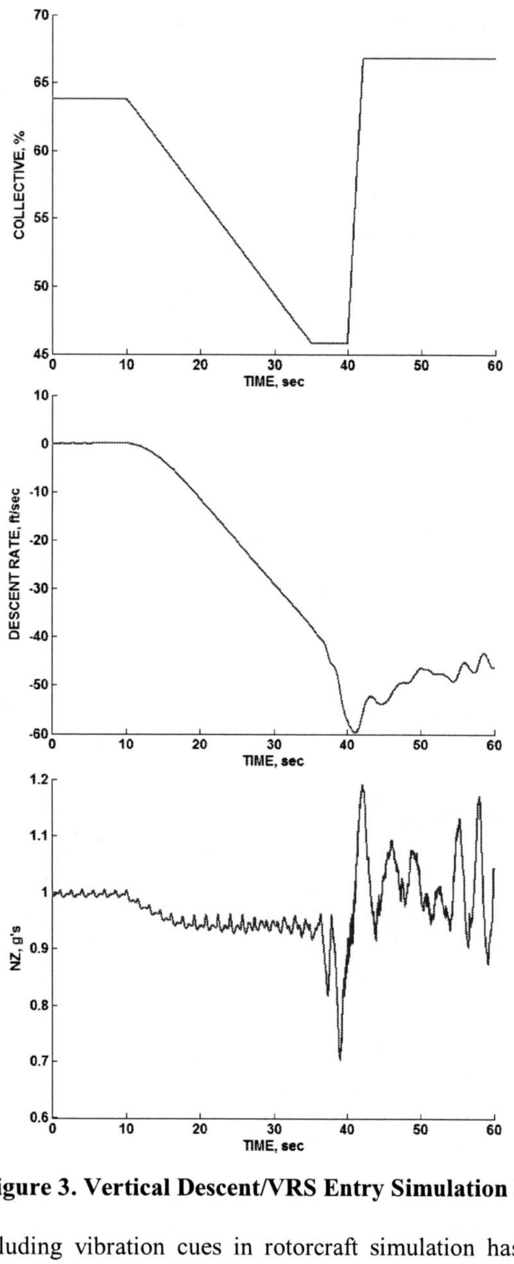

Figure 3. Vertical Descent/VRS Entry Simulation Data

Including vibration cues in rotorcraft simulation has been shown previously to improve simulator fidelity ratings (Ref. 28). Furthermore, improved fidelity may be achieved if the vibration cues are derived from modeling of the underlying physical mechanisms that affect the vibratory loads rather than adopting an approach that treats the vibration cue to the pilot as a simulator special effect. While prediction of pilot seat and cabin vibrations is technically challenging and involves many phenomenological mechanisms including rotor blade aerodynamics/stall, blade motion and flexibility, and fuselage structural response, one important mechanism is the rotor wake and its behavior between flight regime transitions. For example, flight through translational lift conditions (V=20-30 knots) results in a rapid increase in the seat/cabin vibration level, which occurs partly due to the transition of the hover/low speed skewed helical wake structure to rolled-up super vortices in forward flight. Use of a free wake model can provide an indication of these flight regime transitions due to the rotor wake, which can be used to support vibration cue generation to a seat shaker or vibration platform in rotorcraft trainer.

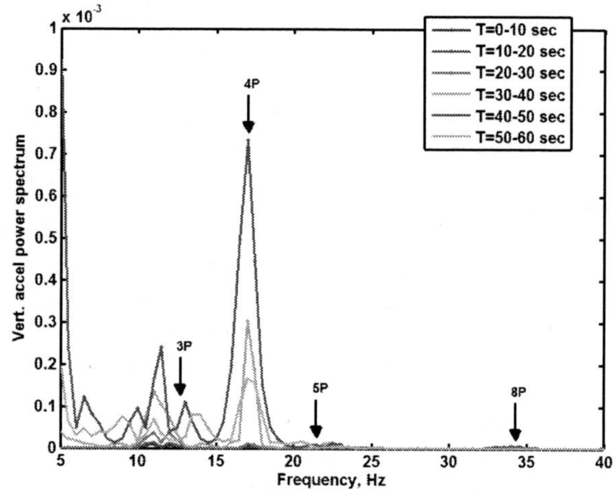

Figure 4. Vertical Acceleration Autospectra for 10-sec Windows from Simulation Data of Figure 3

CHARM Vibration Modeling Assessment

An investigation was conducted to assess the capabilities of CHARM when combined with a blade element rotor model for vibratory loading prediction in a real-time flight dynamics simulation. This investigation used the NAVAIR ExampleHelo/CHARM simulation to assess the prediction of fixed-frame vibration response characteristics to provide a basis for tactile cueing in a pilot trainer. The investigation compared the prediction of accelerations at the center of gravity and pilot station using variations of the inflow model including both baseline dynamic inflow models and CHARM with high resolution and real-time settings.

For results generated in this study, the simulation was trimmed for a target velocity in level flight. Following trimming of the airframe, the simulation was run for 10 seconds and data recorded for spectral analysis. Both stability augmentation system (SAS) and flight path stabilization (FPS) flight control loops were enabled during the simulation to assist in maintaining the target airspeed and aircraft attitude. Small deviations from trim were observed, in particular for low speed conditions with the CHARM model enabled, although these deviations were not significant.

Autospectra for the three helicopter acceleration components at the aircraft center of gravity (AX, AY, AZ) are shown in Figure 5 and Figure 6, corresponding to 30-kt and 160-kt level flight conditions respectively. Well defined spectral peaks can be seen in all responses near 17 Hz and 34 Hz, which correspond to the 4/rev and 8/rev vibratory loading excitation. For the 30-kt condition, the baseline inflow model under-predicts the CHARM vertical acceleration vibration response. This result occurs since vibratory loads in this flight regime are primarily influenced by the wake geometry transitioning from a helical structure to a pair of rolled-up super vortices. CHARM with real-time settings also tends to predict a higher 4/rev vertical response magnitude than the higher resolution model. For the 160-kt condition, the difference between the baseline and CHARM inflow models is less significant. This result is reasonable since vibrations in this flight regime are driven primarily by blade stall, and a common stall model was used by both baseline and CHARM flight dynamics simulations.

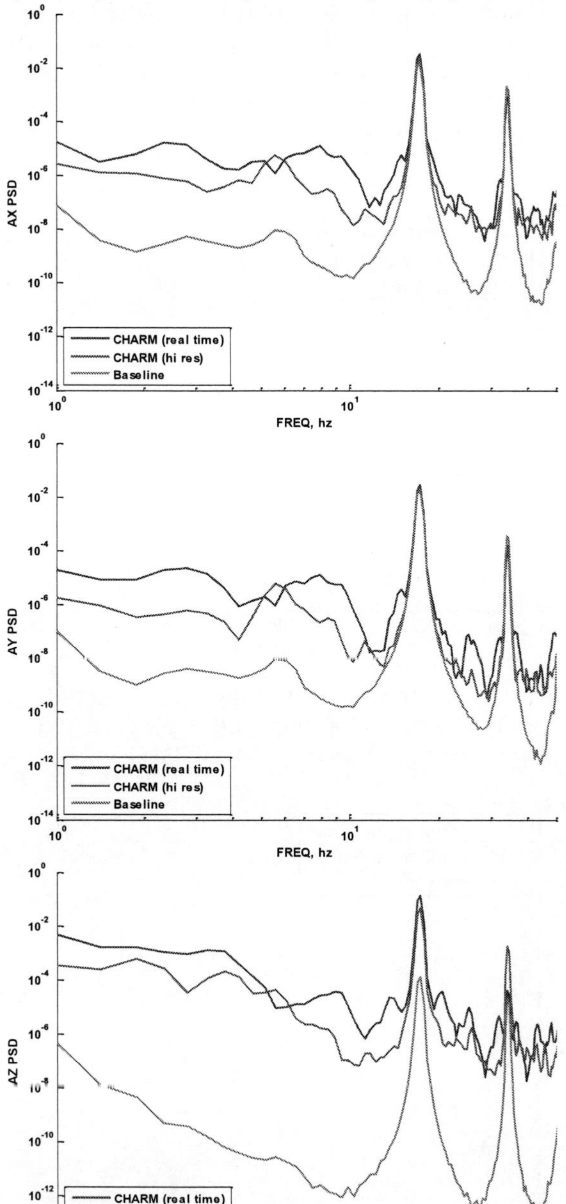

Figure 5. Effect of Inflow Model of Predicted Vibration Response, V = 30 kts

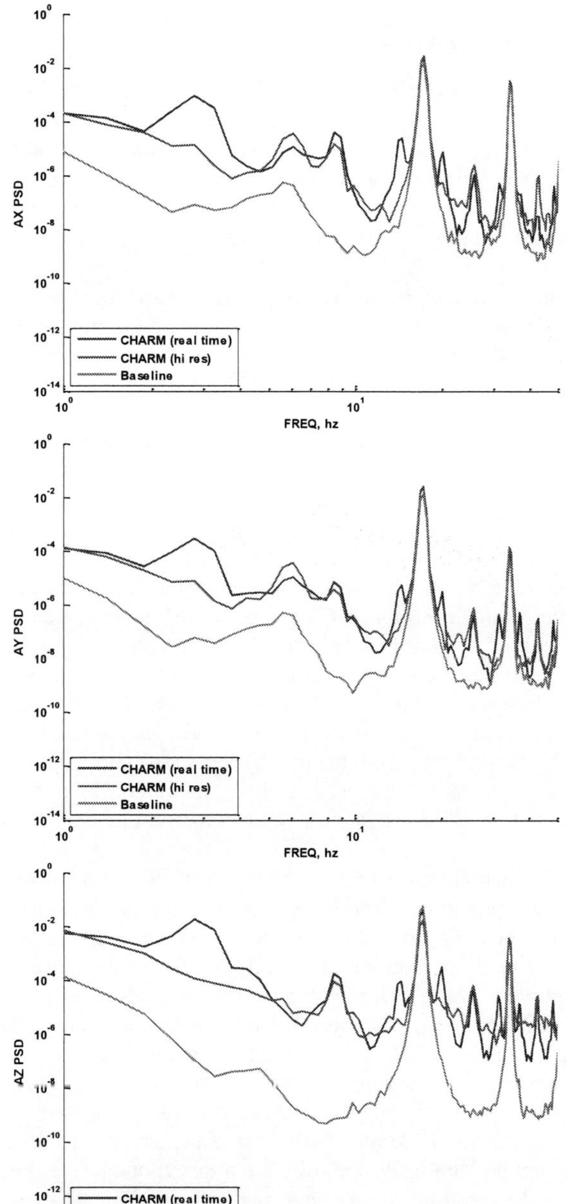

Figure 6. Effect of Inflow Model of Predicted Vibration Response, V = 160 kts

Assessments also were made for vibratory accelerations at the pilot station. Note that the flight dynamics model neglects some physical mechanisms that contribute to the prediction of the cockpit and cabin vibration levels, such as fuselage flexibility. Relative changes in the vibratory loads due to flight regime transitions, however, appear to be captured by the real-time model. Thus, it is expected that the predicted accelerations from the CHARM model can be used to generate a pilot vibration cue that may be beneficial in assessing model fidelity in the final pilot evaluation.

Vibration Cue Generation

Additional investigation was performed to identify an approach for applying the simulation acceleration response to drive a seat shaker or vibration platform in a rotorcraft flight trainer. The primary focus was identification of a signal processing method for generating a pilot cue from the predicted simulation acceleration output.

The primary objective of the signal processing scheme was to isolate the N/rev vibration response while removing low-frequency components associated with helicopter maneuvering. Two filtering methods were examined: a band-pass filter centered on the 4/rev frequency and a high-pass washout filter. The band-pass filter was examined first using simulation data recorded from previous batch simulation. A second-order Butterworth filter with pass band center frequency of 17.2 Hz (4/rev frequency for the H-60) and band width of 1 Hz was applied to the simulation for different level flight conditions. Simulation data were filtered in forward and reverse directions to yield zero phase shift in the filtered time history and fourth-order roll-off. The filtered signals were then used to determine an RMS vibration level. Note that the first and last second of the filtered time history was discarded prior to calculating the RMS magnitude to minimize biasing the result due to the filter starting transients. This approach was verified by comparing to integrating under the autospectra PSD and found to give consistent results.

Figure 7 illustrates the 4/rev vertical vibratory acceleration RMS amplitude at the aircraft center of gravity as a function of flight speed. Figure 8 illustrates the corresponding results for the vertical acceleration at the pilot seat. Both Figure 7 and Figure 8 illustrate the effect of the inflow model (baseline dynamic inflow, CHARM with real-time and high-resolution settings). The CHARM model captures the increase in vibrations in the translational lift transition, which then decrease as forward speed is increased until rising due to blade stall above 150-knots. The baseline dynamic inflow model captures the blade stall effect but does not capture the low-speed transition. This trend qualitatively is consistent with published vibration data and expected trends based on comments from pilot evaluations conducted previously in the NAVAIR MFS MMH facility (Ref. 11).

Note that the pilot station vibration acceleration predicted by CHARM (Figure 8) contains a different trend with forward speed, which generally increases except for a plateau in the translational lift regime. Differences between aircraft and pilot acceleration results can be attributed to the aircraft pitch acceleration, which includes excitation of the fuselage response due to N/rev aerodynamic perturbations at the empennage from the rotor wake. The baseline model response also is unrepresentative. Since the pilot seat vibratory response will be affected by fuselage structural modes, which are not modeled in the flight simulation, and the response at the aircraft center of gravity is qualitatively consistent with expected trends, it was concluded that a seat shaker or vibration platform should be driven by the simulation acceleration at the center of gravity.

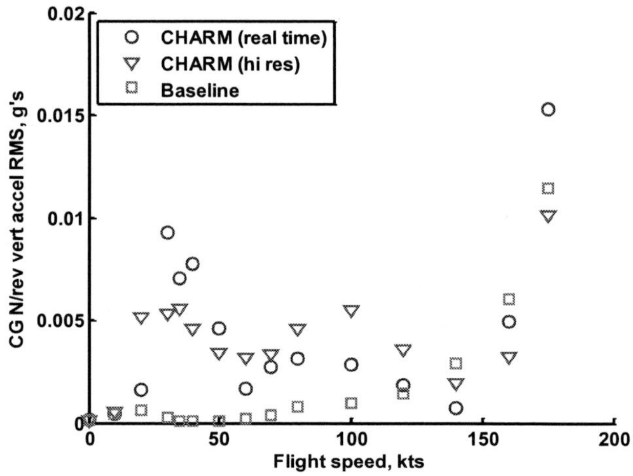

Figure 7. Aircraft N/rev RMS Vibration Magnitude versus Flight Speed with Rotor Inflow Model Effect

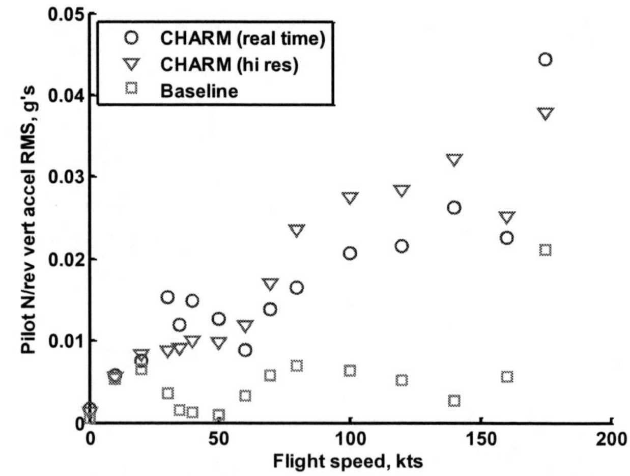

Figure 8. Pilot Station N/rev RMS Vibration Magnitude versus Flight Speed with Rotor Inflow Model Effect

To verify the vibration acceleration cueing and signal processing approach, a demonstration was developed using an in-house piloted simulation that included two tactile transducers to provide vibratory cues. These tactile transducers (Clark Synthesis TST209) were commercial components used in home theater applications. A custom software interface was developed to convert the simulation acceleration output to an audio signal for driving the tactile transducer seat shakers. It was found that the band-pass filtering scheme generated a vibration cue that appeared artificial due to the bandwidth of the tactile transducers. Experimentation indicated that a second-order high-pass washout filter with 12 Hz break frequency applied to the vertical acceleration provided better results. The seat shaker driver software interface included an interface to vary the overall gain at run-time to allow for tuning of the vibration cue. Aside from this overall gain tuning parameter, the seat shaker response was driven by the physical model of the flight dynamics simulation, and therefore provided appropriate vibration cues as a function of flight condition (e.g., translational lift transition, VRS entry). This verification test provided confidence that a suitable vibration cue could be generated in the fleet trainer pilot evaluation.

FLEET TRAINER EVALUATION

Facility

Evaluation of the CHARM free wake model was performed in a naval rotorcraft fleet trainer based on recommendations from evaluations previously conducted in the MFS MMH facility (Ref. 11). The evaluation used the Naval Air Station North Island (NASNI) MH-60R Tactical Operation Flight Trainer TOFT2 facility. This facility provided improved visual and motion cues including vibration cueing using a motion seat as compared to the MMH trainer. Note that the TOFT2 trainer originally was developed by NAVAIR, and familiarity with the existing hardware/software allowed modifications required for this effort to be implemented in a more straightforward manner.

One limitation of the TOFT2 facility was that the server hosting the airframe model, which was based on a rotor disk formulation, was insufficient to run the blade element model with CHARM. A new airframe server was procured to host the CHARM-based flight model for this evaluation. The new airframe server was hosted on a Dell Precision T7810 desktop with dual Intel Xeon E5-2667 v3 (3.2 GHz) processors, which provided 16 computational cores (32 virtual cores). Hardware interface cards (multi-port Ethernet, reflective memory) were installed to support the TOFT2 trainer architecture. In addition to running the airframe model, wake visualization software based on the X-Plane commercial flight simulation also was installed to support development testing.

Airframe Model

The airframe flight model was developed from the FVMS models library blade element model, which was used in previous pilot evaluations tests and engineering studies (i.e., the ExampleHelo model). The FVMS models library included the CHARM free wake model with coupled rotor wake-ship airwake interactions (referred to as the CHARM/SI module). The CHARM model requires separate tabulated data for the ship airwake derived from CFD analysis without the helicopter present. For the fleet trainer evaluation, airwake models were provided by NAVAIR for a DDG-81 class ship. Airwake data were available for 25-kt relative wind speed and three relative wind angles (bow, 30-deg starboard, 45-deg starboard). A separate utility was developed to re-sample the NAVAIR data files to the grid structure needed for the CHARM coupled rotor/airwake model, in addition to scaling the data to any alternate relative wind speed if desired.

To assess the effect of the CHARM free wake model on the simulation, a baseline flight dynamics model was developed using the FVMS blade element model with rotor induced velocity effects from a conventional nonlinear dynamic inflow model. For simulation of shipboard operations using the baseline model, the ship airwake model for the original TOFT2 airframe simulation (referred to as the "SHAD model") was used. It was found during development prior to the pilot evaluation that the definition of the relative wind angle for the SHAD model was reversed between starboard and port directions. Due to time constraints in the trainer, the relative orientation of the NAVAIR DDG-81 airwake models used with CHARM was reversed rather than identifying and resolving the issue with the SHAD model.

Vibration Acceleration Cueing

The MH-60R TOFT2 motion seat software interface provides two methods for vibration acceleration cueing: harmonic N/rev excitation based on rotor speed and direct feedback of the airframe acceleration components. For normal training operations, the rotor speed harmonic excitation interface is used due to the airframe disk model formulation. For this pilot evaluation, direct feedback of airframe acceleration components was used.

The TOFT2 motion seat responds to steady accelerations and N/rev excitation so that it was necessary to refine the acceleration filter to avoid objectionable vibration cues. This development was needed since the frequency response of the TOFT2 motion seat was substantially different than the tactile transducers used in preliminary in-house testing. Furthermore, little to no information of the motion seat response characteristics was available so refinements to the acceleration filter were made empirically. These developments were made at the TOFT2 facility in the days preceding the pilot evaluation test based on engineering judgment. It was discovered early during the pilot evaluation test, however, that the initial acceleration filter produced sub-harmonic excitation (below N/rev). This sub-harmonic excitation would appear in flight conditions between 60 and 80 knots. Ultimately, a narrow band-pass acceleration filter centered on N/rev (approximately 17 Hz) was implemented. Pilot feedback indicated that this filtering scheme produced more representative results but the amplitude was generally smaller than desired. Note that increasing the filtered acceleration amplitude with a constant gain improved the vibration cue to the point when signal saturation was observed and the clipped signal resulted in an objectionable cue. Unfortunately an ideal solution was not obtained without direct knowledge of the motion seat hardware interface and frequency response characteristics.

Test Approach

A formal test plan was developed by the evaluation team (pilot and test engineer), which outlined background and scope of the test, test methods and procedures, evaluation criteria, and program and risk management. Specific test conditions included:

- Hover IGE/OGE performance
- Level flight performance
- Blade stall onset
- Land based operations – normal take-off/approach, steep approach, vertical landing from hover, sideward/rearward flight, running take-off/landing, crosswind take-off/landing
- Vertical descent / vortex ring state entry
- Autorotation
- Shipboard operations – visual approach to ship, shipboard hover/reposition, shipboard departures

For each test condition, both baseline (dynamic inflow) and CHARM/SI models were evaluated in a blind manner (i.e., Model A versus Model B). Testing was conducted over three days. On the final day, some test conditions were repeated to examine sensitivity of model parameters.

The remainder of this section provides a summary of the major findings from the formal pilot evaluation test, in particular those findings that were expected to be affected by the rotor inflow model. Statements made herein reflect the authors' interpretation of comments and feedback from the test team during the evaluation and debrief.

Rotor Torque / Performance

Accurate rotor torque indications clearly were of great importance to the pilots during pilot evaluations in the MFS MMH facility and for this evaluation. The baseline MH-60R TOFT2 model did not have deficiencies in rotor torque predictions, but likely required model tuning to capture the correct trends. Thus, it was not expected that the CHARM/SI model would provide an improvement, but the objective of this test was to produce representative behavior without excessive model tuning.

Torque indications were recorded for several different aircraft configurations, flight conditions and density altitudes, which are summarized in Table 1. Tabulated indicated rotor torque values, normalized by NATOPS values, are provided for the baseline (dynamic inflow) and CHARM simulations. Test points included in and out of ground effect (IGE, OGE) conditions at different density altitudes with IGE cases performed with radar altimeter reading of 10 feet above ground level (AGL). Results from forward flight performance testing are summarized in Table 2. Both models yielded good results with a slight improvement noted with CHARM as compared to the baseline (blade element model with dynamic inflow).

Table 1. Normalized Hover Performance Results

Gross Weight (lbs)	Density Alt (ft)	IGE / OGE	Baseline	CHARM
22000	-2000	IGE	1.02	1.04
	-2000	OGE	0.94	1.00
	0	IGE	1.00	1.02
	0	OGE	0.93	0.96
	2000	IGE	1.00	1.01
	2000	OGE	0.90	0.93
19000	-2000	IGE	1.01	1.07
	-2000	OGE	0.94	1.00
	0	IGE	1.01	1.05
	0	OGE	0.94	0.99
	2000	IGE	0.99	1.01
	2000	OGE	0.93	0.97
16000	-2000	IGE	1.00	1.06
	-2000	OGE	0.94	1.04
	0	IGE	1.00	1.06
	0	OGE	0.96	1.01
	2000	IGE	1.00	1.06
	2000	OGE	0.96	1.00
	8000	IGE	0.98	1.03
	8000	OGE	0.93	0.97
	10000	IGE	1.00	1.02
	10000	OGE	0.92	0.94
	12000	IGE	1.00	1.03
	12000	OGE	0.91	0.93

Table 2. Normalized Level Flight Performance Results

Pressure Alt (ft)	OAT (°C)	Gross Weight (lbs)	Airspeed (KIAS)	Baseline	CHARM
8000	-20	18000	60	1.11	1.06
	-20	18000	120	1.19	1.19
	-20	18000	Vh	0.91	0.93
	-20	22000	60	1.16	1.06
	0	18000	60	1.11	1.06
	0	22000	60	1.17	1.09
	30	18000	60	1.12	1.02
0	-20	18000	60	1.04	0.98
	-20	18000	120	1.13	1.13
	-20	18000	Vh	0.94	0.94
	0	18000	60	1.06	0.98
	0	18000	120	1.09	1.09
	0	18000	Vh	0.97	0.95
	30	18000	60	1.07	1.07
	30	18000	120	1.04	1.04
	30	18000	Vh	0.97	0.99

For the CHARM/SI module, it is necessary to truncate the free wake extent to support real-time operation, which typically causes the indicated torque to be slightly under-predicted in hover OGE conditions. A modeling correction has been implemented within the CHARM/SI module to account for wake truncation effects on the rotor downwash prediction. This correction has been developed to phase out with advance ratio and height above the ground (Ref. 29). Some minor adjustment to this model correction was required based on initial testing prior to the pilot evaluation execution, but the resulting model was found to give representative behavior over the test conditions investigated (see Table 1 and Table 2). Additional investigation in the future should be made to assess if the model correction adjustments needed for the MH-60R TOFT2 trainer are unique to this implementation or require further model improvement.

Blade Stall

Evaluation of the entry into flight conditions resulting in blade stall was performed. It was anticipated that blade stall would be characterized by a pitch-up and roll response, in addition to an increase in N/rev vibrations. This behavior was not observed with either baseline or CHARM models. This result was not surprising since the rotor inflow and wake model do not significantly impact aerodynamic stall on the rotor blades. Further work to improve blade stall modeling in real-time simulation is recommended.

Translational Lift

As part of land-based operations testing, flight through translational lift conditions was assessed. The expectation was to observe an increase in the vibration level with the frequency content predominantly near N/rev. This behavior was not present in the baseline model. With the CHARM model enabled, an increase in vibrations was noted but the levels were substantially smaller than desired. This observation likely occurred due to the limitations of the motion seat hardware/software interface as discussed previously. Adjustments made to the airframe acceleration filter during the pilot evaluation provided improvements but did not produce a satisfactory level of vibration cueing.

Unlike blade stall, increased vibrations during translational lift were present in the CHARM model, as shown in the previous section of this paper. With better insight and more control over the motion seat hardware/software interface, the CHARM model should provide improved vibration cueing. This aspect of CHARM implementation within a rotorcraft trainer should be explored more in the future.

Vertical Descent / VRS Entry

Assessment of vertical descent/VRS entry characteristics was an objective of the fleet trainer evaluation test matrix. Based on feedback from earlier pilot evaluation tests, it was anticipated that the availability of pilot vibration cues would assist the assessment of vertical descent and VRS entry. With adequate vibration cues, it was anticipated that pilot control strategy would be more representative during descent since the vibratory acceleration would increase as VRS conditions were approached.

As observed in previous validation work and pilot testing, the CHARM/SI model provided more accurate modeling of VRS conditions when compared to the baseline dynamic inflow model, which produced non-representative behavior. In this test, the CHARM model captured both the increased vibration and degradation of handling qualities as the aircraft entered VRS conditions, which was not the case for the baseline model. The pilot, however, did not observe an increase in descent rate when collective is increased while in VRS conditions, which was the behavior he expected. During separate investigation of predicted VRS response characteristics using CHARM, it was found that the aircraft descent rate decreased slightly but remained relatively unchanged when collective was increased when in VRS conditions (see Figure 3). It should be noted that an increased descent rate following an increase in collective is not always observed in VRS (e.g., Ref. 30). During this pilot evaluation, the pilot found that the vertical descent could be arrested by increasing collective, which was not representative.

It should be noted that several formal and informal pilot evaluations were performed with the CHARM/SI model during this effort, and more positive feedback was received during informal evaluations. In particular, the CHARM model captured the loss of control and handling qualities degradation during VRS conditions. Increase in the descent rate following a collective increase during vertical descent was observed less consistently. This behavior may be more sensitive to the time/rotor flow state at which the collective input is applied. Proper recovery from VRS conditions, i.e., reducing collective and application of forward cyclic to increase forward speed, was captured by the CHARM model. Thus, the CHARM model may be applicable to the simulation of VRS entry and recovery, and additional research may be needed to understand current model limitations.

Autorotation

Several autorotation landings were performed. Both the baseline model and the CHARM model were representative of autorotation performance.

Shipboard Operations

As noted earlier, the CHARM/SI model captures ship airwake/rotor wake interactions while the baseline SHAD model does not. Of interest is the fact that in this particular pilot evaluation of a MH-60R landing on a DDG-81, both baseline airwake (SHAD) and CHARM/SI models provided representative simulation of the shipboard environment. Although both models were considered sufficient for training applications, there were some minor improvements observed with the CHARM model.

First, when translating through the outer range of the baseline SHAD model, a small transient (described as a slight "burble") was apparent. This transient occurred as the far field influence of the model was applied to aerodynamics model and could be mitigated by phasing in the airwake effect to the simulation. Due to its vortex-based rather than grid-based formulation, the airwake influence is always present in the CHARM/SI model implementation regardless of spatial separation between the ship and aircraft.

Another observation was that the baseline model produced a non-representative transient as the aircraft translated across the ship deck edge. This transient, described as a "sudden jump", was more pronounced for lateral reposition tasks with little to no wind. The behavior was caused by the abrupt change in height above terrain, which affected the response due to switching between IGE and OGE conditions in the aerodynamic model. The CHARM model eliminated this non-representative behavior since both ship deck and ground (sea) surfaces were modeled, in addition to aerodynamic interactions between the rotor wake and surface boundaries. Note that this scenario (operations from a stationary ship without winds) is somewhat atypical, and airwake turbulence tends to dominate the response. This conclusion, however, is based on tests performed to date for operations from a small deck ship, and other platforms with larger vertical separation between landing spots and sea surface may result in different behavior.

CONCLUSIONS

A rotorcraft induced velocity software module, referred to as the CHARM/SI module, was developed that provides increased fidelity and less dependency on empirical data for modeling rotorcraft operations in real-time piloted simulations and trainers. The CHARM/SI module provides particular benefit in flight regimes where rotor/wake interactions dominate flight dynamics, such as vortex ring state (VRS) and shipboard operations. The core technology builds on free-vortex rotor wake and fast panel models and unique methods for coupling ship airwake and helicopter rotor wake models to capture physical interactions between the rotorcraft and ship airwake flow fields in real-time piloted helicopter training environments. The software module is designed for operation within an existing rotorcraft flight dynamics simulation and provides an induced velocity model for capturing normal flight conditions, rotor-fuselage-empennage interactions, vertical descent/vortex ring state, full/partial ground effect for moving ground planes (ship decks), and ship airwake effects, including rotor wake/ship airwake interactions.

The software module was integrated into the MH-60 Multi-Mission Helicopter (MMH) pilot-in-the-loop rotorcraft simulation (trainer) at Manned Flight Simulator (MFS) in Patuxent River, MD and tested in several pilot evaluations. Issues identified in these pilot evaluations were addressed and then the software module was integrated into the MH-60R TOFT2 fleet trainer at the Naval Air Station North Island where a final pilot evaluation was performed. This implementation represented the first time a military fleet helicopter trainer modeled the influence of the helicopter wake on the ship airwake during shipboard operations, as well as the first time a fleet trainer used a real-time free-wake model during piloted flight.

Though the CHARM/SI model provided improved fidelity in several areas, the improvement was not great enough to warrant a pilot recommendation for replacing the existing aerodynamic models. For shipboard operations, no significant deficiency was identified in the baseline trainer model that warranted replacing. A fleet trainer pilot evaluation, however, did identify other areas where the existing trainer model was deficient that could be addressed in future work using the new model, in particular, seat vibration cueing, flight dynamics and vibration associated with blade stall, and recovery from vortex ring state. To address these issues, CDI recommends as future research (1) collaboration with a motion seat manufacturer to properly model seat vibration cueing; (2) use of the CDI CHARM Rotor Module model of the rotor blade aerodynamics and dynamics including a method to properly capture blade stall without the need for empirical modeling; and (3) further research on vortex ring state to determine the specific wake characteristics that result in increased descent rate when collective is pulled in VRS.

The CHARM/SI model is designed to provide high fidelity training in shipboard operations with a unique ability to model the mutual interactions of the ship air wake and rotor wake. In this pilot evaluation, the baseline fleet trainer model was adequate in all shipboard operation scenarios tested and therefore there was no need for upgrading to the CHARM/SI model. If a deficiency in shipboard operations training is identified in the future for either this or another trainer, another pilot evaluation would be recommended.

ACKNOWLEDGMENTS

This work was supported through an SBIR Phase II.5 Accelerated Transition program for the U.S. Navy (Contract No. N68335-12-C-041), Dr. David Findlay serving as technical monitor. The authors also would like to acknowledge the dedicated support and insightful comments of Robert Clauser, Alyssa Zimmerman, Steven Puffenbarger and LT Ryan Brack of NAVAIR in executing the TOFT2 evaluation.

REFERENCES

1. Polsky, S. A., Imber, R. D., Czerwiec, R. M. and Ghee, T. A., "A Computational and Experimental Determination of the Air Flow around the Landing Deck of a US Navy Destroyer (DDG): Part II," AIAA Paper 2007-4484, Proceedings of the AIAA Applied Aerodynamics Conference, Miami, FL, June 2007.
2. Polsky, S. A., "A Computational Study of Unsteady Ship Airwake," AIAA Paper 2002-1022, Proceedings of the AIAA Aerospace Sciences Meeting, Reno, NV, January 2002.
3. Liu, J. and Long, L., "Higher Order Accurate Ship Airwake Predicitons for the Helicopter/Ship Interface Problem," Proceedings of the Americhan Helicopter Society 54th Annual Forum, Washington, DC, May 1998.
4. Bogstad, M. C., Habashi, W. G., Akel, I., Ait-Ali-Yahia, D., Giannias, N. and Longo, V., "Computational Fluid Dynamics Based Advanced Ship Airwake Database for Helicopter Flight Simulators," *Journal of Aircraft*, Vol. 39, (5), September-October 2002, pp. 830-838.
5. Bunnell, J. W., "An Integrated Time-varying Airwake in a UH-60 Black Hawk Shipboard Landing Simulation," AIAA 2001-4065, Proceedings of the AIAA Modeling and Simulation Technologies Conference and Exhibit, Montreal, Canada, August 2001.
6. Keller, J. D., Whitehouse, G. R., Boschitsch, A., Nadal, J., Jeffords, J. and Quire, M., "Computational Fluid Dynamics for Flight Simulator Ship Airwake Modeling," Proceedings of the Interservice/Industry Training, Simulation, and Education Conference (I/ITSEC) 2007, Orlando, FL, December 2007.
7. Forsythe, J. R., Lynch, C. E., Polsky, S. A. and Spalart, P., "Coupled Flight Simulator and CFD Calculations of Ship Airwake using HPCMP CREATETM-AV Kestrel," AIAA Paper 2015-0556, Proceedings of the 53rd AIAA Aerospace Sciences Meeting, Kissimmee, FL, January 2015.
8. Arunajatesan, S., Shipman, J. and Sinha, N., "Numerical Modeling of Coupled VSTOL-Ship Airwake Flowfields," Proceedings of the ASME 2004 Heat Transfer/Fluids Engineering Summer Conference, Charlotte, NC, July 2004.
9. Alpman, E., Long, L. N., Bridges, D. O. and Horn, J. F., "Fully-Coupled Simulations of the Rotorcraft / Ship Dynamic Interface," Proceedings of the American Helicopter Society 63rd Annual Forum, Virginia Beach, VA, May 2007.
10. Keller, J. D., Wachspress, D. A., Hoffler, J. C., Kachman, N. J. and Nichols, J., "Application of a Real Time Free Wake Induced Velocity Model in a Naval Rotorcraft Flight Trainer," Proceedings of the American Helicopter Society 67th Annual Forum, Virginia Beach, VA, May 2011.
11. Keller, J. D., Wachspress, D. A. and Hoffler, J. C., "Real Time Free Wake and Ship Airwake Model for Rotorcraft Flight Training Application," Proceedings of the 71st Annual American Helicopter Society Forum, Virginia Beach, VA, May 2015.
12. Wachspress, D. A., Keller, J. D., Quackenbush, T. R., Whitehouse, G. R. and Yu, K., "High Fidelity Rotor Aerodynamic Module for Real Time Rotorcraft Flight Simulation," Proceedings of the American Helicopter Society 64th Annual Forum, Montreal, Canada, April 2008.
13. Dreier, M. E., *Introduction to Helicopter and Tiltrotor Flight Simulation*, AIAA Education Series, Reston, VA, 2007.
14. Peters, D. A. and HaQuang, N., "Dynamic Inflow for Practical Applications," *Journal of the American Helicopter Society*, Vol. 33, (4), October 1988, pp. 64-68.
15. Peters, D. A. and He, C. J., "Finite State Induced Flow Models Part II: Three-Dimensional Rotor Disk," *Journal of Aircraft*, Vol. 32, (2), March-April 1995, pp. 323-333.
16. Pitt, D. M. and Peters, D. A., "Theoretical Prediction of Dynamic Inflow Derivatives," *Vertica*, Vol. 5, (1), 1981, pp. 21-34.
17. Smith, S. J. and Torgler, B., "Leveraging Technologies to Reverse Engineer a Helicopter for Simulator Development," Proceedings of the Interservice/Industry

Training, Simulation, and Education Conference (I/ITSEC), Orlando, FL, November 2012.

18. Spira, D., Myrand-Lapierre, V. and Soucy, O., "Reducing Blade Element Model Configuration Data Requirements Using System Identification and Optimization," Proceedings of the American Helicopter Society 68th Annual Forum, Fort Worth, TX, May 2012.

19. Johnson, W. J., "Rotorcraft Aeromechanics Applications of a Comprehensive Analysis," Proceedings of the HeliJapan 98 Gifu, Japan, April 1998.

20. Wachspress, D. A., Quackenbush, T. R. and Boschitsch, A. H., "First-Principles, Free-Vortex Wake Analysis for Helicopters and Tiltrotors," Proceedings of the American Helicopter Society 59th Annual Forum, Phoenix, AZ, May 2003.

21. Quackenbush, T. R., Wachspress, D. A., Keller, J. D., Boschitsch, A. H., Wasileski, B. J. and Lawrence, T. H., "Full Vehicle Flight Simulation with Real Time Free Wake Methods," Proceedings of the AHS Aerodynamics, Acoustics, and Test and Evaluation Technical Specialists Meeting, San Francisco, CA, January 2002.

22. Horn, J., Bridges, D., Wachspress, D. and Rani, S., "Implementation of a Free-Vortex Wake Model in Real-Time Simulation of Rotorcraft," *AIAA Journal of Aerospace Computing, Information, and Communications*, Vol. 3, (3), March 2006.

23. Kothmann, B. D., Lu, Y., DeBrun, E. and Horn, J., "Perspectives on Rotorcraft Aerodynamic Modeling for Flight Dynamics Applications," Proceedings of the AHS 4th Decennial Specialist's Conference on Aeromechanics, San Francisco, CA, January 2004.

24. Spoldi, S. and Ruckel, P., "High Fidelity Helicopter Simulation using Free Wake, Lifting Line Tail, and Blade Element Tail Rotor Models," Proceedings of the American Helicopter Society 59th Annual Forum, Phoenix, AZ, May 2003.

25. Quackenbush, T. R., Wachspress, D. A., Boschitsch, A. H. and Keller, J. D., "Real Time Free Wake Methods for Rotorcraft Simulation and Design," CDI Report No. 01-11 prepared under SBIR Contract No. NAS2-99053, 2001.

26. Wachspress, D. A., Whitehouse, G. R., Keller, J. D., Yu, K., Gilmore, P., Dorsett, M. and McClure, K., "A High Fidelity Brownout Model for Flight Simulations and Trainers," Proceedings of the American Helicopter Society 65th Annual Forum, Grapevine, TX, May 2009.

27. Polsky, S., Wilkinson, C., Nichols, J., Ayers, D., Mercado-Perez, J. and Davis, T., "Development and Application of the SAFEDI Tool for Virtual Dynamic Interface Ship Airwake Analysis," Proceedings of the 54th AIAA Aerospace Sciences Meeting, San Diego, CA, January 2016.

28. Miller, D. G., Kocher, E., Hughes, D. F., Grugan, R. A., Taylor, J. E. and Bender, J. S., "Complementary Moving Base and Dynamic Motion Seat Cueing for Rotorcraft Simulation," AIAA 2009-6249, 2009.

29. Keller, J. D. and Wachspress, D. A., "High Fidelity Real-Time Rotor Wake Module with Shipboard Interactions," CDI Report Number 17-03 prepared under NAVAIR contract N68335-12-C-0418, 2017.

30. Taghizad, A., Binet, L., Jimenez, J. and Heutze, D., "Experimental and Theoretical Investigations to Develop a Model of Rotor Aerodynamics Adapted to Steep Descents," Proceedings of the AHS 58th Annual Forum, Montreal, Canada, June 2002.

AUTHOR INDEX

American Helicopter Society International
2701 Prosperity Ave, Suite 210
Fairfax, VA 22031

ISBN 978-1-5108-4407-0

CURRAN ASSOCIATES INC.
proceedings.com